NATIONAL ACADEMIES

Sciences
Engineering
Medicine

NATIONAL
ACADEMIES
PRESS
Washington, DC

Guidance on PFAS Exposure, Testing, and Clinical Follow-Up

Committee on the Guidance on
PFAS Testing and Health Outcomes

Board on Environmental Studies
and Toxicology

Division on Earth and Life Studies

Board on Population Health and
Public Health Practice

Health and Medicine Division

Consensus Study Report

THE NATIONAL ACADEMIES PRESS 500 Fifth Street, NW Washington, DC 20001

This project was funded with federal funds from the Centers for Disease Control and Prevention under contract number 200-2011-38807, task order number 75D30121F00099; National Institutes of Environmental Health Sciences, National Institutes of Health, and U.S. Department of Health and Human Services, under Contract No. HHSN2632018000291, task order number 75N98020F00012. Any opinions, findings, conclusions, or recommendations expressed in this publication do not necessarily reflect the views of any organization or agency that provided support for the project.

International Standard Book Number-13: 978-0-309-48244-8
International Standard Book Number-10: 0-309-48244-5
Digital Object Identifier: https://doi.org/10.17226/26156
Library of Congress Control Number: 2022943858

This publication is available from the National Academies Press, 500 Fifth Street, NW, Keck 360, Washington, DC 20001; (800) 624-6242 or (202) 334-3313; http://www.nap.edu.

Suggested citation: National Academies of Sciences, Engineering, and Medicine. 2022. *Guidance on PFAS Exposure, Testing, and Clinical Follow-Up*. Washington, DC: The National Academies Press. https://doi.org/10.17226/26156.

The **National Academy of Sciences** was established in 1863 by an Act of Congress, signed by President Lincoln, as a private, nongovernmental institution to advise the nation on issues related to science and technology. Members are elected by their peers for outstanding contributions to research. Dr. Marcia McNutt is president.

The **National Academy of Engineering** was established in 1964 under the charter of the National Academy of Sciences to bring the practices of engineering to advising the nation. Members are elected by their peers for extraordinary contributions to engineering. Dr. John L. Anderson is president.

The **National Academy of Medicine** (formerly the Institute of Medicine) was established in 1970 under the charter of the National Academy of Sciences to advise the nation on medical and health issues. Members are elected by their peers for distinguished contributions to medicine and health. Dr. Victor J. Dzau is president.

The three Academies work together as the **National Academies of Sciences, Engineering, and Medicine** to provide independent, objective analysis and advice to the nation and conduct other activities to solve complex problems and inform public policy decisions. The National Academies also encourage education and research, recognize outstanding contributions to knowledge, and increase public understanding in matters of science, engineering, and medicine.

Learn more about the National Academies of Sciences, Engineering, and Medicine at **www.nationalacademies.org**.

Consensus Study Reports published by the National Academies of Sciences, Engineering, and Medicine document the evidence-based consensus on the study's statement of task by an authoring committee of experts. Reports typically include findings, conclusions, and recommendations based on information gathered by the committee and the committee's deliberations. Each report has been subjected to a rigorous and independent peer-review process and it represents the position of the National Academies on the statement of task.

Proceedings published by the National Academies of Sciences, Engineering, and Medicine chronicle the presentations and discussions at a workshop, symposium, or other event convened by the National Academies. The statements and opinions contained in proceedings are those of the participants and are not endorsed by other participants, the planning committee, or the National Academies.

Rapid Expert Consultations published by the National Academies of Sciences, Engineering, and Medicine are authored by subject-matter experts on narrowly focused topics that can be supported by a body of evidence. The discussions contained in rapid expert consultations are considered those of the authors and do not contain policy recommendations. Rapid expert consultations are reviewed by the institution before release.

For information about other products and activities of the National Academies, please visit www.nationalacademies.org/about/whatwedo.

---

[1] Deceased January 2022.

Reviewers

This Consensus Study Report was reviewed in draft form by individuals chosen for their diverse perspectives and technical expertise. The purpose of this independent review is to provide candid and critical comments that will assist the National Academies of Sciences, Engineering, and Medicine in making each published report as sound as possible and to ensure that it meets the institutional standards for quality, objectivity, evidence, and responsiveness to the study charge. The review comments and draft manuscript remain confidential to protect the integrity of the deliberative process.

We thank the following individuals for their review of this report:

Andrea Amico, Testing for Pease
Dean Baker, University of California, Irvine
Linda Birnbaum, National Institute of Environmental Health Sciences and National Toxicology
    Program (retired)
Julia Brody, Silent Spring Institute
Courtney Carignan, Michigan State University
Ellen Chang, Center for Health Sciences at Exponent, Inc.
Nicholas Chartres, University of California, San Francisco
Alan Ducatman, West Virginia University School of Public Health and School of Medicine
Baruch Fischhoff, Howard Heinz University
Philippe Grandjean, Harvard University
Alex John London, Carnegie Mellon University
Donald Mattison, University of South Carolina
Joseph V. Rodricks, Ramboll
Carlos Santos-Burgoa, The George Washington University
Susan L. Santos, Veterans Biomedical Research Institute
Annie St-Amand, Health Canada
Heather M. Stapleton, Duke University
John Washington, U.S. Environmental Protection Agency

Although the reviewers listed above provided many constructive comments and suggestions, they were not asked to endorse the conclusions or recommendations of this report, nor did they see the final draft before its release. The review of this report was overseen by **Joshua Sharfstein**, Johns Hopkins Bloomberg School of Public Health, and **Susan Brantley**, The Pennsylvania State University. They were responsible for making certain that an independent examination of this report was carried out in accordance with the standards of the National Academies and that all review comments were carefully considered. Responsibility for the final content rests entirely with the authoring committee and the National Academies.

Acknowledgments

Many people were critical in helping the committee accomplish its charge. The committee gratefully acknowledges the work of the community liaisons who provided insight and viewpoints pertinent to planning for our public meetings. In particular, we thank Andrea Amico, Emily Donovan, Patrick Elder, Ayesha Khan, Kristen Mello, and Laura Olah, who provided helpful insights on the perspectives and experiences of their communities that informed our community engagement methods. This engagement ensured that our public meetings would include the relevant perspectives, which allowed us to learn about the social, environmental, public health, and medical challenges central to our charge. In addition, we found the information and perspectives provided by the presentations and discussions at our public meetings immensely helpful in informing our deliberations (see Appendix C).

The committee's work was enhanced by the technical expertise; writing contributions; data evaluation, visualization, and extraction; and other support provided by Cary Haver, Jordan Kuiper, Judy Lakind, Kate Marquess, Melissa Miller, Josh Naiman, Anna Ruth Robuck, and Lauren Tobias, who served as consultants. We would also like to acknowledge the U.S. Environmental Protection Agency's Office of Water and Office of Research and Development for sharing publicly available abstracted data from some of the epidemiologic studies included in our literature review, as well as Scott M. Bartell and Nicholas Cuvelier from the University of California, Irvine, for calculating percentiles of per- and polyfluoroalkyl (PFAS) exposure.

Importantly, the committee heard from a number of individuals who shared their personal stories about and experiences with PFAS exposure, testing, and clinical follow-up. These discussions helped ground our work in the lived experiences of the complex issues that had to be tackled in this report, and we are extremely grateful for their courage in sharing those experiences in a public forum.

The committee thanks the staff of the National Academies of Sciences, Engineering, and Medicine who contributed to producing this report, especially the extraordinary, creative, and tireless study staff: Kaley Beins, Elizabeth Boyle, Clifford Duke, Kathryn Guyton, Rose Marie Martinez, Alexandra McKay, Marilee Shelton-Davenport, and Alexis Wojtowicz. Thanks as well go to other staff in the Division on Earth and Life Studies who provided additional support, including Tamara Dawson, Eric Edkin, Elizabeth Eide, Lauren Everett, Nancy Huddleston, Radiah Rose, and Maggie Walser. This project also received important assistance from Megan Lowry (Office of News and Public Information) and Matthew Anderson (Office of Financial Administration). Valuable research assistance was provided by Christopher Lao-Scott, senior research librarian in the National Academies Research Center. Finally, a thank you is extended to Rona Briere and Allison Boman, who assisted the committee with editing the report.

Preface

The creation of the U.S. Environmental Protection Agency (EPA) in 1970 signaled public recognition of industrial impacts on the environment. The government subsequently acknowledged that three elements essential for life—air, water, and food—could be contaminated by industrial activity and threaten human health. In 1980, two steps were taken in response to the government's concern: passage of the Comprehensive Environmental Response, Compensation, and Liability Act and creation of the Agency for Toxic Substances and Disease Registry (ATSDR).

The mission of ATSDR is to prevent or mitigate the adverse impacts on human health and diminished quality of life resulting from exposure to hazardous substances in the environment. The earliest contaminants of concern included pesticides; heavy metals from mining; asbestos; munitions and their manufacturing by-products (including radioactive substances); petrochemicals, including solvents; and products and by-products associated with oil and gas extraction, refinement, and use. Over time, additional industrial products with significant potential to affect the population's health have been identified, including the class of chemicals known as per- and polyfluoroalkyl substances, or PFAS. PFAS have useful properties, such as oil and water repellency, temperature resistance, and friction reduction. For decades, they have been used in numerous applications and products, such as firefighting; chrome-plating; lubricants; insecticides; and coatings and treatments for such surfaces as carpeting, packaging, and cookware. As a result of the production and use of PFAS, many sites across the country are contaminated with PFAS, which in turn can result in contamination of soils and drinking water.

ATSDR faces a critical challenge in protecting people from the potential health impacts of PFAS exposures. Data from the National Health and Nutrition Examination Survey show that nearly 100 percent of people in the United States are exposed to at least one PFAS, but at what level of exposure do harms to human health occur? What PFAS-associated health outcomes might benefit from clinical follow-up or care? Would there be any benefit in testing people to know their PFAS exposure level? What clinical follow-up can help protect people from PFAS-associated harms?

Answering these questions requires bridging approaches used for chemical hazard assessments, such as those carried out by the EPA; public and community health benefits, such as those laid out in the Community Guide to Preventive Services; and medical care health benefit assessments, such as those of the U.S. Preventive Services Task Force. Chemical hazard assessments are conducted to determine whether a chemical exposure causes harm; public health assessments address the impact on and interventions to mitigate threats to the health of a population or community; and medical care health benefit assessments evaluate the effects of medical interventions, including their beneficial effects and potential adverse outcomes. One challenge with blending these different approaches is that in medical care health benefit assessments, the "gold standard" for informing clinical risk/benefit decisions for medical interventions is the randomized controlled trial (RCT). However, RCTs are typically impossible for chemical hazard assessments, for both ethical and practical reasons. For example, most environmental chemicals are not developed to improve human health, making intentional exposures in an RCT unethical. Chemical exposures also vary and may or may not be significant depending on the agent's toxicity, which sometimes makes controlled trials infeasible. And public health assessments, in which comparative studies are still required to support evidence-based recommendations, usually require many years of data and are challenged by the long-standing and often-lamented separation of health care and public health.

In 2010, researchers Stephen Rappaport and Martyn Smith reported that "70 to 90% of disease risks are probably due to differences in environments" and made the case for a more comprehensive approach to evaluating environmental exposures in order to understand the causes of and contributors to chronic disease.[2] Such an approach has, as in the case of lead, and will, for chemicals such as PFAS, ultimately depend on breaking down the barriers between environmental public health and the clinical care setting. These two health sectors have had some limited success in bridging the gap for infectious disease outbreaks and epidemics. Identifying environmental exposures, measuring exposure levels in patients, and providing indicated medical follow-up are elements of a critical frontier that could and should bring the two disciplines closer together to improve the health of those in the nation's communities.

Another challenge for the study committee was the critical need to include community voices in the study process as an important and credible source of evidence to inform guidance recommendations. To meet that challenge, this study included the testimony of more than 30 people who live in or work with a community impacted by PFAS contamination. Community members provided the committee with much needed data based on their lived experiences with PFAS contamination, and moved the committee's work from an academic exercise to a personal reality. The committee used the presentations of community members to inform frameworks within the report and to gain an understanding of the social context that the committee's recommendations will inform.

Atmospheric chemist Susan Solomon has suggested that successfully addressing environmental challenges requires making the problem personal, perceptible, and practical. The voices of affected individuals in contaminated communities make the PFAS issue personal, while the scientists researching the associations with human health make the impacts of PFAS exposure perceptible. In this report, the committee has endeavored to provide practical recommendations that can aid policy makers, state and federal environmental and public health agencies, clinicians, and concerned individuals in addressing this important health problem.

Bruce N. Calonge, *Chair*
Committee on the Guidance on PFAS Testing and
Health Outcomes

---

[2] Rappaport, S., and M. Smith. 2010. *Science* 330(6003):460–461. https://doi.org/10.1126/science.1192603.

Contents

APPENDIXES

BOXES, FIGURES, AND TABLES

BOXES

FIGURES

TABLES

Contents

Acronyms and Abbreviations

AAFP	American Academy of Family Physicians
AAP	American Academy of Pediatrics
ACOG	American College of Obstetricians and Gynecologists
AHRQ	Agency for Healthcare Research and Quality
ALT	alanine aminotransferase
ANHE	Alliance of Nurses for Health Environments
ASTHO	Association of State and Territorial Health Officials
ATSDR	Agency for Toxic Substances and Disease Registry
BE	biomonitoring equivalent
BLL	blood lead level
BMI	body mass index
CAS	Chemical Abstract Services
CDC	Centers for Disease Control and Prevention
CERCLA	Comprehensive Environmental Response, Compensation, and Liability Act
CI	confidence interval
CMS	Centers for Medicare & Medicaid Services
CPG	clinical practice guideline
CPT	Current Procedural Terminology
DNA	deoxyribonucleic acid
DOD	U.S. Department of Defense
EFSA	European Food Safety Authority
EPA	U.S. Environmental Protection Agency
EVIDEM	Evidence and Value in Decision Making
EWG	Environmental Working Group
FDA	U.S. Food and Drug Administration
FOSA	perfluorooctane sulfonamide
GRADE	Grading of Recommendations, Assessment, Development and Evaluations
HBM	human biomonitoring
HDL	high-density lipoprotein
HHS	U.S. Department of Health and Human Services
HMB	German Human Biomonitoring Commission
IARC	International Agency for Research on Cancer
IOM	Institute of Medicine
LDL	low-density lipoprotein
LOD	limit of detection

MCL	maximum contaminant level
MeFOSAA	methylperfluorooctane sulfonamidoacetic acid
NAS	National Academy of Sciences
NHANES	National Health and Nutrition Examination Survey
NIEHS	National Institute of Environmental Health Sciences
NIOSH	National Institute for Occupational Safety and Health
NIST-SRM	National Institute of Standards and Technology Standard Reference Material
NRC	National Research Council
NTP	National Toxicology Program
OECD	Organisation for Economic Co-operation and Development
OR	odds ratio
OSHA	Occupational Safety and Health Administration
PBPK	physiologically based pharmacokinetic
PEHSU	Pediatric Environmental Health Specialty Unit
PFAS-REACH	PFAS Research, Education, and Action for Community Health
PFBA	perfluorobutanoic acid
PFBS	perfluorobutane sulfonic acid
PFCA	perfluorinated aliphatic carboxylic acid
PFDA	perfluorodecanoic acid
PFDoDA	perfluorododecanoic acid
PFHpA	perfluoroheptanoic acid
PFHpS	perfluoroheptanesulfonic acid
PFHxA	perfluorohexanoic acid
PFHxS	perfluorohexane sulfonic acid
PFNA	perfluorononanoic acid
PFOA	perfluorooctanoic acid
PFOS	perfluorooctane sulfonic acid
PFUnA	perfluoroundecanoic acid
PFUnDA	perfluoroundecanoic acid
QA	Quality Assurance
QC	Quality Control
RfD	reference dose
SD	standard deviation
SSEHRI	Social Science Environmental Health Research Institute
TSCA	Toxic Substances Control Act
USDA	U.S. Department of Agriculture
USPSTF	U.S. Preventive Services Task Force
VHA	Veterans Health Administration
WHO	World Health Organization

Summary[1]

Perfluoroalkyl and polyfluoroalkyl substances (PFAS) are a class of chemicals that includes more than 12,000[2] different compounds with various chemical properties. PFAS are commonly used in thousands of products, from nonstick cookware to firefighting foams and protective gear, because they have desirable chemical properties that impart oil and water repellency, friction reduction, and temperature resistance. PFAS as a class have a wide variety of distinct chemical properties and toxicities; for example, some PFAS can bioaccumulate and persist in the human body and the environment, while others transform relatively quickly. The PFAS that do transform, however, will become one or more other PFAS because the carbon–fluorine bond they contain does not break naturally. It is for this reason that PFAS are termed "forever chemicals."

STUDY CONTEXT

Public concern about the impact of PFAS contamination on human health and the environment began in the late 1990s when perfluorooctanoic acid (PFOA) water contamination was identified in Parkersburg, West Virginia. As a result, 3M, a primary PFAS manufacturer initiated a voluntary phase-out of some PFAS (PFOA, perfluorooctanesulfonic acid [PFOS], and perfluorohexanesulfonic acid [PFHxS]). The contamination in Parkersburg also led to a class action lawsuit that identified several health effects related to PFAS exposure and led to the establishment of a medical monitoring program in 2013. Shortly thereafter, the U.S. Environmental Protection Agency (EPA) began requiring all community water systems serving more than 10,000 people to test for certain PFAS, which led to more communities learning that their water was contaminated. In 2016, researchers found that the drinking water supply for Wilmington, North Carolina, was contaminated with a chemical called "GenX," a PFOA replacement. This finding led to public concern about the potential health effects of replacement PFAS (see Figure S-1).

Organizations such as the International Agency for Research on Cancer (IARC), the Agency for Toxic Substances and Disease Registry (ATSDR), and the EPA have linked exposure to PFAS (particularly PFOA and PFOS) to multiple cancers, thyroid dysfunction, small changes in birthweight, and high cholesterol. Gaining a complete picture of the threat can be difficult, however, because of the chemical and toxicological differences among individual PFAS and uncertainty about the exposure level at which their adverse effects may occur. In addition, many of the chronic diseases associated with PFAS exposure have myriad causes.

An estimated 2,854 U.S. locations (in all 50 states and two territories) have some level of PFAS contamination (see Figure S-2). Although not all of the contamination represents exceedances of health advisories, the pervasiveness of the contamination is alarming. Furthermore, almost 100 percent of the U.S. population is exposed to at least one PFAS. Although exposures to the phased-out PFAS have been decreasing (see Figure S-3), people are still exposed to those PFAS from site contamination, occupational uses of stored products, and breakdown of PFAS polymer products that are found in homes. Carpeting, for example, is often treated with fluorotelomer-based polymers that can biodegrade to form phased-out PFAS, such as PFOA. Exposures also occur to the PFAS chemicals used to replace those that have been phased out. Although the harms of the replacement PFAS are less well understood, they may have

---

[1] No references are included in this Summary. References to the content herein are provided in the respective chapters of the main text.

[2] EPA Comptox Dashboard (https://comptox.epa.gov/dashboard/chemical-lists/pfasmaster [accessed May 25, 2022]).

comparable or more serious toxicity than the PFAS they have replaced. The state of New Jersey, for example, recently set a groundwater standard for a replacement PFAS that is an order of magnitude lower than drinking water standards for other PFAS. The New Jersey maximum contaminant levels of perfluorononanoic acid (PFNA), PFOA, and PFOS in drinking water are 13, 14, and 13 nanograms per liter (ng/L), respectively,[3] while the groundwater standard for the replacement PFAS, chloroperfluoropolyether carboxylates, is 2 ng/L.[4]

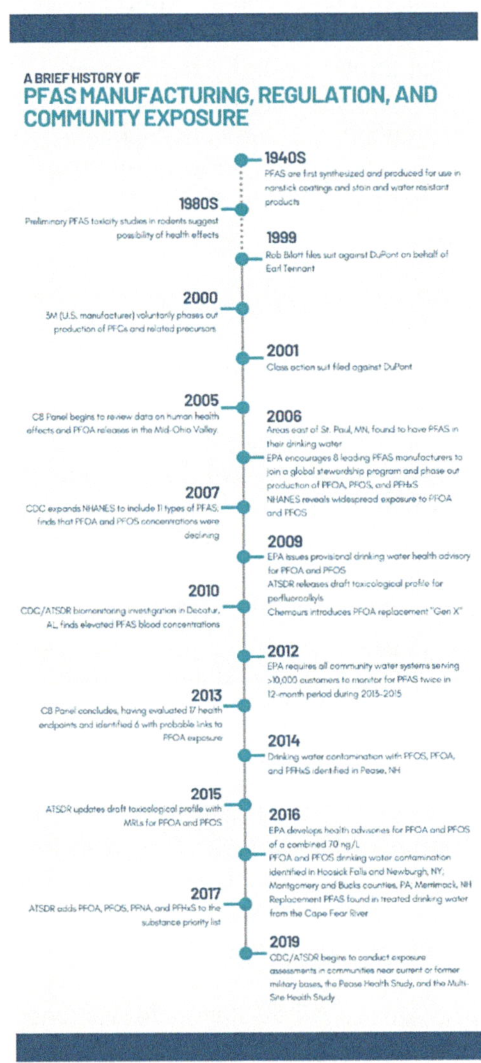

FIGURE S-1 Brief history of PFAS manufacturing, regulation, and community exposure.
NOTE: ATSDR = Agency for Toxic Substances and Disease Registry; CDC = Centers for Disease Control and Prevention; EPA = U.S. Environmental Protection Agency; MRL = minimal risk limit; ng/L = nanograms per liter; NHANES = National Health and Nutrition Examination Survey; PFAS = per- and polyfluoroalkyl substances; PFC = perfluorochemicals; PFHxS = perfluorohexane sulfonic acid; PFNA = perfluorononanoic acid; PFOA = perfluorooctanoic acid; PFOS = perfluorooctane sulfonic acid.
SOURCE: Committee generated based on slides included by Patrick N. Breysse in a presentation to the committee on February 4, 2021.

---

[3] See https://www.nj.gov/health/ceohs/documents/pfas_drinking%20water.pdf (accessed June 8, 2022).
[4] See https://www.nj.gov/dep/standards/ClPFPECA_Standard.pdf (accessed June 8, 2022).

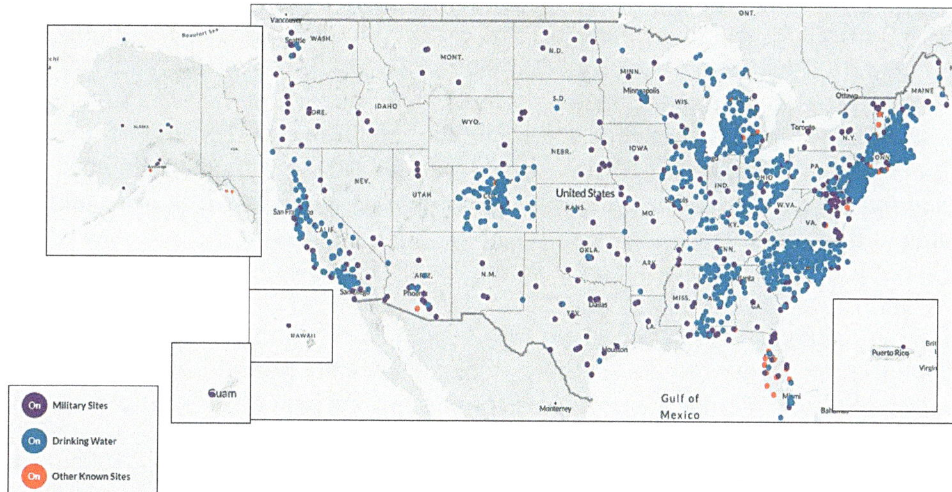

FIGURE S-2 PFAS contamination across the United States.
NOTES: According to the Environmental Working Group (EWG), "locations represented on the map are approximate and intended to portray the general area of a contamination site or a community water system. Locations were mapped using the best data available from official records, including data provided by tests of public drinking water systems, the Safe Drinking Water Information System and the Department of Defense report *Addressing Perfluorooctane Sulfonate (PFOS) and Perfluorooctanoic Acid (PFOA)*, and Department of Defense public records, among others. Data on contaminated industrial and military sites was current as of October 2021." Furthermore, "EWG has worked to ensure the accuracy of the information provided in this map. The map is dynamic. This contaminant site, results, suspected sources and other information in the database may change based on evolving science, new information or other factors. Please be advised that this information frequently relies on data obtained from many sources, and accordingly, EWG cannot guarantee the accuracy of the information provided or any analysis based thereon."
SOURCE: See https://www.ewg.org/interactive-maps/pfas_contamination/map (accessed May 11, 2022). Copyright © Environmental Working Group, www.ewg.org. Reproduced with permission.

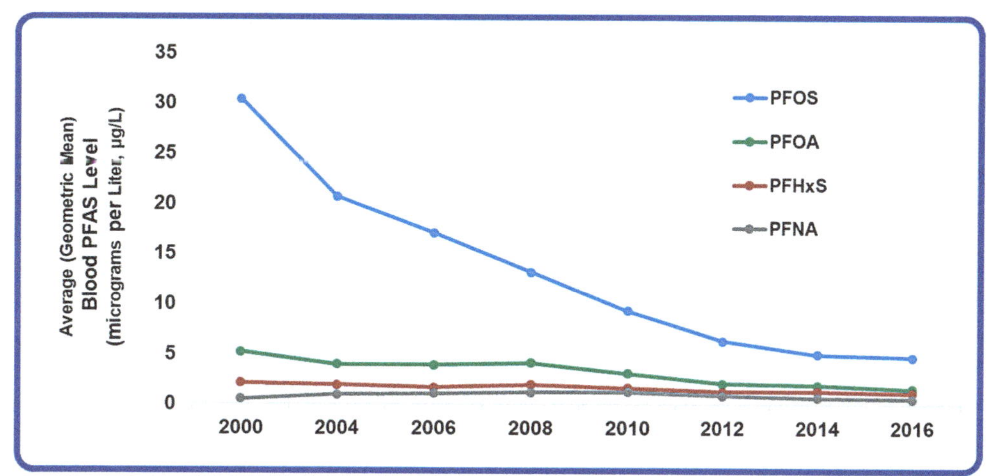

FIGURE S-3 Blood (serum) levels of PFAS, United States, 2000–2016.
NOTE: Average = geometric mean.
SOURCE: Patrick N. Breysse's presentation to the committee on February 4, 2021. DATA SOURCE: Centers for Disease Control and Prevention (2019, January). *Fourth Report on Human Exposure to Environmental Chemicals*, updated tables. Atlanta, GA: U.S. Department of Health and Human Services.

The people who live, work, and play in environments where PFAS contamination exceeds standards do not know how to protect themselves from the health risks of exposure. Many have been exposed to PFAS for decades and may continue to be exposed in their homes or workplaces. Some members of communities that have discovered that their exposures exceed health advisory levels are calling for a medical program to prevent, lead to early detection of, or treat diseases related to the health risks they may face. Developing such a program is challenging, however, because of the uncertainties about the health effects of PFAS and potential harms from additional medical treatments; moreover, many of these diseases are not preventable or even treatable, and many clinicians lack information about what they can and should do for these patients.

To help clinicians[5] respond to patient concerns about PFAS exposure, ATSDR published *PFAS: An Overview of the Science and Guidance for Clinicians on Per- and Polyfluoroalkyl Substances.*[6] This guidance summarizes general information about PFAS and PFAS health studies and suggests answers to example patient questions. However, the ATSDR guidance does not provide specific recommendations on when to test for PFAS, how to order the tests, how to interpret the results, or what clinical follow-up based on PFAS exposure might look like. Interpretation of PFAS blood or urine results is challenging because the specific level of exposure at which harms may occur is unknown, and the science on the potential health effects of exposure to PFAS is advancing quickly, making it difficult to provide advice to clinicians about what follow-up tests might be helpful.

STUDY CHARGE AND APPROACH

ATSDR and the National Institute of Environmental Health Sciences (NIEHS) asked the National Academies of Sciences, Engineering, and Medicine to form an ad hoc committee to advise on PFAS testing and clinical care for patients exposed to PFAS. The Statement of Task asked the committee to

- develop principles for biological testing and clinical evaluation, given substantial scientific uncertainty about the health effects or the value of such measures in informing care,
- review the human health literature for the health effects of PFAS, and
- characterize human exposure pathways and develop principles for exposure reduction.

The Statement of Task asked the committee to recommend

- options and considerations to guide decision making for PFAS testing in a patient's blood or urine,
- PFAS concentrations that could inform clinical care of exposed patients, and
- appropriate patient follow-up and care specific to PFAS-associated health endpoints for those patients known or suspected to be exposed to PFAS.

The committee also was asked to provide advice on changes to ATSDR's clinical guidance. The committee was not asked for community prevention guidance or advice on policies that would reduce exposure to PFAS.

Figure S-4 summarizes the committee's approach to the Statement of Task. A critical component of the approach was community engagement because people who live with potentially harmful exposures have knowledge from experiential learning that precedes scientific findings. The study's community engagement consisted of a panel of community liaisons, three town halls (summarized in Appendix B),

---

[5] The committee uses "clinician" throughout this report to refer to "a healthcare professional qualified in the clinical practice of medicine. Clinicians may be physicians, nurses, pharmacists, or other allied health professionals" as defined by the Centers for Medicare & Medicaid Services (https://www.cms.gov/Medicare/Quality-Initiatives-Patient-Assessment-Instruments/MMS/QMY-Clinicians [accessed June 14, 2022]).

[6] See https://www.atsdr.cdc.gov/pfas/docs/clinical-guidance-12-20-2019.pdf (accessed June 14, 2022).

community speakers at every public committee meeting, open sign-up for public testimony at every public meeting, and encouragement for written testimony throughout the data collection phase of the study. The following sections provide descriptions of the remaining components of the study approach and the committee's associated recommendations.

COMMITTEE'S PRINCIPLES FOR DECISION MAKING UNDER UNCERTAINTY

Providing clear advice to clinicians on clinical follow-up for PFAS-exposed patients is challenging. The questions in the Statement of Task all have some degree of uncertainty. First, clinicians need to know what the health consequences of PFAS exposure are. If everyone is exposed, what exposure level warrants follow-up? What actions can be taken to reduce exposure or prevent disease? What are the harms of those actions? Answering all of these questions requires making ethical judgments. Building on the work of other experts and evidence-to-decision frameworks, the committee developed five principles to guide decision making under uncertainty for use throughout this report and by ATSDR when updating the guidance for clinicians: proportionality, justice, autonomy, feasibility, and adaptability (see Box S-1). In the clinical setting, these principles converge under the principle of autonomy; thus, shared, informed decision making between clinician and patient is the practical way to incorporate the principles into a clinical encounter.

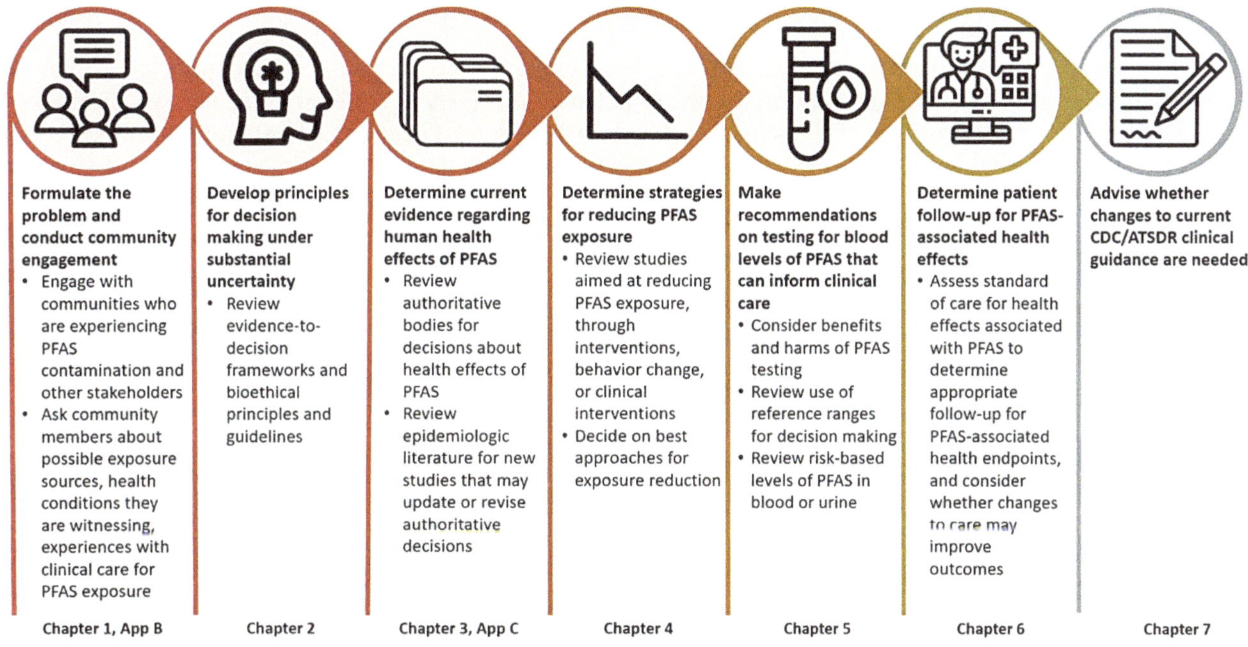

FIGURE S-4 The committee's approach to the Statement of Task and the chapters and appendixes where the topics are discussed.
NOTE: ATSDR = Agency of Toxic Substances and Disease Registry; CDC = Centers for Disease Control and Prevention; PFAS = per- and polyfluoroalkyl substances.

BOX S-1
Principles for Decision Making Under Uncertainty Used in This Report

Proportionality: Decisions should balance plausible harms and benefits proportionally, meaning that the harms and benefits are considered together and weighed based on a qualitative assessment of their potential likelihoods and magnitudes.

Justice: Decisions should be informed by an emphasis on promoting justice, including by balancing benefits and harms fairly across the population of at-risk individuals, advancing health equity, and respecting human rights. In addition, justice requires consideration of sociohistorical context, stakeholders, existing structural inequalities, and issues of agency (the power a community has to advocate for itself in conflicts).

Autonomy: Decisions should be based on informed decision making by individuals and reflect respect for their values.

Feasibility: Decisions should take into account resource availability, including follow-up services.

Adaptability: Decisions should respond to new information about harms, benefits, and other relevant considerations (e.g., health equity and feasibility).

POTENTIAL HEALTH EFFECTS OF PFAS

To recommend PFAS levels that should inform clinical care, the committee conducted a literature review to determine health effects or conditions associated with PFAS. The Statement of Task limited this review to human studies of those PFAS included in the Centers for Disease Control and Prevention's (CDC's) *National Report on Human Exposure to Environmental Chemicals* (see Table S-1). Other PFAS may also cause harm, though they do not all have the same biological persistence and toxicity. Most people are exposed to mixtures of PFAS such that specific effects are difficult to disentangle. Considering these issues, and recognizing that some PFAS are infrequently measured, the committee provided one strength-of-evidence determination for all PFAS for each health effect, recognizing that providing one conclusion across PFAS may not account for the distinct physical, chemical, and toxicological properties of each type of PFAS.

TABLE S-1 PFAS Species Currently Included in the Centers for Disease Control and Prevention's *National Report on Human Exposure to Environmental Chemicals*

Abbreviated Name	Full Name	CAS Registry No.
MeFOSAA	Methylperfluorooctane sulfonamidoacetic acid	2355-31-9
PFHxS	Perfluorohexanesulfonic acid perfluorohexane sulfonic acid	355-46-4
n-PFOA (linear isomer), Sb-PFOA (branched isomers)	Perfluorooctanoic acid	335-67-1*
PFDA	Perfluorodecanoic acid	335-76-2
PFUnDA	Perfluoroundecanoic acid	2058-94-8
n-PFOS (linear isomer), Sm-PFOS (branched isomers)	Perfluorooctanesulfonic acid	1763-23-1*
PFNA	Perfluorononanoic acid	375-95-1

NOTES: CAS = Chemical Abstracts Service. * = CAS number refers to linear isomer only. Previous survey years have also included perfluorobutane sulfonic acid (PFBS), perfluoropentanoic acid (PFpA), perfluorododecanoic acid (PFDoDA), perfluorooctane sulfonamide (FOSA), and 2-(N-ethyl-perfluorooctane sulfonamido)acetate (EtFOSAA), according to Patrick N. Breysse's presentation to the committee on February 4, 2021.

The committee began its review by identifying decisions of other authoritative bodies regarding PFAS exposure and any human health effects. The committee then considered more recent human studies, both systematic reviews and published epidemiologic research articles, that could inform updates to those decisions. This approach improved efficiency while minimizing the risk of excluding scientific findings that could inform the committee's conclusions.

The committee considered animal studies discussed in ATSDR's *Toxicological Profile for Perfluoroalkyls* and in systematic reviews to aid in its interpretation of the human studies. The committee synthesized the available evidence into four categories of association used by other National Academies committees: sufficient evidence of an association, limited or suggestive evidence of an association, inadequate or insufficient evidence of an association, and limited or suggestive evidence of no association (see Figure S-5). All associations between PFAS and a health outcome were considered inadequate or insufficient by default unless available evidence from all aspects of the committee's review warranted placing the evidence in another category of association.

Category of Association	What Does It Mean?
Sufficient Evidence of an Association	• Based on strong evidence, there is high confidence that there is an association between exposure to PFAS and the health outcome. It is unlikely that the association is due to chance or bias.
Limited or Suggestive Evidence of an Association	• Based on limited evidence, there is moderate confidence that there is an association between exposure to PFAS and the health outcome. It is possible that the association is due to chance or bias.
Inadequate or Insufficient Evidence to Determine an Association	• Based on inconsistent evidence, a lack of evidence, or evidence of insufficient quality of an association between exposure to PFAS and the health outcome, no conclusion can be made about a potential association.
Limited or Suggestive Evidence of No Association	• Based on at least limited evidence, there is at least moderate confidence that there is NO association between PFAS and the health outcome.

FIGURE S-5 Categories of association used in this report.
NOTES: The categories of association only describe how strong the evidence is between PFAS and the health outcome. The risk of developing an outcome from exposure to PFAS for things in the same category can vastly differ and are dependent on whether an individual has other risk factors for developing the outcome.

The committee found sufficient evidence of an association for the following diseases and health outcomes:

- **decreased antibody response (in adults and children),**
- **dyslipidemia (in adults and children),**
- **decreased infant and fetal growth, and**
- **increased risk of kidney cancer (in adults).**

The committee found limited or suggestive evidence of an association for the following diseases and health outcomes:

- **increased risk of breast cancer (in adults),**
- **liver enzyme alterations (in adults and children),**

- **increased risk of pregnancy-induced hypertension (gestational hypertension and preeclampsia),**
- **increased risk of testicular cancer (in adults),**
- **thyroid disease and dysfunction (in adults), and**
- **increased risk of ulcerative colitis (in adults).**

The committee observed gaps in the evidence, rendering the evidence inadequate or insufficient, for many health effects including the following:

- immune effects other than reduced antibody response, and ulcerative colitis;
- cardiovascular outcomes other than dyslipidemia;
- developmental outcomes other than small reductions in birthweight;
- cancers other than kidney, breast, and testicular;
- reproductive effects other than hypertensive disorders of pregnancy;
- hepatic effects other than liver enzyme levels;
- endocrine disorders other than those involving thyroid hormone levels;
- respiratory effects;
- hematological effects;
- musculoskeletal effects, such as effects on bone mineral density;
- renal effects, such as renal disease; and
- neurological effects.

For some outcome categories, the research spanned many different tests or measures of effect, all of which assessed slightly different health outcomes, making the evidence difficult to synthesize and use to draw strong conclusions. The committee found this issue most apparent when reviewing the literature on neurodevelopmental effects, such as effects on learning and behavior. Another research issue is that most studies reviewed by the committee were not conducted among people known to have high exposures to PFAS, indicating a gap in understanding the effects of PFAS among those highly exposed.

PFAS EXPOSURE REDUCTION

Some people may be interested in reducing their exposure to PFAS. The primary route to PFAS in nonoccupational settings is likely ingestion, which may include drinking contaminated water; eating seafood from contaminated water; or consuming other contaminated foods, such as vegetables, game, or dairy products. PFAS are often used in materials that come in contact with food, such as microwave popcorn bags or packaging used for fast foods or processed foods. Exposure may also occur when dust containing PFAS is ingested. Inhalation is the most common route in occupational settings, and is a route of exposure for people living near fluorochemical plants or incinerators. PFAS transfer to the fetus during pregnancy and in early life during lactation. Dermal exposure is understudied.

If PFAS are in drinking water, switching to consumption of water lower in PFAS will reduce exposure. In general, however, it is difficult to reduce exposure to PFAS through personal behavior modifications.

For clinicians, based on its review of the evidence on PFAS exposure reduction, the committee makes the following recommendations:

Recommendation 4-1[7]: Clinicians advising patients on PFAS exposure reduction should begin with a conversation aimed at first determining how they might be exposed to PFAS

---

[7] The committee's recommendations are numbered according to the chapter of the main text in which they appear.

(sometimes called an environmental exposure assessment) and what exposures they are interested in reducing. This exposure assessment should include questions about current occupational exposures to PFAS (such as work with fluorochemicals or firefighting) and exposures to PFAS through the environment. Known environmental exposures to PFAS include living in a community with PFAS-contaminated drinking water, living near industries that use fluorochemicals, serving in the military, and consuming fish and game from areas with known or potential contamination.

Recommendation 4-2: If patients may be exposed occupationally, such as by working with fluorochemicals or as a firefighter, clinicians should consult with occupational health and safety professionals knowledgeable about the workplace practices to determine the most feasible ways to reduce that exposure.

Recommendation 4-3: Clinicians should advise patients with elevated PFAS in their drinking water that they can filter their water to reduce their exposure. Drinking water filters are rated by NSF International, an independent organization that develops public health standards for products. The NSF database can be searched online for PFOA to find filters that reduce the PFAS in drinking water included in the committee's charge. Individuals who cannot filter their water can use another source of water for drinking.

Recommendation 4-4: In areas with known PFAS contamination, clinicians should advise patients that PFAS can be present in fish, wildlife, meat, and dairy products and direct them to any local consumption advisories.

There are fewer evidence-based, exposure-reduction recommendations for patients without known sources of exposure:

Recommendation 4-5: Clinicians should direct patients interested in learning more about PFAS to authoritative sources of information on how PFAS exposure occurs and what mitigating actions they can take. Authoritative sources include the Pediatric Environmental Health Specialty Units (PEHSUs), the Agency for Toxic Substances and Disease Registry (ATSDR), and the U.S. Environmental Protection Agency (EPA).

Recommendation 4-6: When clinicians are counseling parents of infants on PFAS exposure, they should discuss infant feeding and steps that can be taken to lower sources of PFAS exposure. The benefits of breastfeeding are well known; the American Academy of Pediatrics, the American Academy of Family Physicians, and the American College of Obstetricians and Gynecologists support and recommend breastfeeding for infants, with rare exceptions. Clinicians should explain that PFAS can pass through breast milk from a mother to her baby. PFAS may also be present in other foods, such as the water used to reconstitute formula and infant food, and potentially in packaged formula and baby food. It is not yet clear what types and levels of exposure to PFAS are of concern for child health and development.

Additionally, there is a critical need for more data to understand PFAS exposure among breastfed infants:

Recommendation 4-7: Federal environmental health agencies should conduct research to evaluate PFAS transfer to and concentrations in breast milk and formula to generate data that can help parents and clinicians make shared, informed decisions about breastfeeding.

PFAS TESTING AND CONCENTRATIONS TO INFORM
CLINICAL CARE OF EXPOSED PATIENTS

Decisions about PFAS testing require shared informed decision making between patient and clinician. Clinicians should explain that exposure biomonitoring may provide important information about an individual's exposure levels that might guide clinical follow-up. At the same time, this information cannot indicate or predict the likelihood that an individual will end up with a particular condition. Allowing people the opportunity to determine whether they will undergo PFAS testing shows respect for patient values. Discussions about PFAS testing should always include information about how PFAS exposure occurs, potential health effects of PFAS, limitations of PFAS testing, and the benefits and harms of the testing (see Box S-2).

BOX S-2
Potential Harms and Benefits of PFAS Testing

Potential Harms
- Fear of blood draw
- Small risk of injury or infection at draw site
- Difficulties in interpreting results
- Stress or concern about the health effects of exposure
- Decreased property values resulting from identifying property contamination
- Social isolation
- Clinical consequences from medical follow-up as a result of exposure

Potential Benefits
- Increased awareness of exposure so it can be reduced
- Empowerment of communities to respond to contamination
- Relief from the stress of not knowing one's exposure level
- Identification of the potential risk for health conditions associated with PFAS exposure, informing subsequent preventive care
- Help in monitoring whether efforts to reduce exposure are working through the conduct of baseline and follow-up tests

---

The German Human Biomonitoring Commission has risk-based guidance levels for two PFAS chemicals—PFOS and PFOA. The European Food Safety Authority has one guidance level for the sum of PFOA, PFOS, PFHxS, and PFNA. No individual values are available for PFHxS and PFNA, and the committee could find no values for methylperfluorooctane sulfonamidoacetic acid (MeFOSAA), PFDA, and perfluoroundecanoic acid (PFUnDA). Based on its review of the evidence for PFAS testing, the committee recommends:

Recommendation 5-1: As communities with PFAS exposure are identified, government entities (e.g., Centers for Disease Control and Prevention [CDC]/Agency for Toxic Substances and Disease Registry [ATSDR], public health departments) should support clinicians with educational materials about PFAS testing so that they can discuss testing with their patients. These educational materials should include the following information:

- How people can be exposed to PFAS: Exposure routes include occupational exposures and work with fluorochemicals or as a firefighter; consumption of contaminated drinking water in communities that obtain their water from sources near commercial airports, military bases, fluorochemical manufacturing plants, wastewater treatment plants, landfills, or incinerators where PFAS-containing waste may have been disposed of or farms where sewage sludge may have been used; and consumption of contaminated fish or game if fishing or hunting occurs in contaminated areas. Individuals living near fluorochemical plants may also be exposed via inhalation of air emissions.
- Potential health effects of PFAS exposure and strategies for reducing exposure.
- Limitations of PFAS blood testing: PFAS blood testing does not identify the sources of exposure or predict future health outcomes; it only assesses body burden at the time of sample collection. For example, a person with low blood levels today may have had higher levels in the past.
- The benefits and harms of PFAS testing.

Recommendation 5-2: Clinicians should offer PFAS testing to patients likely to have a history of elevated exposure. In all discussions of PFAS testing, clinicians should describe the potential benefits and harms of the testing and the potential clinical consequences (such as additional follow-up), related social implications, and limitations of the testing so that patient and clinician can make a shared, informed decision. Patients who are likely to have a history of elevated exposure to PFAS include those who have

- had occupational exposure to PFAS (such as those who have worked with fluorochemicals or served as a firefighter);
- lived in communities where environmental and public health authorities (Centers for Disease Control and Prevention [CDC], Agency for Toxic Substances and Disease Registry [ATSDR], U.S. Environmental Protection Agency [EPA], state and local environmental or health authorities), or academic researchers have documented PFAS contamination; or
- lived in areas where PFAS contamination may have occurred, such as near facilities that use or have used fluorochemicals, commercial airports, military bases, wastewater treatment plants, farms where sewage sludge may have been used, or landfills or incinerators that have received PFAS-containing waste.

Recommendation 5-3: Clinicians should use serum or plasma concentrations of the sum of PFAS* to inform clinical care of exposed patients, using the following guidelines for interpretation:

- Adverse health effects related to PFAS exposure are not expected at less than 2 nanograms per milliliter (ng/mL).
- There is a potential for adverse effects, especially in sensitive populations, between 2 and 20 ng/mL.
- There is an increased risk of adverse effects above 20 ng/mL.

*** Simple additive sum of MeFOSAA, PFHxS, PFOA (linear and branched isomers), PFDA, PFUnDA, PFOS (linear and branched isomers), and PFNA in serum or plasma. Caution is warranted when using capillary blood measurements as levels may differ from serum or plasma levels.**

The cutoff levels should be updated as additional information becomes available. The committee also noted that children younger than 12 years are not routinely included in the *National Report on Human Exposure to Environmental Chemicals*, nor are pregnant people included in large numbers. More reference- and risk-based values are needed, for other PFAS and other biological matrices, but given the expansiveness of the class, this gap can best be addressed with relative potency factor approaches rather than the development of risk-based levels for each PFAS.

Recommendation 5-4: The National Health and Nutrition Examination Survey should begin collecting and sharing more data on children younger than 12 years of age and pregnant people to generate reference populations for those groups.

Testing for PFAS, although expensive, offers an opportunity to identify people who may need to reduce their PFAS exposure and are at increased risk of certain health outcomes. It is important to recognize, however, that race, age, and other social and demographic characteristics already have disadvantaged many patients with respect to accessing clinical preventive services, such as PFAS testing. The disadvantage would be compounded as PFAS testing services should be linked to counseling on steps to mitigate exposure and its impacts. Therefore, encouraging testing primarily among people with relatively stable access to care could have the unintended effect of aggravating disparities in exposure to PFAS absent a funded, national PFAS testing program with a counseling component.

APPLICATION OF THE COMMITTEE'S PRINCIPLES TO PATIENT FOLLOW-UP FOR PFAS-ASSOCIATED HEALTH EFFECTS

Many health outcomes or conditions that the committee found to be associated with PFAS exposure are common in the general population. All have multiple known risk factors. The committee categorized the strength of the evidence for an association between PFAS and various health outcomes, and concluded that all conditions with an adequately supported association should be considered for patient follow-up. The committee then used its established cutoff levels to determine appropriate follow-up based on PFAS exposure level. Risks from PFAS likely increase with exposure, and PFAS levels of 3 ng/mL and 19 ng/mL do not represent the same risk even though they are listed in the same category. Clinical providers should use judgment and shared, informed decision making in making follow-up decisions based on PFAS exposure and other risk factors. Figure S-6 suggests that clinicians engage in shared decision making with their patients regarding follow-up care for PFAS-associated health endpoints.

Recommendation 6-1: Clinicians should treat patients with serum PFAS concentration below 2 nanograms per milliliter (ng/mL) with the usual standard of care.

Recommendation 6-2: For patients with serum PFAS concentration of 2 nanograms per milliliter (2 ng/mL) or higher and less than 20 ng/mL, clinicians should encourage PFAS exposure reduction if a source of exposure is identified, especially for pregnant persons. Within the usual standard of care clinicians should:

- **Prioritize screening for dyslipidemia with a lipid panel (once between 9 and 11 years of age, and once every 4 to 6 years over age 20) as recommended by the American Academy of Pediatrics (AAP) and the American Heart Association (AHA).**
- **Screen for hypertensive disorders of pregnancy at all prenatal visits per the American College of Obstetricians and Gynecologists (ACOG).**

- **Screen for breast cancer based on clinical practice guidelines based on age and other risk factors such as those recommended by the U.S. Preventive Services Task Force (USPSTF).**

≥20 (ng/mL) PFAS*

Encourage PFAS exposure reduction if a source of exposure is identified, especially for pregnant persons.

In addition to the usual standard of care, clinicians should:

- Prioritize screening for dyslipidemia with a lipid panel (for patients over age 2) following American Academy of Pediatrics (AAP) recommendations for high-risk children and American Heart Association (AHA) guidance for high-risk adults.
- At all well visits:
  - Conduct thyroid function testing (for patients over age 18) with serum thyroid stimulating hormone (TSH),
  - Assess for signs and symptoms of kidney cancer (for patients over age 45), including with urinalysis, and
  - For patients over age 15, assess for signs and symptoms of testicular cancer and ulcerative colitis.

2–<20 (ng/mL) PFAS*

Encourage PFAS exposure reduction if a source has been identified, especially for pregnant persons.

Within the usual standard of care clinicians should:

- Prioritize screening for dyslipidemia with a lipid panel (once between 9 and 11 years of age, and once every 4 to 6 years over age 20) as recommended by the AAP and AHA.
- Screen for hypertensive disorders of pregnancy at all prenatal visits per the American College of Obstetricians and Gynecologists (ACOG).
- Screen for breast cancer based on clinical practice guidelines based on age and other risk factors such as those recommended by U.S. Preventive Services Task Force (USPSTF).

<2 (ng/mL) PFAS*

Provide usual standard of care

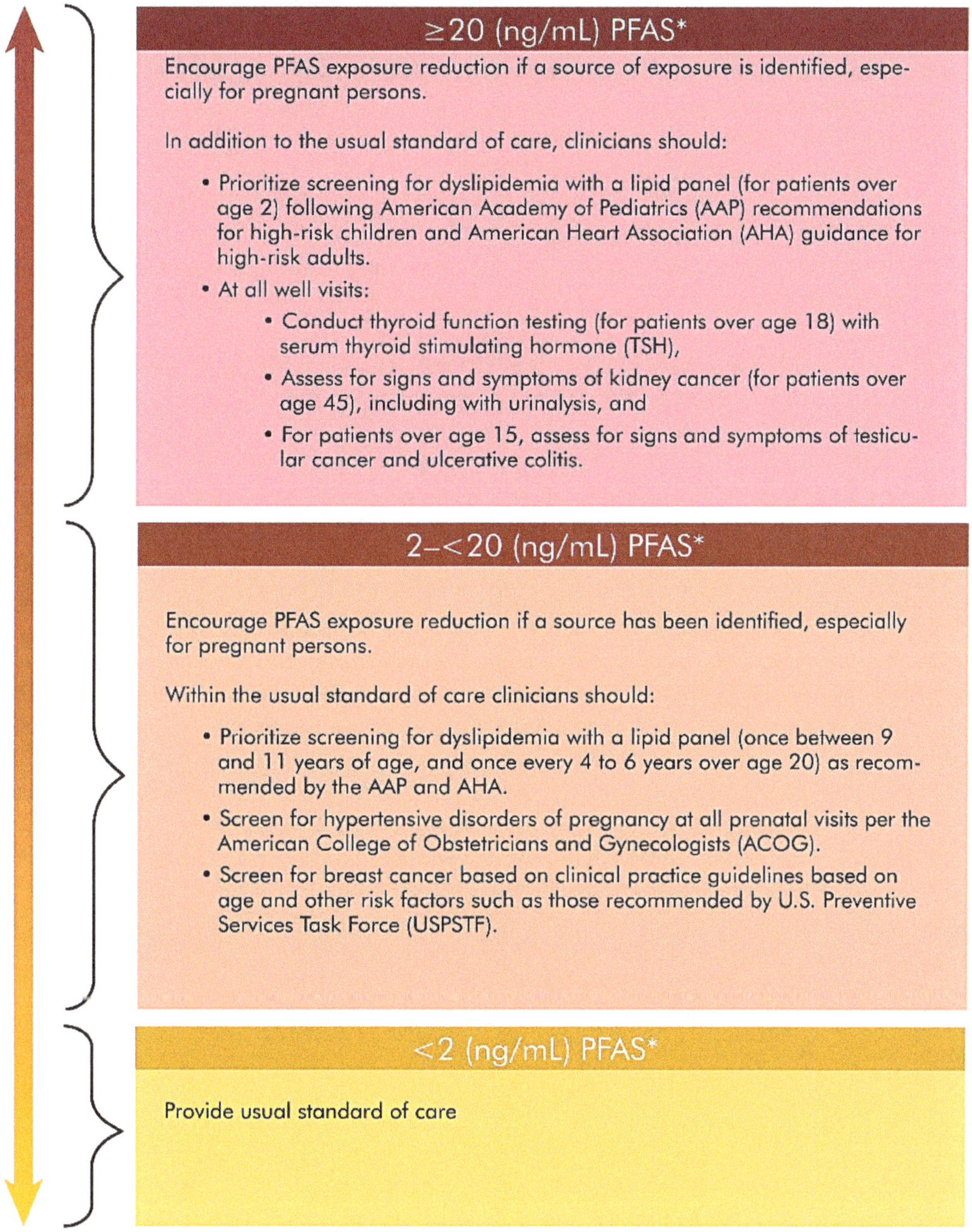

* Simple additive sum of MeFOSAA, PFHxS, PFOA (linear and branched isomers), PFDA, PFUnDA, PFOS (linear and branched isomers), and PFNA in serum or plasma

FIGURE S-6 Clinical guidance for follow-up with patients after PFAS testing.
NOTE: MeFOSAA = methylperfluorooctane sulfonamidoacetic acid; PFDA = perfluorodecanoic acid; PFHxS = perfluorohexane sulfonic acid, PFNA = perfluorononanoic acid; PFOA = perfluorooctanoic acid; PFOS = perfluorooctanesulfonic acid; PFUnDA = perfluoroundecanoic acid.

Recommendation 6-3: For patients with serum PFAS concentration of 20 nanograms per milliliter (ng/mL) or higher, clinicians should encourage PFAS exposure reduction if a source of exposure is identified, especially for pregnant persons. In addition to the usual standard of care, clinicians should:

- **Prioritize screening for dyslipidemia with a lipid panel (for patients over age 2) following American Academy of Pediatrics (AAP) guidelines for high-risk children and American Heart Association (AHA) guidance for high-risk adults.**
- **At all well visits:**
  - **conduct thyroid function testing (for patients over age 18) with serum thyroid stimulating hormone (TSH),**
  - **assess for signs and symptoms of kidney cancer (for patients over 45), including with urinalysis, and**
  - **for patients over 15, assess for signs and symptoms of testicular cancer and ulcerative colitis.**

APPLYING THE COMMITTEE'S EXPOSURE, TESTING, AND CLINICAL FOLLOW-UP RECOMMENDATIONS

The committee created a flow chart summarizing PFAS education, exposure assessment, and clinical follow-up (see Figure S-7). In communities where PFAS exposure has been identified, ATSDR and other government entities should support local clinicians with educational materials about PFAS exposure and testing. Clinicians should then determine whether a particular patient is likely to have a history of elevated exposure to PFAS. If so, the clinician should offer PFAS testing and make a shared, informed decision on that testing. If testing is chosen, the labs should be ordered (Test Code 39307 Current Procedural Terminology [CPT] code 82542). Test results should be interpreted by summing the concentrations of MeFOSAA, PFHxS, PFOA (linear and branched isomers), PFDA, PFUnDA, PFOS (linear and branched isomers), and PFNA. The laboratory may not report results for all PFAS considered by the committee or may include different PFAS in its panel. In that case, the sum of PFAS should include only the PFAS in the analyte list that the committee considered. For example, if the lab tests for PFOA, PFOS, PFHxS, PFNA, and perfluorobutane sulfonic acid (PFBS), the summation should include PFOA, PFOS, PFHxS, and PFNA. Differing analytes lists may cause some variation in response. Still, as long as PFOA, PFOS, PFHxS, and PFNA are included in the analyte list, the results may not vary too greatly as these four analytes are most commonly detected in the United States. If any analyte is below the limit of detection, the clinician should calculate the analyte limit of detection divided by the square root of 2 and use this value in the summation. The sum thus derived should be compared against Figure S-6 to determine an appropriate clinical follow-up plan based on shared, informed decision making between patient and clinician.

REVISING ATSDR'S PFAS CLINICAL GUIDANCE

The committee recommends several changes to ATSDR's guidance to ensure consistency with the findings, conclusions, and recommendations presented in this report and improve the guidance's writing, design, dissemination, and implementation:

Recommendation 7-1: The Agency for Toxic Substances and Disease Registry (ATSDR) should update its PFAS clinical guidance to make it more succinct and accord with the review of PFAS-associated health effects, exposure reduction considerations, PFAS testing recommendations and interpretation, and recommendations for clinical follow-up presented in this report. When describing the health effects of PFAS, ATSDR should avoid using terms typically used to

categorize toxicants, such as "endocrine disrupter" or "neurotoxin," because they are vague and not necessarily clinically meaningful. When discussing the strength of the association between PFAS and a health outcome, ATSDR should use standard categories of association (such as sufficient evidence of an association, limited or suggestive evidence of an association, inadequate or insufficient evidence of an association, and limited or suggestive evidence of no association).

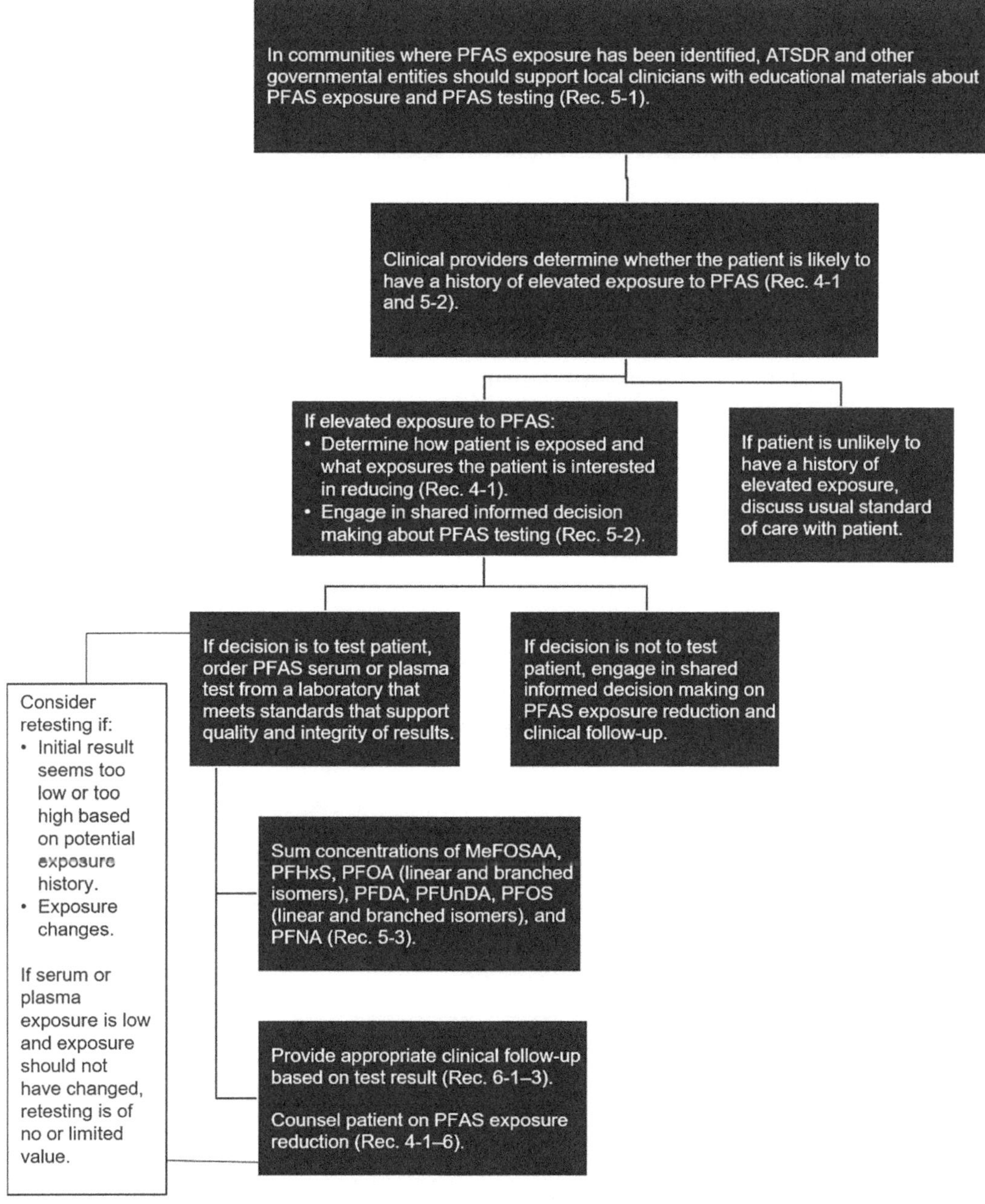

FIGURE S-7 Flow chart showing how the committee's recommendations work together in a clinical setting. NOTE: ATSDR = Agency for Toxic Substances and Disease Registry; MeFOSAA = methylperfluorooctane sulfonamidoacetic acid; PFDA = perfluorodecanoic acid; PFHxS = perfluorohexane sulfonic acid, PFNA = perfluorononanoic acid; PFOA = perfluorooctanoic acid; PFOS = perfluorooctanesulfonic acid; PFUnDA = perfluoroundecanoic acid.

Recommendation 7-2: The Agency for Toxic Substances and Disease Registry (ATSDR) should incorporate a reader-centered approach when developing its guidance, with the knowledge that many different audiences will turn to its clinical guidance document to prepare for discussions with their clinicians. ATSDR should also solicit feedback on the guidance from a variety of stakeholders, such as community groups, practicing clinicians, and medical associations. In addition, ATSDR should encourage clinicians to use evidence-based organizational health literacy strategies to support shared, informed decision making; patient-centered care; cultural humility; and accessible language when communicating with patients about potential health risks.

Evidence on the health effects of PFAS should be updated every 2 years, and the clinical guidance should be updated at least every 5 years. In that process, ATSDR should strive to ensure that its clinical guidance follows criteria for making guidelines trustworthy when possible. Figure S-8 provides an overview of how the process used by this committee could be applied by ATSDR to improve its clinical guidance.

Recommendation 7-3: The Agency for Toxic Substances and Disease Registry (ATSDR) should develop a process for updating its PFAS guidance that adheres to criteria for making guidelines trustworthy, such as being based on a thorough, transparent, unbiased review of the evidence and being developed by a knowledgeable panel of experts free from strong biases and conflicts of interest. A review of the evidence on the health effects of PFAS should be completed by an authoritative neutral party every 2 years, and the clinical guidance should be updated every 5 years or sooner if warranted by the evidence on the health effects of PFAS. Clinicians and members of communities with elevated PFAS exposure should be engaged to inform the problem and review updated guidance.

FIGURE S-8 Suggested framework for updating the Agency for Toxic Substances and Disease Registry's clinical guidance based on new evidence.

IMPLEMENTING THE COMMITTEE'S RECOMMENDATIONS
TO IMPROVE PUBLIC HEALTH

Public health requires the use of multifaceted approaches to emerging health issues. In environmental health—the subset of public health focused on environmental factors—mitigation of potential harms associated with chemical exposures is often complicated because there is no exposure surveillance system for most chemicals. The committee's recommendations will best protect the public health if they are part of a national effort focused on increased biomonitoring, exposure surveillance, and education of clinicians and public health professionals on environmental health issues:

Recommendation 8-1: Laboratories conducting PFAS testing of serum or plasma should report the results to state public health authorities, following the respective states' statutes and reporting regulations. This reporting would improve PFAS exposure surveillance; it could be linked with the Centers for Disease Control and Prevention's (CDC's) environmental public health tracking network and help build capacity for improvements in the state-based national biomonitoring network.

The people and communities with high exposures to PFAS need to be identified. As the committee looks forward, it sees a pressing need for a robust environmental health infrastructure to continue to respond to PFAS and address other complex emerging and persistent environmental challenges.

1

Introduction

More than a decade ago, the United Nations General Assembly passed Resolution 64/292, recognizing that safe and clean drinking water is a human right and is essential to realizing all human rights. Yet today, drinking water in thousands of communities across the United States is contaminated with chemicals known as per- and polyfluoroalkyl substances (PFAS) (Hu et al., 2016) (see Box 1-1). PFAS do not occur in nature and are manufactured for a wide range of purposes, from nonstick cookware, to stain-resistant fabrics and carpets, to firefighting foams and protective gear. PFAS constitute a large class of up to 12,000 different chemicals (EPA, 2020). They are commonly used in thousands of products because they have desirable chemical properties that impart oil and water repellency, friction reduction, and temperature resistance (ITRC, 2017). PFAS as a class have a wide variety of distinct chemical properties and toxicities; for example, some PFAS can accumulate and persist in the human body and the environment, while others transform relatively quickly. The PFAS that do transform, however, will become one or more other PFAS, because the carbon–fluorine bond they contain does not break naturally. It is for this reason that PFAS are termed "forever chemicals."

In the most rudimentary sense, PFAS can be thought of as either polymer or nonpolymer. Nonpolymer PFAS, such as perfluroalkyl acids (PFAAs), are more commonly detected in the environment and are often considered highly persistent and mobile (Blum et al., 2015; ITRC, 2021). By contrast, many polymer PFAS, such as polytetrafluoroethylene (PTFE), are insoluble, less bioavailable, and often considered less of a direct concern, with respect to human or ecological health (Lohmann et al., 2020; ITRC, 2021). Nonetheless, the production of fluoropolymers requires other PFAS chemicals, such as PFAAs. For example, the nonpolymer PFAS perfluorooctanoic acid (PFOA) was used as a surfactant in the emulsion polymerization of fluoropolymers.[1]

BOX 1-1
What Are PFAS?

What are PFAS? A consensus definition does not exist. Buck and colleagues (2011) define PFAS as fluorinated substances that "contain 1 or more C atoms on which all the H substituents (present in the nonfluorinated analogues from which they are notionally derived) have been replaced by F atoms, in such a manner that they contain the perfluoroalkyl moiety CnF2n+1–" (p. 513).

The Organisation for Economic Co-operation and Development (OECD) defines PFAS as "fluorinated substances that contain at least one fully fluorinated methyl or methylene carbon atom (without any H/Cl/Br/I atom attached to it), that is, with a few noted exceptions, any chemical with at least a perfluorinated methyl group (–CF3) or a perfluorinated methylene group (–CF2–) is a PFAS" (OECD, 2021, p. 7).

continued

---

[1] See https://www.acs.org/content/acs/en/molecule-of-the-week/archive/p/perfluorooctanoic-acid.html (accessed June 22, 2022).

BOX 1-1 Continued

The U.S. Environmental Protection Agency's (EPA's) (2020) CompTox Chemicals Dashboard includes all substances that contain a specific set of substructural elements. The Dashboard says that "there is no precisely clear definition of what constitutes a PFAS substance" and PFAS lists include "partially fluorinated substances, polymers, and ill-defined reaction products" (EPA, 2020, para. 1).

What are PFAS used for? PFAS have desirable chemical properties such as oil and water repellency, temperature resistance, and friction reduction. Since the late 1940s, PFAS have been used in numerous applications, such as firefighting, chrome-plating, lubricants, and insecticides, as well as coating and treating of such surfaces as carpeting, packaging, and cookware (ATSDR, 2021).

How long do PFAS stay in the human body following exposure? PFAS levels in people's bodies will persist unless exposure ceases, and will continue to persist even after exposure ends. Half-life estimates based on repeated serum measurements range from days (e.g., PFBA) to years (e.g., PFOA, PFOS). It is generally assumed that it takes five half-lives to eliminate PFAS after exposure has ceased (ATSDR, 2021).

What are the potential hazards of PFAS? Health effects associated with PFAS include altered immune function, elevated cholesterol, thyroid disease, hypertension during pregnancy, testicular cancer, and kidney cancer (ATSDR, 2021; Fenton et al., 2021; Steenland et al., 2020).

SOCIOHISTORICAL TIMELINE OF PFAS

Public concern about the impact of PFAS contamination on human health and the environment began in the late 1990s when PFOA water contamination was identified in Parkersburg, West Virginia.[2] In response, 3M, a primary PFAS manufacturer, initiated a voluntary phase-out of some PFAS (PFOA, perfluorooctanesulfonic acid [PFOS], and perfluorohexanesulfonic acid [PFHxS]]) (ITRC, 2017). The contamination in Parkersburg also led to a class action lawsuit that identified several health effects related to PFAS exposure and subsequently to the establishment of a medical monitoring program (see Box 1-2). Shortly thereafter, the U.S. Environmental Protection Agency (EPA) began requiring all community water systems serving more than 10,000 people to test for certain PFAS, which led to more communities learning that their water was contaminated.[3] In 2016, researchers found that the drinking water supply for Wilmington, North Carolina, was contaminated with a chemical called "GenX," a PFOA replacement.[4] This finding led to public concern about the potential health effects of replacement PFAS (see Figure 1-1).

An estimated 2,854 U.S. locations (in all 50 states and two territories) have some level of PFAS contamination (see Figure 1-2). Although not all of the contamination represents exceedances of health advisories, the pervasiveness of the contamination is alarming. Furthermore, almost 100 percent of the U.S. population is exposed to at least one PFAS. Although exposures to the phased-out PFAS have been decreasing (see Figure 1-3), people are still exposed to those PFAS from site contamination, occupational uses of stored products, and breakdown of PFAS polymer products that are found in homes. Carpeting, for example, is often treated with fluorotelomer-based polymers that can biodegrade to form phased-out PFAS, such as PFOA (Washington and Jenkins, 2015). Exposures also occur to the PFAS chemicals used to replace those that have been phased out. Although the harms of the replacement PFAS are less well understood, they may have comparable or more serious toxicity than the PFAS they have replaced (Kwiatkowski et al., 2020). The state of New Jersey, for example, recently set a groundwater standard for

---

[2] See https://pfasproject.com/parkersburg-west-virginia (accessed June 16, 2022).

[3] See https://www.epa.gov/sites/default/files/2016-05/documents/ucmr3-factsheet-list1.pdf (accessed June 16, 2022).

[4] See https://news.ncsu.edu/2018/04/finding-genx (accessed June 16, 2022).

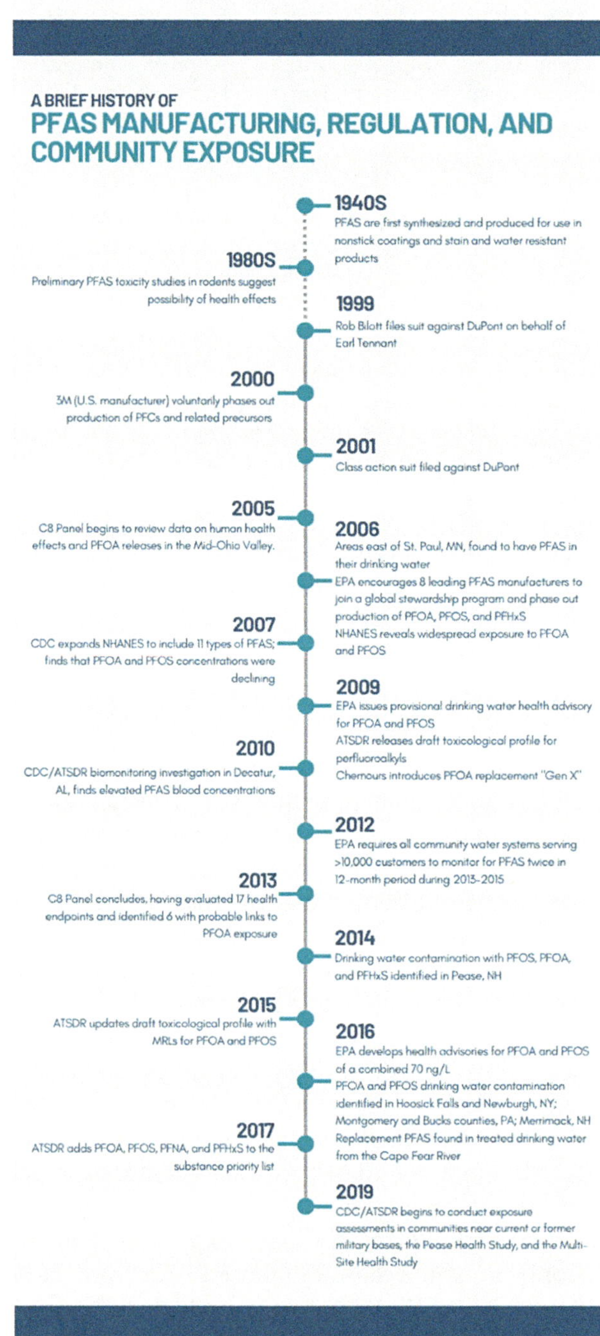

FIGURE 1-1 Brief history of PFAS manufacturing, regulation, and community exposure.
NOTE: ATSDR = Agency for Toxic Substances and Disease Registry; CDC = Centers for Disease Control and Prevention; EPA = U.S. Environmental Protection Agency; MRL = minimal risk limit; ng/L = nanograms per liter; NHANES = National Health and Nutrition Examination Survey; PFAS = per- and polyfluoroalkyl substances; PFC = perfluorochemicals; PFHxS = perfluorohexane sulfonic acid; PFNA = perfluorononanoic acid; PFOA = perfluorooctanoic acid; PFOS = perfluorooctane sulfonic acid.
SOURCE: Committee generated based on slides included by Patrick N. Breysse in a presentation to the committee on February 4, 2021.

a replacement PFAS (chloroperfluoropolyether carboxylates) that is an order of magnitude lower than drinking water standards for other PFAS. The New Jersey maximum contaminant levels of perfluorononanoic acid (PFNA), PFOA, and PFOS in drinking water are 13, 14, and 13 nanograms per liter (ng/L), respectively,[5] while the groundwater standard for the replacement PFAS, chloroperfluoropolyether carboxylates, is 2 ng/L.[6]

A reason PFAS contamination became well known only recently, after decades of their use, is that they can be difficult to detect. The physical and chemical properties that make PFAS persistent and mobile in the environment also make them particularly challenging to analyze (Guelfo et al., 2021). Analytical methods sensitive enough to detect environmentally relevant concentrations became widely available in the early 2010s. Although analyte lists continue to expand, currently available methods still allow identification of only a small fraction of the thousands of PFAS that have reportedly been created and used since the 1950s. As existing analytical methods improve, additional PFAS and new release sites will likely be identified (De Silva et al., 2021; Guelfo et al., 2021).

Estimated in 2,854 sites in 50 states and two territories

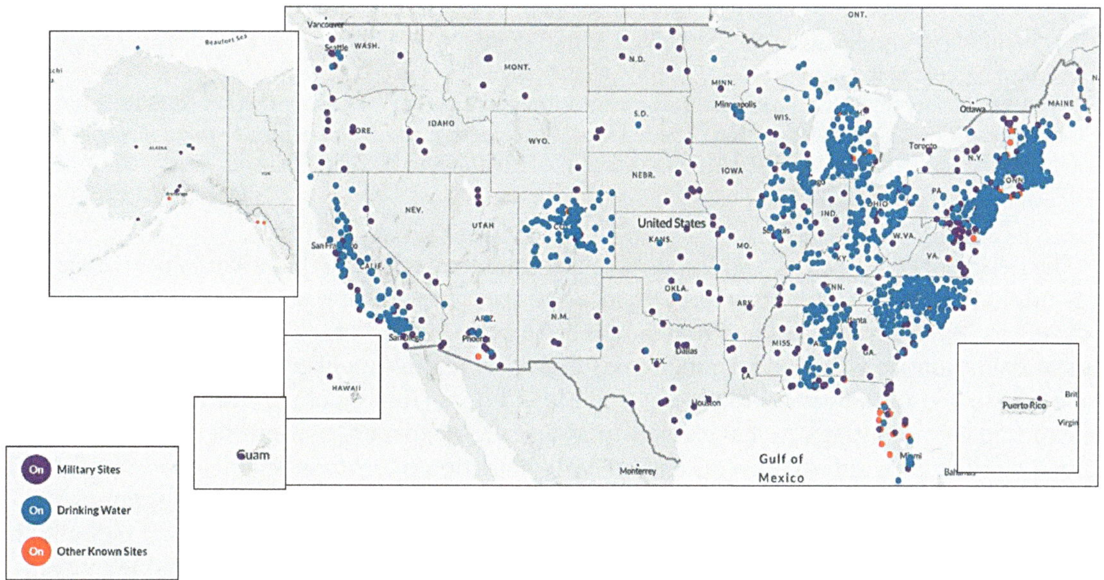

FIGURE 1-2 PFAS contamination across the United States.
NOTES: According to the Environmental Working Group (EWG), "locations represented on the map are approximate and intended to portray the general area of a contamination site or a community water system. Locations were mapped using the best data available from official records, including data provided by tests of public drinking water systems, the Safe Drinking Water Information System and the Department of Defense report *Addressing Perfluorooctane Sulfonate (PFOS) and Perfluorooctanoic Acid (PFOA)*, and Department of Defense public records, among others. Data on contaminated industrial and military sites was current as of October 2021." Furthermore, "EWG has worked to ensure the accuracy of the information provided in this map. The map is dynamic. This contaminant site, results, suspected sources and other information in the database may change based on evolving science, new information or other factors. Please be advised that this information frequently relies on data obtained from many sources, and accordingly, EWG cannot guarantee the accuracy of the information provided or any analysis based thereon."
SOURCE: See https://www.ewg.org/interactive-maps/pfas_contamination/map (accessed May 11, 2022). Copyright © Environmental Working Group, www.ewg.org. Reproduced with permission.

---

[5] See https://www.nj.gov/health/ceohs/documents/pfas_drinking%20water.pdf (accessed June 8, 2022).
[6] See https://www.nj.gov/dep/standards/ClPFPECA_Standard.pdf (accessed June 8, 2022).

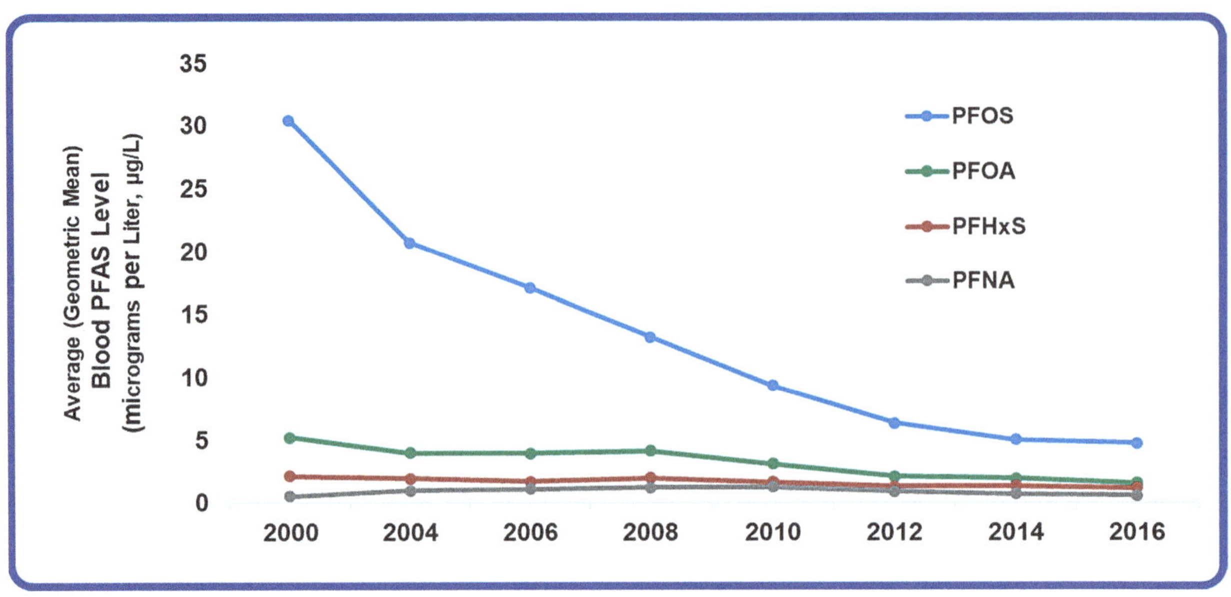

FIGURE 1-3 Blood (serum) levels of PFAS, United States, 2000–2016.
NOTE: Average = geometric mean.
SOURCE: Patrick N. Breysse's presentation to the committee on February 4, 2021. DATA SOURCE:
Centers for Disease Control and Prevention (2019, January). *Fourth Report on Human Exposure to Environmental Chemicals*, updated tables. Atlanta, GA: U.S. Department of Health and Human Services.

The people who live, work, and play in environments where PFAS contamination exceeds standards often do not know how to protect themselves from the health risks of exposure. Many have been exposed to PFAS for decades and may continue to be exposed in their homes or workplaces. Some members of communities with documented exposure report feeling violated and concerned about the health and well-being of their families, friends, and loved ones (Raponi et al., 2021; Rizzuto, 2021). Many in these exposed communities who have discovered that their exposure exceeds health advisory levels are calling for a medical program to prevent, lead to early detection of, or treat any diseases related to the health risks they may face (Raponi et al., 2021; Rizzuto, 2021). Developing such a program is challenging, however, because of the uncertainties about the health effects of PFAS and potential harms from additional medical treatments; moreover, many of these diseases are not preventable or even treatable once exposures have occurred (Wones et al., 2009). Accordingly, clinicians generally do not know how to respond to patients' concerns about PFAS.

ENVIRONMENTAL JUSTICE

Compounding the challenge of responding clinically to PFAS contamination are issues of environmental justice and systemic racism in medical care. It is well established that environmental risks are not distributed uniformly across populations. Race, ethnicity, poverty, age, life stage, and other social factors can place people at disproportionately high risk for diseases with environmental causes as a result of hazardous exposures at increased levels compared to the general population (Gochfeld and Burger, 2011). While environmental justice research specific to PFAS contaminants has been limited, place-based factors that may put individuals at greater risk of exposure (siting of chemical companies, refineries, and industrial sites), coupled with insufficient access to environmental screening, information, and adequate health care, have disproportionate impacts on Black, Hispanic, and Indigenous communities, as well as low-income populations. A special report from *Scientific American* called this a "triple whammy of race, poverty, and environment converging nationwide to create communities near pollution sources where nobody else wants to live" (Kay and Katz, 2012, para. 12).

According to a 2014 paper by the Environmental Justice Health Alliance for Chemical Policy Reform, more than 134 million Americans live within the "vulnerability zones" of industrial facilities that store or use highly hazardous chemicals (Orum et al., 2014). The companies themselves define these vulnerability zones, determined by guidance from the EPA,[7] as areas that could be affected by the release of toxic chemicals. The scope of the research on which this paper was based was broad, but included industries with known PFAS contamination, such as waste management and chemical manufacturing. The authors found that, relative to the U.S. average, communities within the vulnerability zones are disproportionately African American or Latinx; are more likely to live in poverty; and have lower housing values, incomes, and education levels. As the extent of contamination becomes known, housing values have been shown to decrease, and those with the means to move are more likely to do so, leaving behind those with the fewest resources and options (Harclerode et al., 2021).

When considering environmental justice, it is important to think beyond exposure disparities (Mohai et al., 2009); core environmental justice issues relevant to PFAS also include rural health, industrial siting, and access to environmental exposure reduction (Bullard, 1996) and clinical care. A framework is needed that allows consideration of structural factors beyond race, ethnicity, and socioeconomic status and accounts for how individuals and communities are impacted by decisions made at the local, state, and policy levels by government, industry, and health care professionals. Pellow's Environmental Justice Framework accounts for the complexity of relationships and decisions that impact PFAS exposure and associated health outcomes (Pellow, 2004) (see Figure 1-4). Sze and London (2008) advance this argument by incorporating negotiation at the stakeholder level, as well as problem identification at the level of sociohistorical and structural factors, and then solution-oriented approaches reflecting considerations of sustainability and safety.

FIGURE 1-4 Pellow's Environmental Justice Framework.

---

[7] See https://www.epa.gov/rmp/forms/vulnerable-zone-indicator-system (accessed June 16, 2022).

A first consideration in addressing environmental justice with respect to PFAS is understanding the sociohistorical context rather than a particular discrete event that has contributed to the presence of PFAS in certain communities. For example, PFAS contamination did not just randomly occur in rural communities serviced by well water that happened to be near industrial sites. Rather, the locating of certain industrial sites and decisions to dispose of PFAS with limited regard for the surrounding community's access to safe water are rooted in the relationship and history of these industries and communities. The disposal of these chemicals did not occur as single events lacking context, but reflected a pattern of decisions made over time. Understanding the historical and social context influencing how and where PFAS are distributed is an essential part of identifying effective mitigation strategies.

Second, Pellow's Environmental Justice Framework emphasizes the complex roles of the stakeholders involved. In the case of PFAS, environmental inequality can affect many different stakeholders, ranging from industry, to workers, to community members and organizations, to government entities such as local departments of public health. The complexity of these roles means there can exist synergistic and contradictory allegiances that impact what is valued, how resources are accessed and distributed, who has power, and what is considered profitable. With PFAS, industry's power and profit motive may trump a community's access to information and ability to test for and reduce exposure. Likewise, key health information for clinicians is governed by boards that decide about the content of medical education and training. Clinicians' lack of knowledge about environmental exposures such as PFAS, particularly in highly contaminated communities, can be particularly deleterious for the least-advantaged stakeholder groups.

A third consideration is the effects of social inequalities on exposure to PFAS, as relevant social stratification that affects exposure is not always clear or easy to study. Rural and urban disparities, for example, can be analyzed using National Health and Nutrition Examination Study (NHANES) data (with appropriate clearance), and living in a rural versus an urban area can definitely affect PFAS exposure, often through exposure to well water that is common in rural areas. Other, less obviously measurable influences can also drive exposures, however. Food insecurity, for example, even if only temporary, increases subsistence fishing (Quimby et al., 2020), which may cause people to fish for food in contaminated lakes or rivers. These structural and social factors put people at risk of exposure and can contribute to inequality. In short, even though relatively affluent areas can be heavily contaminated with PFAS, and non-Hispanic Whites are exposed to some of the highest PFAS concentrations (see below), solving the larger problem of PFAS contamination will still require attention to persistent social inequities. Otherwise, action to mitigate PFAS exposure could aggravate existing (or create new) environmental health disparities.

Fourth is the need to consider the role of agency—specifically, the power a community has to advocate for itself in conflicts. PFAS provide a unique example. PFAS contamination rose to national prominence partly because of exposures among well-educated, high-income, and mainly White communities. With PFAS, as with many social problems, power matters. At the same time, PFAS exposure varies regionally. In some parts of the country, the highest exposures are borne by less advantaged communities, making their relative ability to advocate for testing and responsive health care or to demand effective mitigation from their government somewhat tenuous.

While research on disparities in PFAS exposure is limited, several studies suggest that no differences exist by race, ethnicity, or socioeconomic status (Buekers et al., 2018; Nguyen et al., 2020; Sagiv et al., 2015). If one looks at time-aggregated estimates of PFAS exposure for the U.S. population, non-Hispanic Whites, older adults, and people of higher socioeconomic status have higher concentrations of these chemicals (see Figure 1-5). However, time-aggregated exposures present an incomplete picture. Looking at PFAS exposure by year stratified by race orethnicity suggests that in 2000, before the voluntary phase-out of PFOS andrelated chemicals, those in the racial group other non-Hispanic had the highest exposure to PFOS and PFHxS (see Figure 1-6). The implication of this finding is that widespread policy changes—such as the phase out of certain types of PFAS—greatly reduced exposure disparities. Because there are no national estimates of biomonitoring data for replacement PFAS, however, it is unknown whether the disparities observed with PFOS, PFHxS, and PFNA exist today for other PFAS.

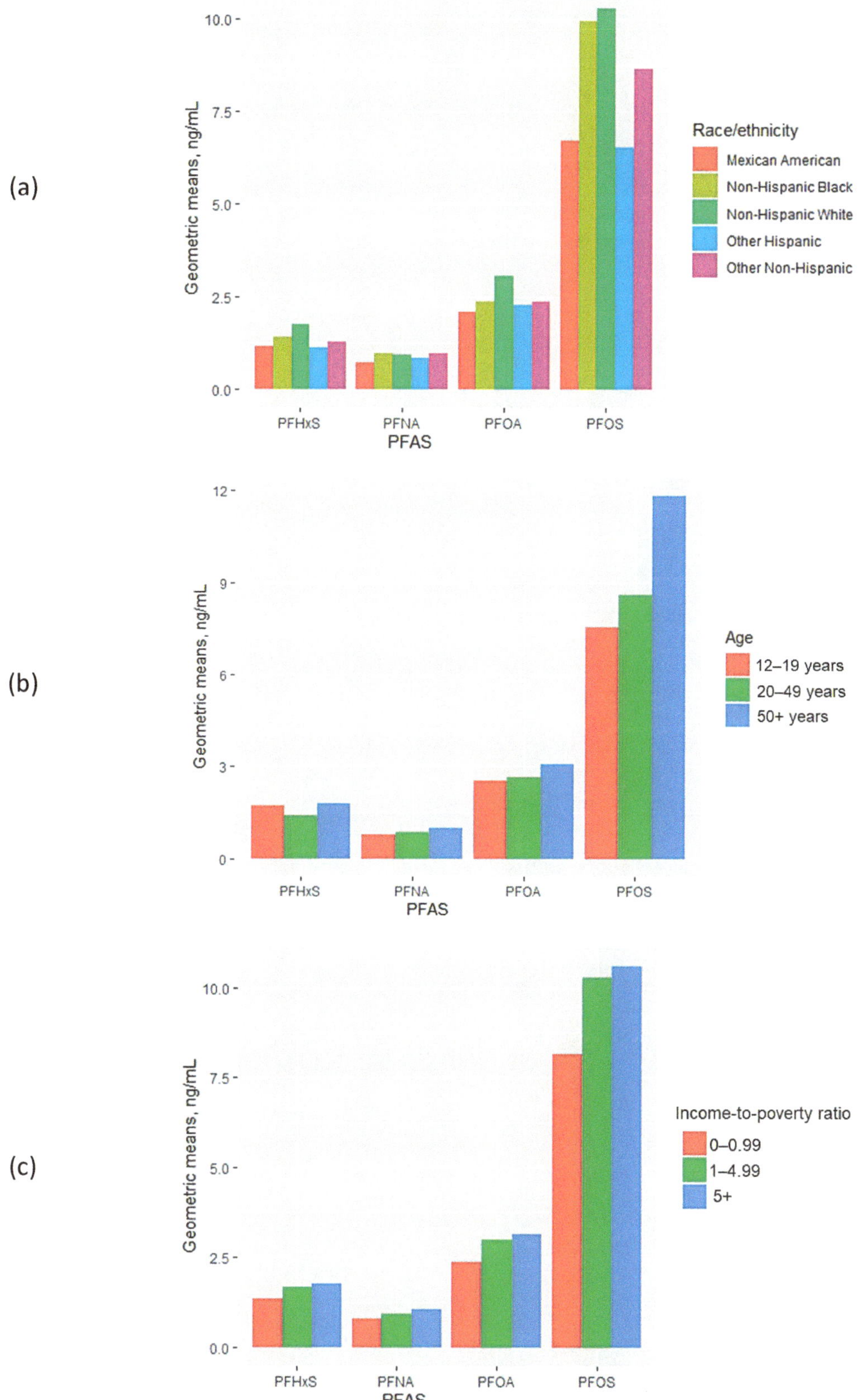

FIGURE 1-5 Serum PFAS concentrations (unadjusted geometric means) from the National Health and Nutrition Examination Survey, 1999–2016, by race/ethnicity (panel a), age (panel b), and income-to-poverty ratio (panel c).

   Guidance on PFAS Exposure, Testing, and Clinical Follow-Up

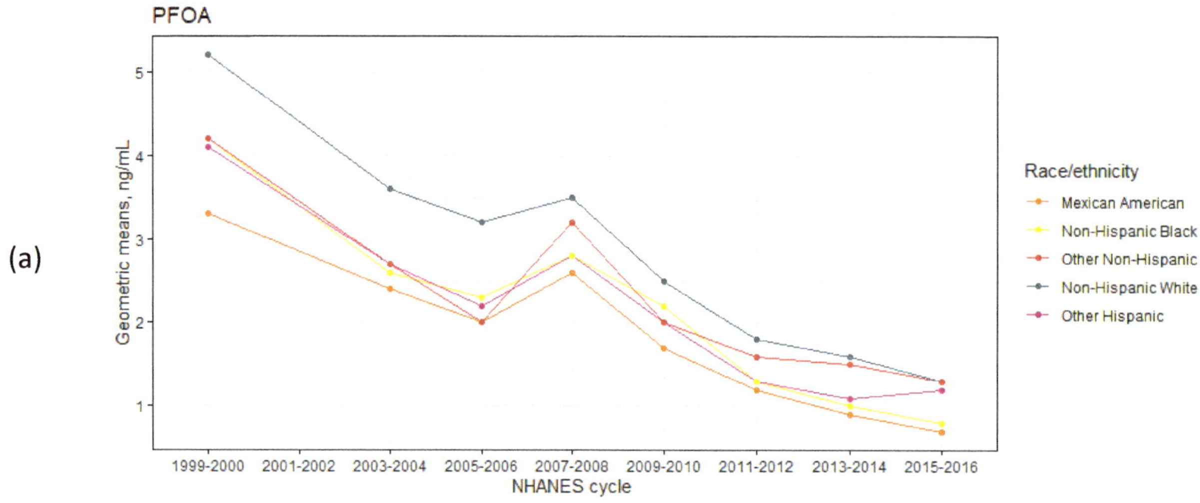

(a)

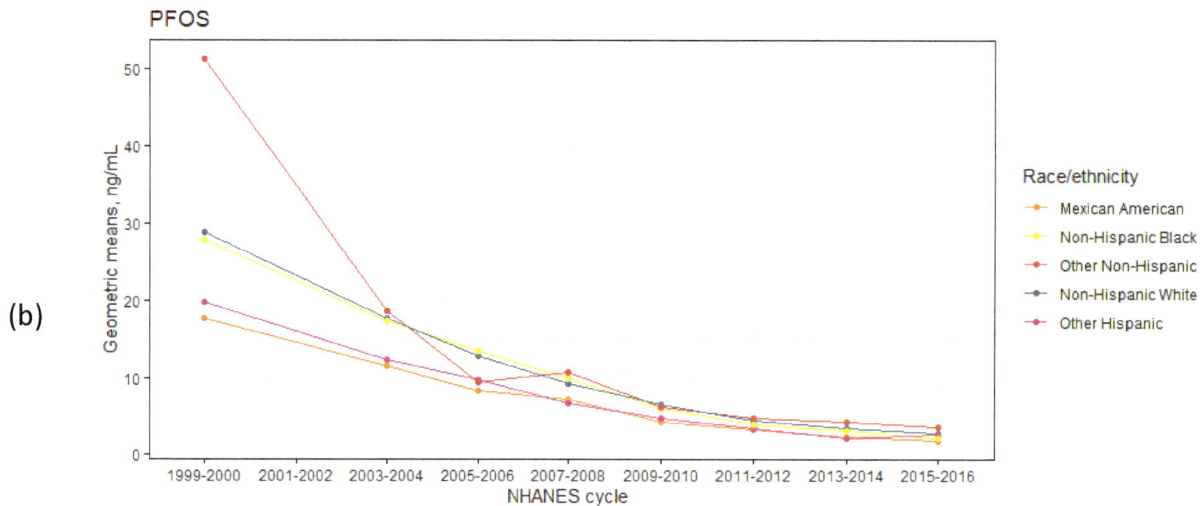

(b)

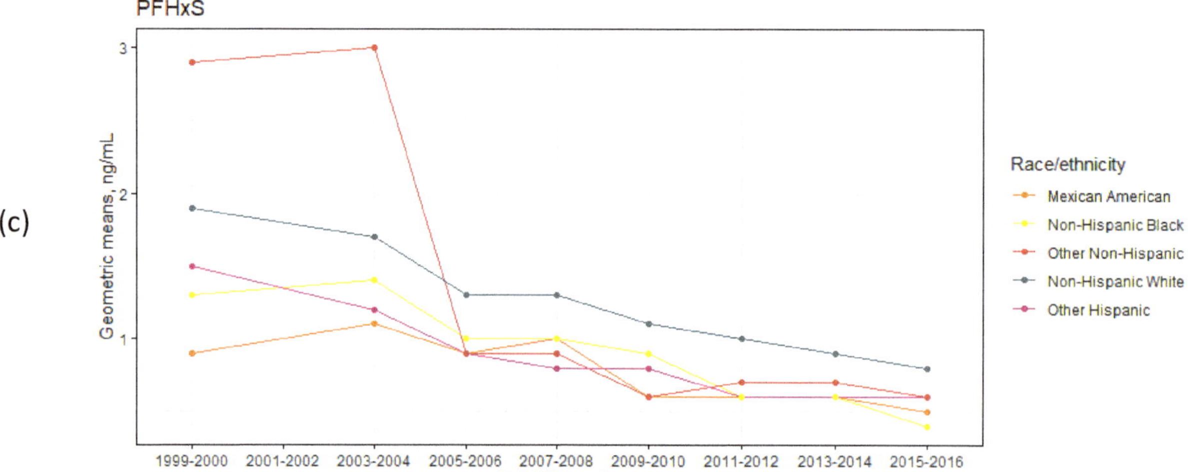

(c)

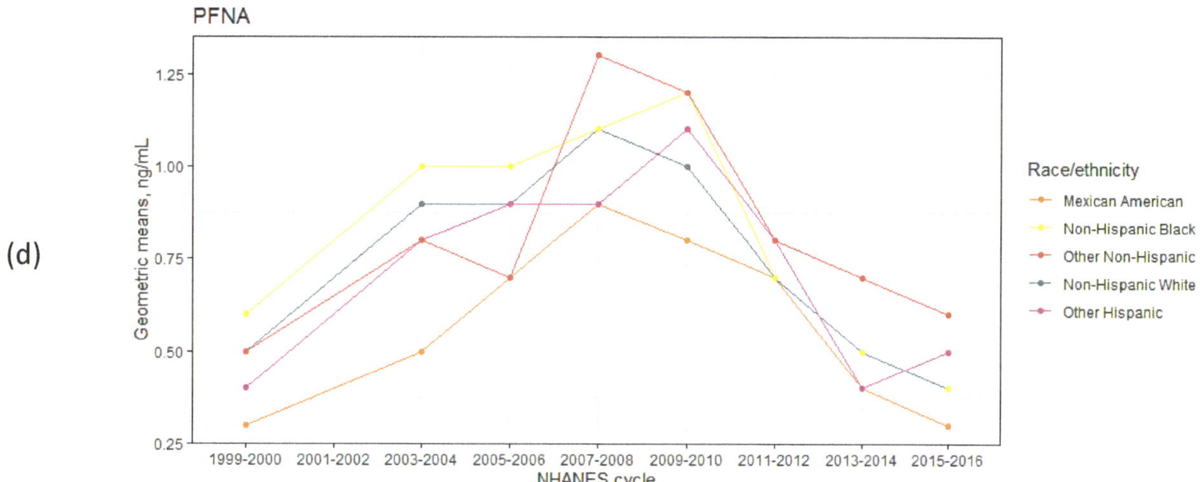

FIGURE 1-6 Serum PFAS concentrations (unadjusted geometric means) from the National Health and Nutrition Examination Survey, 1999–2016, by race/ethnicity, for PFOA (panel a), PFOS (panel b), PFHxS (panel c), and PFNA (panel d).

PFAS CONTAMINATION AND ROUTES OF EXPOSURE

PFAS contamination is global. PFAS have been detected in regions with little human activity, including the atmosphere of remote locations (Shoeib et al., 2010), the Arctic and Antarctic seas (Armitage et al., 2006), and remote soils of every continent (Rankin et al., 2016). Environmental contamination with PFAS occurs in countless ways. Fluorochemical manufacturing sites release PFAS into water as well as into the air, from which they can settle into both soil and water. At military sites and commercial airports PFAS-containing foams are used in training exercises for firefighting. PFAS also can leak from landfills where PFAS-containing wastes are disposed, and can be released into the environment by PFAS-containing wastewater from wastewater treatment plants (Evich et al., 2022; Gomez et al., 2021; Sunderland et al., 2019) (see Figure 1-7). PFAS-treated consumer products have been implicated as sources of exposure in indoor settings (Harrad et al., 2010), and fluorotelomer products used to treat a wide range of consumer household and occupational products have been shown to degrade to form PFAS (Washington and Jenkins, 2015).

PFAS in the environment can contaminate drinking water when the chemicals reach public drinking water systems and private wells (Hu et al., 2016). PFAS can bioaccumulate in fish, shellfish, livestock, dairy, and game animals that contact them through contaminated food or water (De Silva et al., 2021; Death et al., 2021; Domingo and Nadal, 2017). Produce also can be contaminated if it is grown with contaminated drinking water or PFAS-contaminated compost or biosolids (Blaine et al., 2013; Scher et al., 2018).

Certain occupations may lead to increased PFAS exposures. They include, for example, jobs in facilities used to manufacture fluorochemicals or to produce PFAS-containing products, such as textiles or food contact materials. Other jobs with a known increased risk of exposure to PFAS include electroplating, painting, carpet installation and treatment, and jobs that require prolonged work with ski wax; increased exposures also occur among military and civilian firefighters who use PFAS-containing foams in training exercises and wear PFAS-impregnated gear (ATSDR, 2021). In addition, food workers and others in the hospitality industry may have elevated exposure if they handle PFAS-containing food packaging as part of their job (Carnero et al., 2021; Curtzwiler et al., 2021; Schaider et al., 2017).

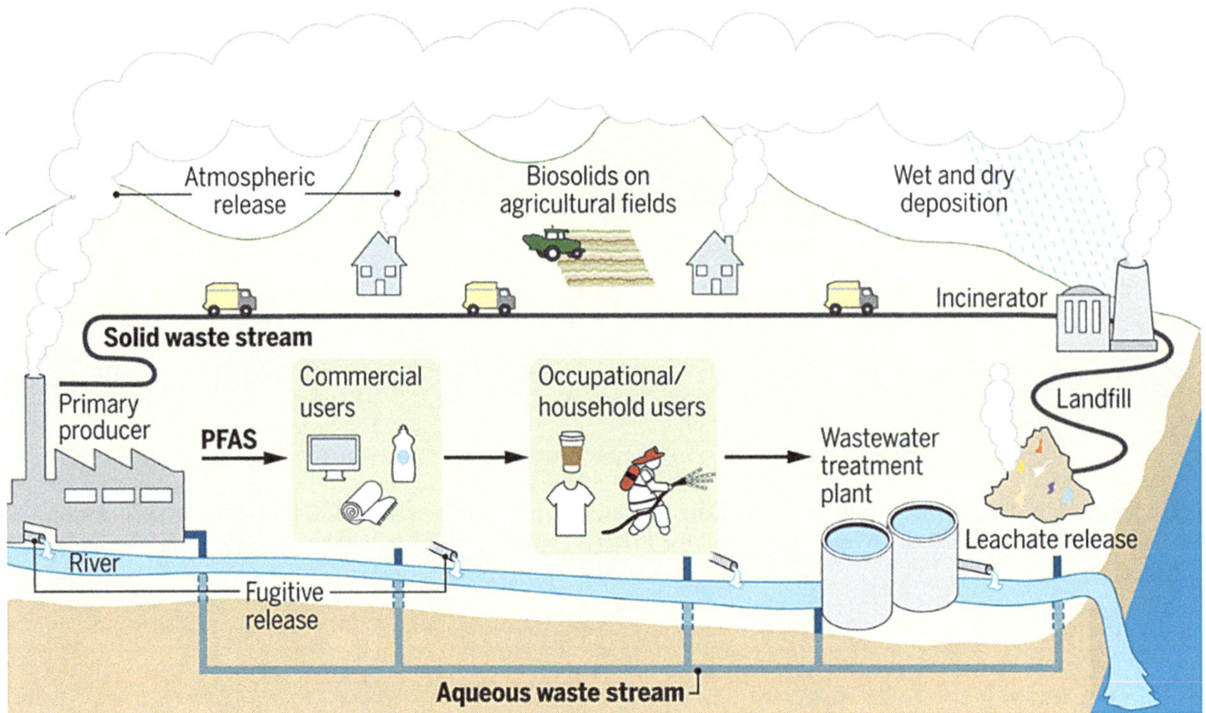

FIGURE 1-7 Examples of how PFAS enter the environment.
SOURCE: Evich et al., 2022.

Ingestion is the most well-studied route of exposure to PFAS in nonoccupational settings (Trudel et al., 2008). PFAS can be ingested by drinking contaminated water or by eating contaminated seafood or other contaminated foods, such as vegetables, game, or dairy products (Bao et al., 2019, 2020; Death et al., 2021; Domingo and Nadal, 2017; Herzke et al., 2013; Li et al., 2019). PFAS are often used in cookware and food contact materials, such as microwave popcorn bags and packaging of fast foods and processed foods (Carnero et al., 2021; Curtzwiler et al., 2021; Schaider et al., 2017). Exposure can also occur through accidental ingestion of PFAS-containing dusts (ATSDR, 2021[8]). PFAS cross the placenta, and PFAS from the mother's body burden can be passed on to her developing fetus (Gao et al., 2019; Manzano-Salgado et al., 2015); PFAS can also pass from mother to child through breast milk (Serrano et al., 2021; Zheng et al., 2021).

Inhalation and transdermal absorption are less well studied. Inhalation is well documented in occupational settings that use aerosolized PFAS (Gilliland, 1992). Volatile PFAS have been detected indoors (Fromme et al., 2015; Morales-McDevitt et al., 2021) and have been associated with persistent PFAS in serum (Fraser et al., 2013). Inhalation near factory emissions and incinerators contributes to exposures in nearby communities (ATSDR, 2021[9]). Inhalation from showering in contaminated water is an active area of research, but there are as yet no data formally evaluating this route.

PFAS are used in thousands of products (e.g., water- and stain-resistant clothing) and personal care products (e.g., sunscreen, makeup, dental floss). They also are used in such products as paint, textiles, firefighting foam, electroplating materials, ammunition, climbing ropes, guitar strings, artificial turf, and soil remediation substances (Glüge et al., 2020). The extent to which the use of such products contributes to human exposures remains unclear, however, because the relative contribution of PFAS exposures from sources other than food or water is not well characterized (DeLuca et al., 2021).

---

[8] This text was altered to correct the references after release of a pre-publication version of the report.

[9] This text was altered to correct the references after release of a pre-publication version of the report.

POLICIES THAT COULD REDUCE EXPOSURE TO PFAS

The EPA's major policy levers for reducing harmful exposures to PFAS are the Toxic Substances Control Act (TSCA), which limits chemicals in commerce; the Safe Drinking Water Act; and the Comprehensive Environmental Response, Compensation, and Liability Act (CERCLA, commonly known as the "Superfund"), which designates hazardous sites for cleanup.

In 2000, the 3M Company (Maplewood, Minnesota), the primary U.S. manufacturer of PFOS, began a voluntary phase-out of perfluorooctanyl chemicals and related precursors, including PFOS, PFHxS, and PFOA[8] (ITRC, 2017). Between 2002 and 2013, the EPA issued four Significant New Use Rules under the TSCA that prevented others from manufacturing the chemicals 3M had voluntarily phased out (Richter et al., 2021) and required notifying the agency prior to any manufacture, use, or import of 271 other PFAS.[9] In 2006, the EPA encouraged eight leading PFAS manufacturers to join a global stewardship program focused on phasing out and ending the production of PFOA. At the same time, the TSCA allows some industries (including photographic imaging, semiconductors, etching, metal plating, and aviation) to continue using banned PFAS at low levels, and these industries are exempted from the provision requiring that the EPA be notified of any new use of banned PFAS (ITRC, 2017). Furthermore, these bans apply only to a handful of chemicals, most of which could easily be substituted for by other PFAS. And other PFAS not yet banned remain in use by industry while undergoing further biologic profiling and toxicity evaluation (Houck et al., 2021).

If PFAS were designated as hazardous substances, CERCLA could be used to lower exposures from contaminated sites, giving the EPA more authority to investigate and remediate those waste sites. The agency has not yet designated PFOS or PFAS as hazardous, however. And the Safe Drinking Water Act would be more protective if the standards were enforceable.[10]

Policy changes that could reduce PFAS exposures may be forthcoming. Recently, the EPA established an EPA Council on PFAS charged with building on the agency's ongoing work to understand better and ultimately reduce the potential risks posed by these chemicals. A PFAS Strategic Roadmap also was released, committing the agency to action. In addition, the EPA released for review by a science advisory board draft documents that will inform an enforceable National Primary Drinking Water Regulation for PFOA and PFOS.[11]

The EPA is not alone in its lack of action. The U.S. Food and Drug Administration (FDA) has delayed limiting PFAS in bottled water,[12] but has revoked regulations that allowed for long-chain PFAS (such as PFOA and PFOS) in food packaging in 2016,[13] and reached an agreement to phase out PFAS substances containing 6:2 fluorotelomer alcohol (Hahn, 2020). The Occupational Safety and Health Administration (OSHA) has set no occupational exposure limits.

The lack of federal standards has left states and communities responsible for creating policies to reduce exposures to these chemicals in the interest of protecting public health (Brennan et al., 2021) There is no consistency in these policies, and not all states have the authority to set standards more protective than those of the federal government (ECOS, 2020). Furthermore, leaving policy making to

---

[8] See https://archive.epa.gov/epapages/newsroom_archive/newsreleases/33aa946e6cb11f35852568e10052 46b4.html (accessed June 16, 2022).

[9] See https://www.epa.gov/assessing-and-managing-chemicals-under-tsca/risk-management-and-polyfluoroalkyl-substances-pfas (accessed June 16, 2022).

[10] See https://www.epa.gov/pfas/epa-actions-address-pfas (accessed June 16, 2022).

[11] See https://www.epa.gov/pfas/pfas-strategic-roadmap-epas-commitments-action-2021-2024 (accessed June 16, 2022).

[12] See https://www.consumerreports.org/bottled-water/fda-delays-setting-limits-on-pfas-in-bottled-water-a8292013869 (accessed June 30, 2022).

[13] See https://www.fda.gov/food/chemical-contaminants-food/authorized-uses-pfas-food-contact-applications (accessed June 16, 2022).

state governments encourages a confusing patchwork of rules and advice. Health inequities thrive in such chaotic circumstances.

PROVIDING CLINICAL ADVICE IN COMMUNITIES EXPOSED TO PFAS

To help clinicians[14] respond to patient concerns about PFAS exposure, the Agency for Toxic Substances and Disease Registry (ATSDR) published *PFAS: An Overview of the Science and Guidance for Clinicians on Per- and Polyfluoroalkyl Substances.*[15] This guidance summarizes general information about PFAS and PFAS health studies and suggests answers to example patient questions. However, the ATSDR guidance does not provide specific recommendations on when to test for PFAS, how to order the tests, how to interpret the results, or what clinical follow-up based on PFAS exposure might look like. Interpretation of PFAS blood or urine results is challenging because the specific level of exposure at which harms may occur is unknown, and the science on the potential health effects of exposure to PFAS is advancing quickly, making it difficult to advise clinicians about what follow-up tests might be helpful.

ATSDR has the legal authority to issue guidance to clinicians in its "Criteria for Determining the Appropriateness of a Medical Monitoring Program under Comprehensive Environmental Response, Compensation, and Liability Act (CERCLA),"[16] promulgated in 1995 (42 USC 9604[i][9]). Under the criteria, the purpose of medical monitoring is to detect individuals with exposures to hazardous substances and refer them to medical care for further evaluation and treatment. The criteria recommends actions that clinicians can take, such as early detection, treatment, or other interventions that interrupt "the progress to symptomatic disease, improve the prognosis of disease, improve the quality of life of the affected individual, or address diseases that are amenable to primary prevention" (42 USC 9604[i][14], p. 38842). Medical monitoring under the criteria is not a research mechanism, and suggests that other epidemiologic studies be carried out to further investigate the cause–effect relationship between exposures and health outcomes. ATSDR's authority to issue guidance to clinicians comes from a different section in the same law (42 USC 9604[i][14]), which grants the agency the authority to develop education materials on medical surveillance, screening, and methods of diagnosing and treating injury or disease related to exposure to hazardous substances.

The first well-documented case of community water contamination with PFAS resulted in recommendations for medical monitoring (see Box 1-2). Releases of PFOA-contaminated drinking water from the DuPont Washington Works facility near Parkersburg, West Virginia, were detected in the local area and a few nearby communities across the Ohio River. A class action lawsuit was filed—*Jack W. Leach et al. v. E.I. du Pont de Nemours & Company* (no. 01-C-608 W.Va., Wood County Circuit Court, filed April 10, 2002). The lawsuit resulted in a medical monitoring program that included suggested screenings for high cholesterol, thyroid disease, ulcerative colitis, testicular cancer, kidney cancer, pregnancy-induced hypertension, and blood testing for PFOA (Frisbee et al., 2009). The medical monitoring guidance generated by this high-profile case has led some community members to believe that the ATSDR guidance should provide recommendations for clinical follow-up, and advocates in many PFAS-exposed communities are working to establish medical monitoring programs (Rizzuto, 2021).

The absence of recommendations on testing and follow-up in ATSDR's PFAS clinical guidance is aligned with standard medical practice, as clinicians are expected to order tests only when how to interpret and act on the results is known. In addition, reporting of levels of PFAS in blood and urine to patients raises ethical questions, such as whether to report results in the absence of established health guidelines.

---

[14] The committee uses "clinician" throughout this report to refer to "a healthcare professional qualified in the clinical practice of medicine. Clinicians may be physicians, nurses, pharmacists, or other allied health professionals" as defined by the Centers for Medicare & Medicaid Services (https://www.cms.gov/Medicare/Quality-Initiatives-Patient-Assessment-Instruments/MMS/QMY-Clinicians [accessed June 16, 2022], para 2).

[15] See https://www.atsdr.cdc.gov/pfas/docs/clinical-guidance-12-20-2019.pdf (accessed June 16, 2022).

[16] See https://www.govinfo.gov/content/pkg/FR-1995-07-28/pdf/FR-1995-07-28.pdf (accessed June 16, 2022).

BOX 1-2
The C-8 Science Panel and the C-8 Medical Panel

Background: In the early 2000s, contamination of drinking water with PFOA (also known as C-8 for the eight carbons in its chemical structure) was discovered in six water districts in two states near the DuPont Washington Works facility near Parkersburg, West Virginia. As a result, a class action lawsuit, *Jack W. Leach et al. v. E.I. du Pont de Nemours & Company* (no. 01-C-608 W.Va., Wood County Circuit Court, filed April 10, 2002), was filed. The lawsuit resulted in the formation of the C-8 Science Panel to determine "probable link conditions" associated with PFOA-contaminated drinking water. The settlement agreement defined "probable link" to mean that, based on the weight of the available scientific evidence, it is more likely than not that there is a link between exposure to PFOA and a particular human disease (Frisbee et al., 2009). The settlement also called for the formation of a separate medical panel "to develop general guidelines for medical monitoring related to the human diseases for which the science panel delivered a probable link finding and is different from what would otherwise be prescribed" if exposure to PFOA were absent.

Science Panel Findings: The science panel identified pregnancy-induced hypertension (including preeclampsia), kidney cancer, testicular cancer, thyroid disease, ulcerative colitis, and high cholesterol (hypercholesterolemia) as probable linked conditions (C-8 Medical Panel, 2013).

Medical Panel Findings: The medical panel recommended screenings for high cholesterol, thyroid disease, ulcerative colitis, testicular cancer, kidney cancer, pregnancy-induced hypertension, and blood testing for PFOA (C-8 Medical Panel, 2013).

At a workshop on PFAS exposure that occurred more than 1 year before the present study began, however, Andrea Amico, a member of a PFAS-impacted community in Pease, New Hampshire, traveled to Washington, DC, to represent her community's views on PFAS testing. She said that people in her community

> were exposed … without their consent, and now they have to fight tooth and nail to get a blood test result to know how much exposure they had? It just seems incredibly wrong.… We don't have all the answers yet, but not testing them is not the right answer.

Amico also argued for more access to testing despite the knowledge gaps regarding the interpretation of test results. She believes PFAS exposure testing would allow people to compare their levels with those of others in highly exposed communities, which could help them understand their potential health risks (NASEM, 2020). As it stands now, clinicians in communities with known PFAS contamination are left to make their own decisions about whether to test patients for PFAS exposure, which PFAS to test for, how to interpret the results, what health effects are associated with PFAS, and how to determine appropriate follow-up care for exposed patients. Pressed for time in clinic visits, clinicians may dismiss interest in PFAS testing or avoid recommending specific medical follow-up, and they may not even be clear on when or how to order a test.

COMMITTEE'S TASK AND APPROACH

Prompted by the tension between people in PFAS-exposed communities wanting preventive care for their exposure and clinicians not knowing what care should be provided, ATSDR and the National Institute of Environmental Health Sciences (NIEHS) asked the National Academies of Sciences, Engineering, and Medicine to form an ad hoc committee to provide advice on clinical care for patients exposed to PFAS. The committee included experts in epidemiology, toxicology, preventive medicine, pediatrics, nursing, public health, environmental medicine, philosophy, ethics, exposure science, and risk communication (see Appendix A for biographical information on the committee members). The committee's Statement of Task is provided in Box 1-3.

BOX 1-3
Statement of Task

An ad hoc committee appointed by the National Academies of Sciences, Engineering, and Medicine (the National Academies) will consider current evidence regarding human health effects of the most widely studied per- and polyfluoroalkyl substances (PFAS). The National Academies will provide the Centers for Disease Control and Prevention and the Agency for Toxic Substances and Disease Registry (CDC/ATSDR) and the National Institute of Environmental Health Sciences (NIEHS) an objective and authoritative review of current evidence regarding human health effects of those PFAS being monitored in the CDC's National Report on Human Exposure to Environmental Chemicals. The National Academies will also provide recommendations regarding potential changes to CDC/ATSDR PFAS clinical guidance including:

- Options and considerations to guide decision making for PFAS testing in a patient's blood or urine.
- PFAS concentrations that could inform clinical care of exposed patients.
- Appropriate patient follow-up and care specific to PFAS-associated health endpoints for those patients known or suspected to be exposed to PFAS.

This information will be used to inform how communities and individuals exposed to PFAS could be best served by clinicians. Specifically, the committee will undertake the following tasks:

1. Assess the strength of evidence for the spectrum of putative health effects suggested by human studies (including immune response, lipid metabolism, kidney function, thyroid disease, liver disease, glycemic parameters and diabetes, cancer, and fetal and child development) to establish a basis for prioritized clinical surveillance or monitoring of PFAS health effects. This assessment should characterize the likelihood of those health effects occurring (qualitative probability) given real world human exposures and identify the human populations at most risk (consider life stage, health status, exposure level). Data/evidence gaps that contribute to uncertainty about health effects of most concern should be annotated.
2. Develop general principles for clinical evaluation or biological testing given substantial scientific uncertainty about health effects or the value of such measures in informing care. These principles should address reasons for testing (e.g., opportunities to reduce morbidity and mortality), when to test, who to test, how to test, what to test for, risks of testing, and the related social and ethical implications of testing.
3. Review current knowledge about the contribution of PFAS exposure sources (i.e., drinking water, diet, the indoor environment, etc.) to human exposure and develop principles clinicians can use to advise patients on exposure reduction.
4. Advise whether changes to current CDC/ATSDR clinical guidance/recommendations on PFAS blood or urine testing are needed given the committee's general principles and assessment of the associations between PFAS exposure and clinically relevant health outcomes. Ultimately, the goal is to provide guidance on how clinicians can advise patients on PFAS testing and health outcomes that may be associated with PFAS as well as what to advise patients regarding standard medical or preventive care and exposure reduction.
5. Outline a process by which the CDC/ATSDR PFAS clinical guidance can be effectively reviewed and revised over the next decade.

Figure 1-8 provides an overview of the steps in the committee's approach to its charge: problem formulation and engagement with communities, development of principles for decision making under substantial scientific uncertainty, determination of strategies for reducing exposure to PFAS, development of advice for PFAS testing and levels of PFAS in blood or urine that could inform clinical care of exposed patients, determination of the health effects of PFAS, appropriate follow-up care of exposed patients, and advice on whether changes are needed with respect to ATSDR's PFAS clinical guidance.

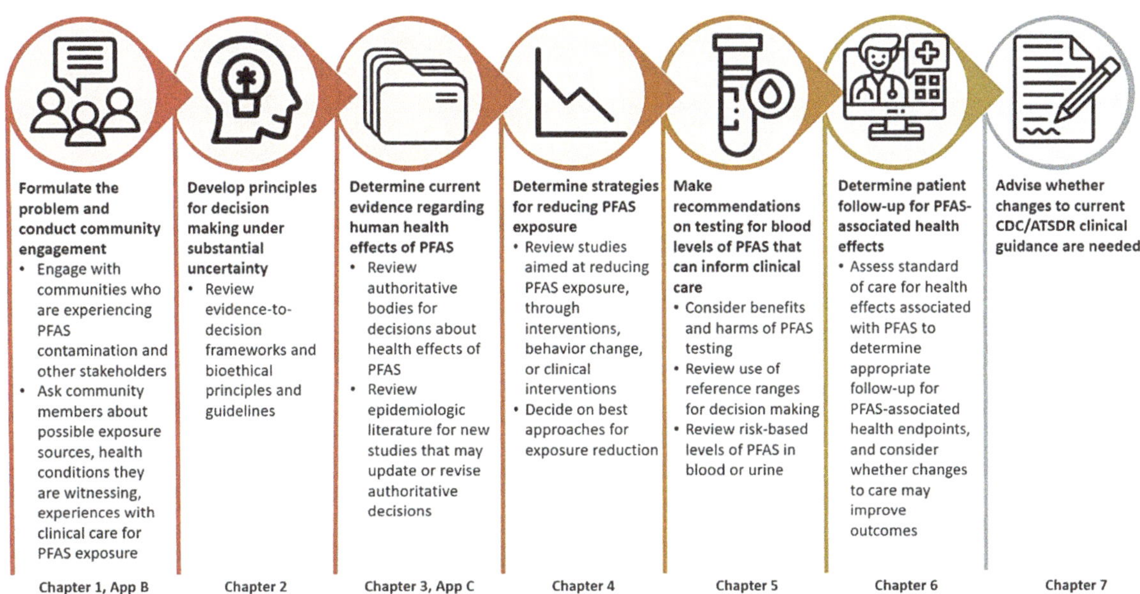

FIGURE 1-8 The committee's approach to the Statement of Task and the chapters and appendixes where the topics are discussed.
NOTE: ATSDR = Agency of Toxic Substances and Disease Registry; CDC = Centers for Disease Control and Prevention; PFAS = per- and polyfluoroalkyl substances.

Problem Formulation and Community Engagement

The committee held six public meetings between February 4 and August 12, 2021 (see Appendix C for the agendas of these public meetings). Leading up to its first meeting, the committee began to receive public input on the study. The testimony provided at the first public meeting revealed that many members of the public feel ignored or dismissed by clinicians and the medical community, and have assumed the role of community scientists, becoming experts in their communities' illnesses and experiences. People who live with potentially harmful exposures can have knowledge that is inaccessible to scientists as their experiential learning provides essential context for scientific findings (Brown, 1992). To incorporate this knowledge into the study, the committee, National Academies staff, and the study sponsors developed a community engagement plan. To aid in this process, the committee considered the Jemez Principles for Democratic Organizing, a set of recommendations developed to serve as a foundation on which diverse coalitions can together make justice-based decisions specifically weighing on inclusivity, as well as letting people speak for themselves (Solís and SPWU, 1997). Ultimately, the committee's community engagement plan consisted of several key elements (see Figure 1-9):

- a panel of community liaisons;
- three 2021 town halls (April 7, Eastern Town Hall; May 6, Middle Town Hall; May 25, Western Town Hall);
- community speakers from PFAS-impacted communities at every public meeting;
- open sign-up for public testimony to the committee at every public meeting; and
- encouragement for providing written testimony to the committee throughout the data-collection phase of the study.

FIGURE 1-9 The committee's approach to community engagement.

The community liaisons were solicited through a nomination process similar to that used for National Academies committees. The community liaisons were intended to be an inclusive group; all nominees who appeared to be genuinely interested or had experience working with PFAS or similar environmental contamination issues were asked to join the group. National Academies staff held several meetings with the liaisons throughout the study process. The first call with the liaisons clearly described the mission of the National Academies, the goals of the study, what the study would not address, and the study timeline.

The role of the community liaisons was to provide input to the committee on behalf of PFAS-impacted communities. The liaisons aided the study process by suggesting speakers, topics, and discussion questions for the public meetings, and by answering questions to inform the report or the committee's work. They also prepared letters to the committee providing consensus opinions related to the Statement of Task and suggested reviewers for the report for consideration by National Academies staff.

Health Effects Associated with PFAS Exposure

The committee carefully considered the purpose of the literature review of putative health effects prescribed in its Statement of Task: to *establish a basis for prioritized clinical surveillance or monitoring of PFAS health effects.* Accordingly, the committee's review was focused on determining a set of health effects that may be associated with PFAS, which could then be used for preventive medicine recommendations and decisions. The Statement of Task limited the review to those PFAS included in the Centers for Disease Control and Prevention's (CDC's) *National Report on Human Exposure to Environmental Chemicals* (see Table 1-1).

Although the committee believes that studies on health effects in humans will most likely be limited to those PFAS in Table 1-1 because they are the most commonly studied in humans, other PFAS may cause harm because of similarities in biological persistence and toxicities (Kwiatkowski et al., 2020). Moreover, while different PFAS have distinct physical, chemical, and toxicological properties, people are exposed to more than a single PFAS. As a result, exposures are often to mixtures of PFAS such that specific effects are difficult to disentangle. Considering these issues and recognizing that some PFAS are less frequently measured than others, the committee ultimately decided to provide one strength-of-evidence determination for all PFAS for each health effect. Further description of the methods used for the committee's review is provided in Appendix D.

TABLE 1-1 PFAS Species Currently Included in the Centers for Disease Control and Prevention's *National Report on Human Exposure to Environmental Chemicals*

Abbreviated Name	Full Name	CAS Registry No.
MeFOSAA	Methylperfluorooctane sulfonamidoacetic acid	2355-31-9
PFHxS	Perfluorohexanesulfonic acid perfluorohexane sulfonic acid	355-46-4
n-PFOA (linear isomer), Sb-PFOA (branched isomers)	Perfluorooctanoic acid	335-67-1*
PFDA	Perfluorodecanoic acid	335-76-2
PFUnDA	Perfluoroundecanoic acid	2058-94-8
n-PFOS (linear isomer), Sm-PFOS (branched isomers)	Perfluorooctanesulfonic acid	1763-23-1*
PFNA	Perfluorononanoic acid	375-95-1

NOTES: CAS = Chemical Abstracts Service. * = CAS number refers to linear isomer only. Previous survey years have also included perfluorobutane sulfonic acid (PFBS), perfluoropentanoic acid (PFpA), perfluorododecanoic acid (PFDoDA), perfluorooctane sulfonamide (FOSA), and 2-(N-ethyl-perfluorooctane sulfonamido)acetate (EtFOSAA), according to Patrick N. Breysse's presentation to the committee on February 4, 2021.

Reducing Exposure to PFAS

This report does not review EPA policy actions, which are beyond the scope of the committee's task. A review of individual-level exposure reduction strategies is presented in Appendix E. Chapter 4 provides an overview of that literature and medical interventions to reduce PFAS body burden. The report reviews these individual-level strategies because the task calls for them; however, the committee does not believe that people should be responsible for protecting themselves from harmful chemicals. Well-established exposure prevention frameworks, such as the hierarchy of controls and the health impact pyramid, dictate that individuals should not be responsible for making such decisions (Frieden, 2010). Systems approaches to exposure reduction, such as setting an enforceable water standard, regulating PFAS from nonessential uses, and cleaning up hazardous waste sites, will be far more effective at reducing population-level exposures (see Figure 1-10).

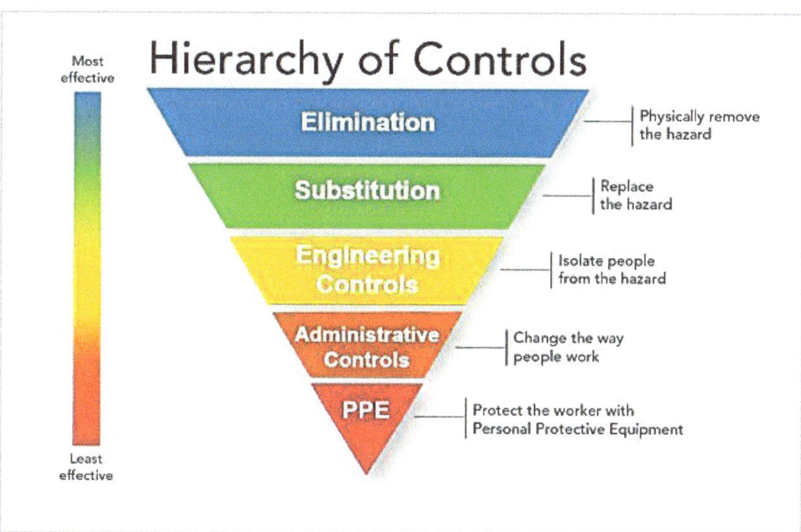

FIGURE 1-10 Hierarchy of controls.
SOURCE: See https://www.cdc.gov/niosh/topics/hierarchy/default.html (accessed June 16, 2022).

Advice on PFAS Testing and Levels That Could Inform Clinical Care

The committee's approach to providing advice on PFAS testing and clinical care included a review of the harms and benefits of testing. To determine the levels of PFAS exposure that could inform clinical care of exposed patients, the committee did not complete dose-response modeling to determine a tolerable risk level. Instead, the committee reviewed various strategies for interpreting PFAS concentrations, such as use of reference ranges for comparison, and levels of risk determined by other agencies.

Follow-Up Care for Patients Known or Suspected to Be Exposed to PFAS

The committee considered relevant for patient follow-up recommendations for each condition that its review found had suggestive or limited suggestive evidence of an association with PFAS exposure. The committee believed that while it was important to categorize the strength of the evidence, all conditions with an association should be considered for patient follow-up since acknowledging the potential risk may make clinicians and patients more likely to prioritize screenings.

ORGANIZATION OF THE REPORT

This report is organized into eight chapters and five appendixes. Chapter 2 presents the committee's general principles for decision making under substantial scientific uncertainty. Chapter 3 presents the committee's assessment of the health effects of PFAS. Chapter 4 reviews strategies for reducing exposure to PFAS. Chapter 5 provides recommendations for PFAS testing and concentrations in blood or urine that could inform clinical care. Chapter 6 presents the committee's specific recommendations for patient follow-up. Chapter 7 addresses suggested changes to the ATSDR PFAS clinical guidance. Finally, Chapter 8 considers the implementation of the committee's recommendations. Appendix A provides biographical information on the committee members, National Academies staff, and the community liaisons; Appendix B provides a summary of the town halls; Appendix C contains the agendas of the committee's public meetings; Appendix D is a summary of the methods used in the committee's literature review; and Appendix E is a white paper describing a review of the PFAS personal intervention literature.

REFERENCES

Armitage, J., I. T. Cousins, R. C. Buck, K. Prevedouros, M. H. Russell, M. MacLeod, and S. H. Korzeniowski. 2006. Modeling global-scale fate and transport of perfluorooctanoate emitted from direct sources. *Environmental Science & Technology* 40(22):6969–6975. https://pubs.acs.org/doi/10.1021/es0614870.
ATSDR (Agency for Toxic Substances and Disease Registry). 2021. *Toxicological profile for perfluoroalkyls*. Atlanta, GA: U.S. Department of Health and Human Services. https://www.atsdr.cdc.gov/toxprofiles/tp200.pdf (accessed July 1, 2022).
Bao, J., W. J. Yu, Y. Liu, X. Wang, Y. H. Jin, and G. H. Dong. 2019. Perfluoroalkyl substances in groundwater and home-produced vegetables and eggs around a fluorochemical industrial park in China. *Ecotoxicology and Environmental Safety* 171:199. https://doi.org/10.1016/j.ecoenv.2018.12.086.
Bao, J., C. L. Li, Y. Liu, X. Wang, W. J. Yu, Z. Q. Liu, L. X. Shao, and Y. H. Jin. 2020. Bioaccumulation of perfluoroalkyl substances in greenhouse vegetables with long-term groundwater irrigation near fluorochemical plants in Fuxin, China. *Environmental Research* 188:109751. https://doi.org/10.1016/j.envres.2020.109751.

Blaine, A. C., C. D. Rich, L. S. Hundal, C. Lau, M. A. Mills, K. M. Harris, and C. P. Higgins. 2013. Uptake of perfluoroalkyl acids into edible crops via land applied biosolids: Field and greenhouse studies. *Environmental Science and Technology* 47:14062. https://doi.org/10.1021/es403094q.

Blum, A., S. Balan, M. Scheringer, X. Trier, G. Goldenman, I. Cousins, M. Diamond, T. Fletcher, C. Higgins, and A. Lindeman. 2015. The Madrid statement on poly- and perfluoroalkyl substances (PFASs). *Environmental Health Perspectives* 123(5):A107–A111. https://doi.org/10.1289/ehp.1509934.

Brennan, N. M., A. T. Evans, M. K. Fritz, S. A. Peak, and H. E. von Holst. 2021. Trends in the regulation of per- and polyfluoroalkyl substances (PFAS): A scoping review. *International Journal of Environmental Research and Public Health* 18(20):10900. https://doi.org/10.3390/ijerph182010900.

Brown, P. 1992. Popular epidemiology and toxic waste contamination: Lay and professional ways of knowing. *Journal of Health and Social Behavior* 33(3):267–281. https://doi.org/10.2307/213 7356.

Buck, R. C., J. Franklin, U. Berger, J. M. Conder, I. T. Cousins, P. D. Voogt, A. A. Jensen, K. Kannan, S. A. Mabury, and S. P. J. van Leeuwen. 2011. Perfluoroalkyl and polyfluoroalkyl substances in the environment: Terminology, classification, and origins. *Integrated Environmental Assessment and Management* 7(4):513–541. https://doi.org/10.1002/ieam.258.

Buekers, J., A. Colles, C. Cornelis, B. Morrens, E. Govarts, and G. Schoeters. 2018. Socio-economic status and health: Evaluation of human biomonitored chemical exposure to per- and polyfluorinated substances across status. *International Journal of Environmental Research and Public Health* 15(12):2818. https://doi.org/10.3390/ijerph15122818.

Bullard, R. D. 1996. Environmental justice: It's more than waste facility siting. *Social Science Quarterly* 77(3):493–499. https://www.jstor.org/stable/42863495.

C-8 Medical Panel. 2013. *C-8 Medical Panel Report.* https://www.hpcbd.com/wp-content/uploads/sites/ 1601601/2021/01/Medical-Panel-Report-2013-05-24.pdf (accessed March 24, 2021).

Carnero, A. R., A. Lestido-Cardama, P. V. Loureiro, L. Barbosa-Pereira, A. R. B. de Quirós, and R. Sendón. 2021. Presence of perfluoroalkyl and polyfluoroalkyl substances (PFAS) in food contact materials (FCM) and its migration to food. *Foods* 10(7). https://doi.org/10.3390/foods10071443.

Curtzwiler, G. W., P. Silva, A. Hall, A. Ivey, and K. Vorst. 2021. Significance of perfluoroalkyl substances (PFAS) in food packaging. *Integrated Environmental Assessment and Management* 17(1):7–12. https://doi.org/10.1002/ieam.4346.

De Silva, A. O., J. M. Armitage, T. A. Bruton, C. Dassuncao, W. Heiger-Bernays, X. C. Hu, A. Kärrman, B. Kelly, C. Ng, A. Robuck, M. Sun, T. F. Webster, and E. M. Sunderland. 2021. PFAS exposure pathways for humans and wildlife: A synthesis of current knowledge and key gaps in understanding. *Environmental Toxicology and Chemistry* 40(3):631–657. https://doi.org/10.1002/ etc.4935.

Death, C., C. Bell, D. Champness, C. Milne, S. Reichman, and T. Hagen. 2021. Per- and polyfluoroalkyl substances (PFAS) in livestock and game species: A review. *Science of the Total Environment* 774:144795. https://doi.org/10.1016/j.scitotenv.2020.144795.

DeLuca, N. M., M. Angrish, A. Wilkins, K. Thayer, and E. A. Cohen Hubal. 2021. Human exposure pathways to poly- and perfluoroalkyl substances (PFAS) from indoor media: A systematic review protocol. *Environment International* 146:106308. https://doi.org/10.1016/j.envint.2020.106308.

Domingo, J. L., and M. Nadal. 2017. Per- and polyfluoroalkyl substances (PFASs) in food and human dietary intake: A review of the recent scientific literature. *Journal of Agricultural and Food Chemistry* 65(3):533–543. https://doi.org/10.1021/acs.jafc.6b04683.

ECOS (Environmental Council of the States). 2020. *Processes and considerations for setting state PFAS standards.* https://www.ecos.org/wp-content/uploads/2020/02/Standards-White-Paper-FINAL-February-2020.pdf (accessed March 25, 2020).

EPA (U.S. Environmental Protection Agency). 2020. *EPA CompTox chemicals dashboard.* https://comptox.epa.gov/dashboard/chemical-lists/pfasmaster (accessed June 16, 2022).

Evich, M. G., M. J. B. Davis, J. P. McCord, B. Acrey, J. A. Awkerman, D. R. U. Knappe, A. B. Lindstrom, T. F. Speth, C. Tebes-Stevens, M. J. Strynar, Z. Wang, E. J. Weber, W. M. Henderson, and J. W. Washington. 2022. Per- and polyfluoroalkyl substances in the environment. *Science* 375(6580):eabg9065. https://doi.org/10.1126/science.abg9065.

Fenton, S. E., A. Ducatman, A. Boobis, J. C. DeWitt, C. Lau, C. Ng, J. S. Smith, and S. M. Roberts. 2021. Per- and polyfluoroalkyl substance toxicity and human health review: Current state of knowledge and strategies for informing future research. *Environmental Toxicology and Chemistry* 40(3):606–630. https://doi.org/10.1002/etc.4890.

Fraser, A. J., T. F. Webster, D. J. Watkins, M. J. Strynar, K. Kato, A. M. Calafat, V. M. Vieira, and M. D. McClean. 2013. Polyfluorinated compounds in dust from homes, offices, and vehicles as predictors of concentrations in office workers' serum. *Environment International* 60:128–136. https://doi.org/10.1016/j.envint.2013.08.012.

Frieden, T. R. 2010. A framework for public health action: The health impact pyramid. *American Journal of Public Health* 100(4):590–595. https://doi.org/10.2105/AJPH.2009.185652.

Frisbee, S. J., A. P. Brooks, Jr., A. Maher, P. Flensborg, S. Arnold, T. Fletcher, K. Steenland, A. Shankar, S. S. Knox, C. Pollard, J. A. Halverson, V. M. Vieira, C. Jin, K. M. Leyden, and A. M. Ducatman. 2009. The C-8 health project: Design, methods, and participants. *Environmental Health Perspectives* 117(12):1873–1882. https://doi.org/10.1289/ehp.0800379.

Fromme, H., S. Dietrich, L. Fembacher, T. Lahrz, and W. Völkel. 2015. Neutral polyfluorinated compounds in indoor air in Germany—The LUPE 4 study. *Chemosphere* 139:572–578. https://doi.org/10.1016/j.chemosphere.2015.07.024.

Gao, K., T. Zhuang, X. Liu, J. Fu, J. Zhang, J. Fu, L. Wang, A. Zhang, Y. Liang, M. Song, and G. Jiang. 2019. Prenatal exposure to per- and polyfluoroalkyl substances (PFASs) and association between the placental transfer efficiencies and dissociation constant of serum proteins-PFAS complexes. *Environmental Science & Technology* 53(1):6529–6538. https://doi.org/10.1021/acs.est.9b00715.

Gilliland, F. D. 1992. *Fluorocarbons and human health: Studies in an occupational cohort.* Thesis. University of Minnesota. https://static.ewg.org/reports/2019/pfa-timeline/1992_1-46-Gilliland-Thesis.pdf (accessed July 1, 2022).

Glüge, J., M. Scheringer, I. T. Cousins, J. C. DeWitt, G. Goldenman, D. Herzke, R. Lohmann, C. A. Ng, X. Trier, and Z. Wang. 2020. An overview of the uses of per- and polyfluoroalkyl substances (PFAS). *Environmental Science: Processes & Impacts* 22(12):2345–2373.

Gochfeld, M., and J. Burger. 2011. Disproportionate exposures in environmental justice and other populations: The importance of outliers. *American Journal of Public Health* 101(Suppl 1):S53–S63. https://doi.org/10.2105/AJPH.2011.300121.

Gomez, J. A., D. Raynes, T. Doriss, L. Andersen, A. C. Capaccio, C. K. Gilbert, R. P. Johnson, and D. C. Royer. 2021. *Man-made chemicals and potential health risks: EPA has completed some regulatory related actions for PFAS.* Washington, DC: Government Accountability Office. https://www.gao.gov/assets/gao-21-37.pdf (accessed July 1, 2022).

Guelfo, J. L., S. Korzeniowski, M. A. Mills, J. Anderson, R. H. Anderson, J. A. Arblaster, J. M. Conder, I. T. Cousins, K. Dasu, and B. J. Henry. 2021. Environmental sources, chemistry, fate, and transport of per- and polyfluoroalkyl substances: State of the science, key knowledge gaps, and recommendations presented at the August 2019 SETAC Focus Topic Meeting. *Environmental Toxicology and Chemistry* 40(12):3234–3260.

Hahn, S. M. 2020. *FDA announces voluntary agreement with manufacturers to phase out certain short-chain PFAS used in food packaging.* Silver Spring, MD: U.S. Food and Drug Administration. https://www.fda.gov/news-events/press-announcements/fda-announces-voluntary-agreement-manufacturers-phase-out-certain-short-chain-pfas-used-food (accessed June 30, 2022).

Harclerode, M., S. Baryluk, H. Lanza, and J. Frangos. 2021. Preparing for effective, adaptive risk communication about per- and polyfluoroalkyl substances in drinking water. *American Water Works Association Water Science* 3(5):e1236. https://doi.org/10.1002/aws2.1236.

Harrad, S., C. A. de Wit, M. Abou-Elwafa Abdallah, C. Bergh, J. A. Bjorklund, A. Covaci, P. P. O. Darnerud, J. de Boer, M. Diamond, and S. Huber. 2010. Indoor contamination with hexabromocyclododecanes, polybrominated diphenyl ethers, and perfluoroalkyl compounds: An important exposure pathway for people? *Environmental Science & Technology* 44(9):3221–3231.

Herzke, D., S. Huber, L. Bervoets, W. D'Hollander, J. Hajslova, J. Pulkrabova, G. Brambilla, S. P. De Filippis, S. Klenow, G. Heinemeyer, and P. De Voogt. 2013. Perfluorinated alkylated substances in vegetables collected in four european countries: Occurrence and human exposure estimations. *Environmental Science and Pollution Research* 20:7930.

Houck, K. A., G. Patlewicz, A. M. Richard, A. J. Williams, M. A. Shobair, M. Smeltz, M. S. Clifton, B. Wetmore, A. Medvedev, and S. Makarov. 2021. Bioactivity profiling of per- and polyfluoroalkyl substances (PFAS) identifies potential toxicity pathways related to molecular structure. *Toxicology* 457:152789.

Hu, X. C., D. Q. Andrews, A. B. Lindstrom, T. A. Bruton, L. A. Schaider, P. Grandjean, R. Lohmann, C. C. Carignan, A. Blum, S. A. Balan, C. P. Higgins, and E. M. Sunderland. 2016. Detection of poly- and perfluoroalkyl substances (PFASs) in U.S. drinking water linked to industrial sites, military fire training areas, and wastewater treatment plants. *Environmental Science and Technology Letters* 3(10):344–350. https://doi.org/10.1021/acs.estlett.6b00260.

ITRC (Interstate Technology and Research Council). 2017. *History and use of per- and polyfluoroalkyl substances (PFAS)*. https://pfas-1.itrcweb.org/wp-content/uploads/2017/11/pfas_fact_sheet_history_and_use__11_13_17.pdf (accessed March 25, 2020).

ITRC. 2021. *Per- and polyfluoroalkyl substances technical and regulatory guidance*. https://pfas-1.itrcweb.org/wp-content/uploads/2022/03/PFAS-Full-PDF-December-2021-Update.pdf (accessed May 25, 2022).

Kay, J., and C. Katz. 2012. Pollution, poverty and people of color: Living with industry. *Scientific American*. http://www.scientificamerican.com/article/pollution-poverty-people-color-living-industry (accessed July 1, 2022).

Kwiatkowski, C. F., D. Q. Andrews, L. S. Birnbaum, T. A. Bruton, J. C. DeWitt, D. R. U. Knappe, M. V. Maffini, M. F. Miller, K. E. Pelch, A. Reade, A. Soehl, X. Trier, M. Venier, C. C. Wagner, Z. Wang, and A. Blum. 2020. Scientific basis for managing PFAS as a chemical class. *Environmental Science & Technology Letters* 7(8):532–543. https://doi.org/10.1021/acs.estlett.0c00255.

Li, P., X. Oyang, Y. Zhao, T. Tu, X. Tian, L. Li, Y. Zhao, J. Li, and Z. Xiao. 2019. Occurrence of perfluorinated compounds in agricultural environment, vegetables, and fruits in regions influenced by a fluorine-chemical industrial park in China. *Chemosphere* 225:659.

Lohmann, R., I. T. Cousins, J. C. DeWitt, J. Glüge, G. Goldenman, D. Herzke, A.B. Lindstrom, M. F. Miller, C. A. Ng, S. Patton, M. Scheringer, X. Trier, and Z. Wang. 2020. Are fluoropolymers really of low concern for human and environmental health and separate from other PFAS? *Environmental Science and Technology* 54(20):12820–12828. https://doi.org/10.1021/acs.est.0c03244.

Manzano-Salgado, C. B., M. Casas, M. J. Lopez-Espinosa, F. Ballester, M. Basterrechea, J. O. Grimalt, A. M. Jiménez, T. Kraus, T. Schettgen, J. Sunyer, and M. Vrijheid. 2015. Transfer of perfluoroalkyl substances from mother to fetus in a Spanish birth cohort. *Environmental Research* 142:471–478. https://doi.org/10.1016/j.envres.2015.07.020.

Mohai, P., D. Pellow, and J. Timmons Roberts. 2009. Environmental justice. *Annual Review of Environment and Resources* 34(1):405–430. https://doi.org/10.1146/annurev-environ-082508-094348.

Morales-McDevitt, M. E., J. Becanova, A. Blum, T. A. Bruton, S. Vojta, M. Woodward, and R. Lohmann. 2021. The air that we breathe: Neutral and volatile PFAS in indoor air. *Environmental Science & Technology Letters* 8(10):897–902. https://doi.org/10.1021/acs.estlett.1c00481.

NASEM (National Academies of Sciences, Engineering, and Medicine). 2020. *Understanding, controlling, and preventing exposure to PFAS: Proceedings of a workshop—in brief.* Washington, DC: The National Academies Press. https://doi.org/10.17226/25856.

Nguyen, V. K., A. Kahana, J. Heidt, K. Polemi, J. Kvasnicka, O. Jolliet, and J. A. Colacino. 2020. A comprehensive analysis of racial disparities in chemical biomarker concentrations in United States women, 1999–2014. *Environment International* 137:105496. https://doi.org/10.1016/j.envint.2020.105496.

OECD (Organisation for Economic Co-operation and Development). 2021. *Reconciling terminology of the universe of per- and polyfluoroalkyl substances: Recommendations and practical guidance.* Series on Risk Management, no. 61. Paris: OECD. https://www.oecd.org/officialdocuments/publicdisplaydocumentpdf/?cote=ENV/CBC/MONO(2021)25&docLanguage=En.

Orum, P., R. Moore, M. Roberts, and J. Sanchez. 2014, May. *Who's in danger? Race, poverty, and chemical disasters.* Brattleboro, VT: Environmental Justice and Health Alliance for Chemical Policy Reform. https://comingcleaninc.org/assets/media/images/Reports/Who%27s%20in%20Danger%20Report%20FINAL.pdf (accessed July 1, 2022).

Pellow, D. 2004. The politics of illegal dumping: An environmental justice framework. *Qualitative Sociology* 27(4):511–525. https://doi.org/10.1023/B:QUAS.0000049245.55208.4b.

Quimby, B., S. E. S. Crook, K. M. Miller, J. Ruiz, and D. Lopez-Carr. 2020. Identifying, defining and exploring angling as urban subsistence: Pier fishing in Santa Barbara, California. *Marine Policy* 121:104197. https://doi.org/10.1016/j.marpol.2020.104197.

Rankin, K., S. A. Mabury, T. M. Jenkins, and J. W. Washington. 2016. A North American and global survey of perfluoroalkyl substances in surface soils: Distribution patterns and mode of occurrence. *Chemosphere* 161:333. https://doi.org/10.1016/j.chemosphere.2016.06.109.

Raponi, I., P. Brown, and A. Cordner. 2021. Improved medical screening in PFAS-impacted communities to identify early disease. *Environmental Health News.* https://www.ehn.org/pfas-testing-2653577444.html (accessed November 22, 2021).

Richter, L., A. Cordner, and P. Brown. 2021. Producing ignorance through regulatory structure: The case of per- and polyfluoroalkyl substances (PFAS). *Sociological Perspectives* 64(4):631–656.

Rizzuto, P. 2021. The doctor will see the PFAS-exposed plaintiff now. *Bloomberg Law: Energy and Environment Report.* https://news.bloomberglaw.com/environment-and-energy/the-doctor-will-see-the-pfas-exposed-plaintiff-now (accessed November 22, 2021).

Sagiv, S. K., S. L. Rifas-Shiman, T. F. Webster, A. M. Mora, M. H. Harris, A. M. Calafat, X. Ye, M. W. Gillman, and E. Oken. 2015. Sociodemographic and perinatal predictors of early pregnancy per- and polyfluoroalkyl substance (PFAS) concentrations. *Environmental Science & Technology* 49(19):11849–11858.

Schaider, L. A., S. A. Balan, A. Blum, D. Q. Andrews, M. J. Strynar, M. E. Dickinson, D. M. Lunderberg, J. R. Lang, and G. F. Peaslee. 2017. Fluorinated compounds in U.S. fast food packaging. *Environmental Science & Technology Letters* 4(3):105–111. https://doi.org/10.1021/acs.estlett.6b00435.

Scher, D. P., J. E. Kelly, C. A. Huset, K. M. Barry, R. W. Hoffbeck, V. L. Yingling, and R. B. Messing. 2018. Occurrence of perfluoroalkyl substances (PFAS) in garden produce at homes with a history of PFAS-contaminated drinking water. *Chemosphere* 196:548.

Serrano, L., L. M. Iribarne-Durán, B. Suárez, F. Artacho-Cordón, F. Vela-Soria, M. Peña-Caballero, J. A. Hurtado, N. Olea, M. F. Fernández, and C. Freire. 2021. Concentrations of perfluoroalkyl substances in donor breast milk in Southern Spain and their potential determinants. *International Journal of Hygiene and Environmental Health* 236:113796. https://doi.org/10.1016/j.ijheh.2021.113796.

Shoeib, M., P. Vlahos, T. Harner, A. Peters, M. Graustein, and J. Narayan. 2010. Survey of polyfluorinated chemicals (PFCs) in the atmosphere over the northeast Atlantic Ocean. *Atmospheric Environment* 44(24):2887–2893.

Solís, R., and SPWU (Southwest Public Workers Union). 1997, April. Jemez principles for democratic organizing. *SouthWest Organizing Project.* https://nation.mothersoutfront.org/jemez_principles_for_democratic_organizing (accessed July 1, 2022).

Steenland, K., T. Fletcher, C. R. Stein, S. M. Bartell, L. Darrow, M. J. Lopez-Espinosa, P. Barry Ryan, and D. A. Savitz. 2020. Review: Evolution of evidence on PFOA and health following the assessments of the C-8 Science Panel. *Environment International* 145. https://doi.org/10.1016/j.envint.2020.106125.

Sunderland, E. M., X. C. Hu, C. Dassuncao, A. K. Tokranov, C. C. Wagner, and J. G. Allen. 2019. A review of the pathways of human exposure to poly- and perfluoroalkyl substances (PFASs) and present understanding of health effects. *Journal of Exposure Science & Environmental Epidemiology* 29:131.

Sze, J., and J. K. London. 2008. Environmental justice at the crossroads. *Sociology Compass* 2(4):1331–1354.

Trudel, D., L. Horowitz, M. Wormuth, M. Scheringer, I. T. Cousins, and K. Hungerbühler. 2008. Estimating consumer exposure to PFOS and PFOA. *Risk Analysis* 28(2):251–269. https://doi.org/10.1111/j.1539-6924.2008.01017.x.

Washington, J. W., and T. M. Jenkins. 2015. Abiotic hydrolysis of fluorotelomer-based polymers as a source of perfluorocarboxylates at the global scale. *Environmental Science & Technology* 49(24):14129–14135. https://doi.org/10.1021/acs.est.5b03686.

Wones, R., S. M. Pinney, J. M. Buckholz, C. Deck-Tebbe, R. Freyberg, and A. Pesce. 2009. Medical monitoring: A beneficial remedy for residents living near an environmental hazard site. *Journal of Occupational and Environmental Medicine/American College of Occupational and Environmental Medicine* 51(12):1374.

Zheng, G., E. Schreder, J. C. Dempsey, N. Uding, V. Chu, G. Andres, S. Sathyanarayana, and A. Salamova. 2021. Per- and polyfluoroalkyl substances (PFAS) in breast milk: Concerning trends for current-use PFAS. *Environmental Science & Technology* 55(11):7510–7520.

2

Principles for Decision Making Under Uncertainty

Given that there is some degree of uncertainty in all science, science-based decision making also involves uncertainty (Fischhoff and Davis, 2014). This uncertainty can arise from many different factors, including lack of available evidence, statistical variability, model uncertainty, and "deep" uncertainty about the fundamental scientific processes relevant to a decision (IOM, 2013). Yet, despite this uncertainty, when evidence exists, it helps predict what may happen when a decision is made (Fischhoff and Davis, 2014). Values also influence how scientific evidence is gathered and interpreted, thereby influencing how uncertainty is characterized and what decisions are made on the basis of that particular interpretation of the evidence (Douglas, 2009; Elliott, 2017; IOM et al., 1995). Thus, it is important to consider uncertainty in the broader context of the decision to be made and the values underlying the decision-making process (IOM, 2013).

In response to uncertainty, some public health professionals may wish to avoid acting as soon as science determines the risks of an exposure for fear that the association is not a true one and that acting too quickly could lead to inappropriate and costly public health measures (Boffetta et al., 2008). On the other hand, some may fear that if society waits for more certain evidence, meaningful action will be delayed (Blair et al., 2009). As a group of scientists with different backgrounds and perspectives, the committee grappled with both of these potential outcomes of its decisions. The committee established the principles in this chapter to help accomplish its charge, but also believes they may be useful to support the Agency for Toxic Substances and Disease Registry (ATSDR) in making decisions about medical follow-up and advice to communities exposed to per- and polyfluoroalkyl substances (PFAS) or other chemicals with uncertain effects.

DEVELOPMENT OF THE COMMITTEE'S PRINCIPLES

In its effort to identify principles for decision making under uncertainty, the committee solicited input from members of PFAS-impacted communities through its virtual town halls (see Chapter 1). The committee also reviewed decision-making frameworks, including a review paper (Norris et al., 2021), the criteria used by the U.S. Preventive Services Task Force (USPSTF) (Krist et al., 2018; Sawaya et al., 2007), GRADE (Grading of Recommendations Assessment, Development and Evaluation) (Moberg et al., 2018), a National Academies evidence framework for decisions regarding genetic testing (NASEM, 2017), the Evidence and Value: Impact on DEcision-Making (EVIDEM) framework (Goetghebeur et al., 2008), the principles used by the C-8 Medical Panel (C-8 Medical Panel, 2013), the ethical principles from the Belmont Report (HHS, 1979), and principles for precautionary reasoning (Resnik, 2021). A brief overview of these sources is included below.

The existing principles and frameworks the committee reviewed were not entirely satisfactory for the purposes of this study. Most of the existing frameworks were not designed to address situations characterized by limited evidence and substantial scientific uncertainty—the contexts the committee was charged to address. The committee understood that when evidence is incomplete, dangers can be associated both with taking action and with failing to act (Douglas, 2009; Elliott and Richards, 2017). Therefore, the committee adapted the principles and frameworks it reviewed to develop a set of principles appropriate for decision making under substantial scientific uncertainty.

42

PRINCIPLES PUT FORWARD BY THE COMMITTEE

The committee decided to use the ethical principles proposed in the Belmont Report (HHS, 1979) and developed by Beauchamp and Childress (2001) as a foundation for its approach. Although these principles (nonmaleficence, beneficence, autonomy, and justice) were not initially proposed for decision making under uncertainty, they provide an ethical starting point for addressing such situations. In addition, they encompass many of the criteria and concepts included in other frameworks, such as considerations of benefits, harms, health equity, and human rights. The committee combined the principles of nonmaleficence and beneficence into a principle of proportionality—drawing on the Resnik (2021) framework for precautionary reasoning in response to uncertainty. The committee also added two other principles included in other decision-making frameworks but not captured by the principles from the Belmont Report. One is the principle of feasibility, found in several of the frameworks reviewed by Norris and colleagues (2021), including GRADE. It captures the importance of considering the capabilities of the current medical system and the ways it might need to develop to respond to decision needs. The other is the principle of adaptability, which builds on the call for review and revision in the National Academies framework for genetic testing and the emphasis in the USPSTF framework on revising recommendations in response to emerging scientific information. Box 2-1 briefly describes the committee's resulting five principles for exposure biomonitoring and patient follow-up under substantial scientific uncertainty about the health effects of PFAS exposure. Each of these principles is discussed in turn below.

Proportionality

The committee's first principle, proportionality, is adapted from Resnik's recommendations for precautionary decision making under uncertainty (Resnik, 2021, p. 81):

> **Proportionality:** Decisions should balance plausible harms and benefits proportionally, meaning that the harms and benefits are considered together and weighed based on a qualitative assessment of their potential likelihoods and magnitudes.

BOX 2-1
Principles for Decision Making Under Uncertainty Used in This Report

Proportionality: Decisions should balance plausible harms and benefits proportionally, meaning that the harms and benefits are considered together and weighed based on a qualitative assessment of their potential likelihoods and magnitudes.

Justice: Decisions should be informed by an emphasis on promoting justice, including by balancing benefits and harms fairly across the population of at-risk individuals, advancing health equity, and respecting human rights. In addition, justice requires consideration of sociohistorical context, stakeholders, existing structural inequalities, and issues of agency (the power a community has to advocate for itself in conflicts).

Autonomy: Decisions should be based on informed decision making by individuals and reflect respect for their values.

Feasibility: Decisions should take into account resource availability, including follow-up services.

Adaptability: Decisions should respond to new information about harms, benefits, and other relevant considerations (e.g., health equity and feasibility).

This principle draws on a common theme that cuts across most frameworks the committee considered: balancing harms and benefits. For example, the USPSTF states that in determining the benefits of a preventive services framework, one should consider evidence about the accuracy of screening, harms of early-intervention treatment and not treating, and the treatment benefits and harms of a particular preventive service. The USPSTF framework also considers the accuracy of a screening test and its relationship to the clinical health effect, and the harms associated with the screening for the health condition and treatment of the condition.

The principle of proportionality addresses harms and benefits in situations of substantial scientific uncertainty. To balance harms and benefits proportionally is to consider them together such that, all else being equal, policies with fewer or less severe harms (assessed in terms of their likelihood and magnitude) can be justified based on lesser benefits relative to policies with increased harms. This principle is particularly relevant for decision making under uncertainty because it can justify taking actions to realize potentially significant benefits even when the evidence for those benefits is limited, as long as the potential harms associated with the actions are minimal.

Proportionality provides an alternative to probability-based decision principles that require numerical ranking of the probabilities or utilities associated with effects. Although applying the principle of proportionality calls for considering the likelihood and magnitude of benefits and harms in some form, it does not require assigning numerical rankings to their likelihoods or magnitudes. Because of the qualitative nature of this principle, reasonable decision makers can potentially disagree about how to apply it in specific situations because judgments are required to weigh harms and benefits. Especially when the benefits associated with two policies are similar, and the likelihood and the magnitude of their harms cannot be estimated precisely, the principle of proportionality may not offer decisive guidance. Instead, this principle provides a general way of reasoning compatible with a range of more specific strategies for balancing harms and benefits under different forms of uncertainty (Douglas, 2009; Workman et al., 2020). It affirms that both harms and benefits need to be considered and that they need to be weighed against one another when making decisions (Resnik, 2021).

For this report, the harms of a decision could include physical effects, psychological effects, and opportunity costs (Harris et al., 2014). The committee does not consider the financial costs of paying for tests or treatment a harm. Costs could be considered under the principle of feasibility, but the committee excluded costs from feasibility as well. The reason for excluding costs is that, although a certain test or treatment may be expensive today, costs typically decrease over time if demand for the test or treatment increases. Many other organizations, such as the USPSTF, do not consider financial costs when formulating recommendations based on assessment of health benefits and harms, and the committee followed this approach. The committee did consider as potential harm the loss of income or financial damage incurred as a result of decisions (e.g., decreased property values caused by a community's learning about chemical contamination).

The principle of proportionality calls for considering benefits and harms that are plausible. The concept of plausibility provides a minimal threshold for determining which benefits and harms should even be considered. Resnik (2021, pp. 80–81) defines a scientific statement as plausible "if it is consistent with well-established scientific facts, hypotheses, laws, models, or theories"—a lower standard of evidence than is typically required for even weak scientific confirmation. Instead, this definition is designed to rule out "armchair speculation" about potential harms and benefits. If a harm or benefit is judged to be plausible, it can be considered in the overall assessment of proportionality. Nevertheless, harms and benefits deemed to have a reasonably low level of evidential support might still be given relatively little weight in proportionality assessment unless their potential magnitudes were estimated to be exceptionally high.

In situations of substantial scientific uncertainty, it is generally unrealistic to expect more than a qualitative assessment of the likelihoods and magnitudes of harms and benefits. This assessment may be informed by evidence, including epidemiological studies, toxicological studies (both animal models and mechanistic studies), modeling, and support from analogy. It should be guided by norms for the responsible acquisition and utilization of evidence, knowledge, and expertise. Still, the appropriate standards of evidence for drawing tentative conclusions about potential likelihoods and magnitudes may be much lower than would apply in other scientific contexts (Douglas, 2009; Elliott and Richards, 2017; Ginsberg et al., 2019).

Justice

Justice is the second principle for decision making in situations of substantial scientific uncertainty:

> **Justice:** Decisions should be informed by an emphasis on promoting justice, including by balancing benefits and harms fairly across the population of at-risk individuals, advancing health equity, and respecting human rights. In addition, justice requires consideration of sociohistorical context, stakeholders, existing structural inequalities, and issues of agency (the power a community has to advocate for itself in conflicts).

Justice is central to the Belmont Report and Beauchamp and Childress's framework for biomedical ethics (Beauchamp and Childress, 2001; HHS, 1979), and elements of justice are included in many of the frameworks discussed in the review by Norris and colleagues (2021). Beauchamp and Childress (2001, p. 226) describe justice as "fair, equitable, and appropriate treatment in light of what is due or owed to persons." Thus, an injustice involves "a wrongful act or omission that denies people benefits to which they have a right or distributes burdens unfairly."

In this report, the focus is on three elements of justice: fairness, equity, and human rights. Resnik (2001) emphasizes that decisions under uncertainty should be handled in a manner that is procedurally and distributively fair: the *process* for making decisions should be fair in the sense of incorporating all relevant stakeholders and employing transparent and accountable procedures, and the *endpoints* of the decisions should also be fair.

In the health care setting, the second element of justice—equity—calls for promoting health equity. Health equity is "the state in which everyone has the opportunity to attain full health potential, and no one is disadvantaged from achieving this potential because of social position or any other socially defined circumstance" (NASEM, 2017, p. 32). Promoting health equity is not the same as achieving similar health outcomes for everyone. Instead, promoting health equity in the context of this report means designing approaches that enable people to lead full, healthy lives regardless of their social circumstances, such as race, socioeconomic status, and geographic location. Health equity is a particularly important element of justice because it establishes the positive responsibility to provide the conditions necessary for people to lead healthy lives regardless of their social circumstances; without this positive responsibility, some ethicists have argued that the principle of justice does not provide significant guidance beyond what is already found in other bioethical principles, such as proportionality or autonomy (London, 2022).

Finally, justice involves respect for human rights. The committee's conception of human rights is drawn from the World Health Organization (WHO)-Integrate evidence-to-decision framework (Rehfuess et al., 2019), which emphasizes rights related to the availability and accessibility of health care and such general rights as nondiscrimination.

In cases of environmental pollution such as PFAS contamination, it is crucial to incorporate the concept of environmental justice into efforts to promote fairness, health equity, and human rights. According to the U.S. Environmental Protection Agency (EPA), "Environmental justice is the fair treatment and meaningful involvement of all people regardless of race, color, national origin, or income, concerning the development, implementation, and enforcement of environmental laws, regulations, and policies. This goal will be achieved when everyone enjoys:

- The same degree of protection from environmental and health hazards, and
- Equal access to the decision-making process to have a healthy environment."[1]

---

[1] See https://www.epa.gov/environmentaljustice (accessed September 17, 2021).

The environmental justice movement in the United States arose in recent decades out of the realization that communities of color, which already faced other social disadvantages, were also exposed to disproportionately high levels of environmental pollution (Bullard, 2018; Shrader-Frechette, 2002). Many places where PFAS contamination has been identified are rural and served by private well water, and they face other environmental and social challenges, such as co-occurrences of other contaminants; occupational exposures; and health inequities, such as less access to health care and a lack of economic resources with which to mitigate exposures. These less-advantaged rural communities may also have structural and agency-related factors that can impact their ability to minimize their exposure, as well as to seek and access adequate health care linked to exposures. Promoting environmental justice requires addressing these inequities, such as by taking steps to eliminate disproportionate lack of access to health care, exposure biomonitoring, patient follow-up for PFAS-associated health effects, and environmental remediation and mitigation. Required as well is addressing the needs of vulnerable populations, such as pregnant and nursing women, young children, and those who are immunocompromised.

Autonomy

The committee's third principle is autonomy:

Autonomy: Decisions should be based on informed decision making by individuals and reflect respect for their values.

The Belmont Report emphasizes the importance of respecting individuals' autonomy and ability to make judgments; it also warns against withholding information that would enable them to judge. Beauchamp and Childress (2001) likewise emphasize that respect for autonomy requires eliminating constraints on individuals' decision making and fostering their self-determination. In practice, this principle has often been operationalized by focusing on obtaining individuals' informed consent to medical treatments or research experiments, but its implications are much broader. In the context of this report, the principle of autonomy calls for health care services to provide individuals with information and collaborate with them to facilitate decisions that accord with their values. This approach is fundamental in situations of substantial scientific uncertainty, when difficult decisions need to be made about how to weigh harms and benefits given limited information about their likelihoods and magnitudes. Different individuals can reasonably approach such situations in different ways. Therefore, the principle of autonomy affirms that patients should play a role in making these decisions.

In addition to fostering decisions that accord with patient values, the principle of autonomy supports other steps to empower patients and their communities. To facilitate patients' self-determination, health care providers and researchers should take steps to report information back to patients in ways that facilitate their future informed decision making. When patients are tested for exposure to potentially toxic substances, it is essential to provide them with (Brody et al., 2014; Morello-Frosch et al., 2009)

- information about their exposure level;
- a description of the potential health impacts, as well as the related signs and symptoms;
- how their exposure level compares with those of others in their area and across the country;
- the potential significance of their exposure level; and
- how they can reduce future risk, either of the exposure itself or its potential health impact(s).

The PFAS Research, Education, and Action for Community Health (PFAS-REACH) project is an example of an effort to provide accessible information to community members about how to interpret exposure biomonitoring results and what follow-up activities could be warranted to address PFAS-associated health effects (Boronow et al., 2017). It is also essential to respect the decision of some individuals not to be tested. Some individuals may not want to have information that increases their uncertainty or could lead to future medical monitoring that might not improve their health outcomes.

The principle of autonomy can also guide actions taken at the community level to help foster decision making by individuals. For example, there is growing recognition that community organizations and advocacy groups can generate important information about public health threats and develop actionable solutions (Corburn, 2005; Elliott, 2017; Wandersman, 2003). Thus, health care providers and public health institutions can foster individuals' autonomy by collaborating with these organizations that can help facilitate informed patient decision making.

Feasibility

The committee's fourth principle for making decisions under substantial scientific uncertainty is feasibility:

> **Feasibility:** Decisions should into take account resource availability, including follow-up services.

Feasibility, included in many evidence-to-decision frameworks reviewed by Norris and colleagues (2021), denotes the ability to conduct testing, clinical evaluation, and follow-up activities. An assessment of feasibility encompasses the infrastructure and resources currently available and whether they are sufficient to achieve the goals of a decision. In the context of PFAS exposure biomonitoring, for example, feasibility could include such considerations as the time and knowledge providers have available to facilitate environmental exposure assessments or their access to environmental occupational health physicians. Feasibility assessments may also include consideration of whether an action or policy is sustainable over time and it entails important legal, ethical, or bureaucratic barriers (Moberg et al., 2018). At the same time, however, it is crucial to recognize that feasibility assessments are relative to a particular point in time, and inputs and outputs can change. In some cases, a particular policy may not be feasible immediately because of a lack of resources or other institutional barriers, but those barriers may reflect past or present injustices that need to be remedied. Thus, for example, a feasibility assessment can guide policy makers to allocate additional resources when exposure biomonitoring or follow-up services are likely to have benefits or are needed to promote a just health care system. Therefore, the principle of feasibility needs to be considered alongside the principle of justice so it can guide future investments and institutional changes instead of detracting from efforts to address inequities and promote just outcomes. Unfortunately, the lack of a coordinated health system in the United States complicates any assessment of feasibility that goes beyond the requirements imposed on public health agencies. Individuals must rely on insurance or self-pay for many follow-up services, and access to high-quality primary care and specialty care services can be fraught with challenges.

Adaptability

The fifth and final principle is adaptability:

> **Adaptability:** Decisions should respond to new information about harms, benefits, and other relevant considerations (e.g., health equity and feasibility).

The National Academies evidence framework for decisions regarding genetic testing (NASEM, 2017) emphasizes that adaptability is important in decision contexts in which scientific information changes rapidly, as is the case with genetic testing. The USPSTF also emphasizes that its evaluation of the benefits and harms of preventive services is a process that needs to be revisited in light of new advances in research, testing, and treatment capabilities (Sawaya et al., 2007). This responsiveness to new information is especially important in situations of substantial scientific uncertainty, such as the decision context surrounding PFAS. In such situations, it is often necessary to develop plans for acting even without decisive information about what course of action is best, and the harms associated with failing to act when warranted can often be more severe than the harms associated with taking actions that ultimately turn out to be unwarranted. When actions are taken without compelling information, however, it is crucial to reexamine regularly the state of the evidence and the consequences of the actions taken to reassess the best course of action. The adaptive management movement in environmental policy exemplifies this commitment to an ongoing process of learning and reevaluation. According to a National Research Council report on managing water resources, "Adaptive management promotes flexible decision-making that can be adjusted in the face of uncertainties as endpoints from management actions and other events become better understood" (NRC, 2004, pp. 1–2).

COMMITTEE'S CONSIDERATIONS IN DEVELOPING ITS PRINCIPLES

Among the committee's considerations in developing its principles for decision making under substantial scientific uncertainty were input from contaminated communities and a number of previously published evidence-to-decision frameworks.

Community Input from Town Halls

During its three town halls, the committee obtained perspectives from people living in or working with communities with PFAS contamination during its three Town Halls (see Appendix B). The committee recognizes that the views expressed at the town halls do not necessarily represent the views of all affected communities or all individuals within those communities; nonetheless, they provided input for the committee to consider. Important themes relevant to the development of principles for decision making under substantial scientific uncertainty included the following:

- Community members shared that they had to educate their clinicians about the exposures in their communities and the potential health effects that could result. They stated further that clinicians do not know how to provide advice on patient follow-up following PFAS exposure, whether PFAS exposure biomonitoring should be provided, or how to interpret PFAS exposure biomonitoring results if shared by the patient.
- Community members expressed the view that PFAS exposure biomonitoring should be available through traditional clinical care to all individuals. In addition, given that PFAS are ubiquitous in the environment and that many of the contamination sites have not been identified, exposure biomonitoring should be more widely available.
- Significant harms were not associated with exposure biomonitoring for PFAS. Community members understand that PFAS exposure biomonitoring may not directly inform their health, but they wish to have knowledge of their exposure levels as a first step.
- Speakers stated that PFAS exposure information could inform health care decisions and patient follow-up related to potential adverse health effects associated with PFAS exposure.

Evidence-to-Decision Frameworks Considered by the Committee

Norris and colleagues recently conducted a review of evidence-to-decision frameworks to inform the development of a new framework for recommending interventions to prevent or mitigate the harmful effects of adverse environmental exposures (Norris et al., 2021). The authors reviewed 18 frameworks and compared the criteria they used to justify decisions. Those criteria included the priority of the problem and several considerations related to benefits and harms, including desirable effects, undesirable effects, the certainty of evidence regarding desirable and undesirable effects, and the balance of effects. Other criteria included values, resources, equity, acceptability, feasibility, and human rights. Briefly, the frameworks reviewed by Norris and colleagues are summarized in Table 2-1.

TABLE 2-1 Evidence-to-Decision Frameworks Reviewed by Norris and Colleagues (2021)

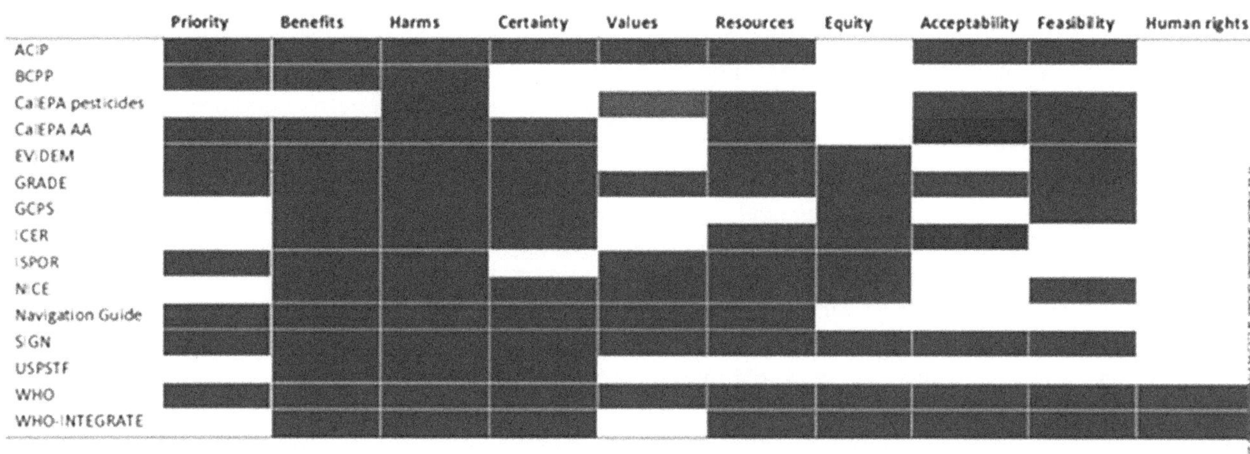

NOTES: ACP = Advisory Committee on Immunization Practices; BCPP = Breast Cancer Prevention Partners; CalEPA = California Environmental Protection Agency; CalEPA AA = California Environmental Protection Agency, Alternative Assessments; EVIDEM = Evidence and Value: Impact on DEcision-Making; GRADE = Grading of Recommendations Assessment, Development and Evaluations; GCPS = Guide to Community Preventive Services; ICER = Institute for Clinical and Economic Review; ISPOR = International Society for Pharmacoeconomics and Outcomes Research; NICE = United Kingdom National Institute for Health and Care Excellence; SIGN = Scottish Intercollegiate Guidelines Network; USPSTF = U.S. Preventive Services Task Force; WHO = World Health Organization.
SOURCE: Norris et al., 2021. Creative Commons CC BY.

GRADE (Grading of Recommendations Assessment, Development, and Evaluations)

GRADE has become the dominant evidence-to-decision framework over the past 15 years, including for clinical decision making and public health (Guyatt et al., 2008; Moberg et al., 2018; Norris et al., 2021). It includes evidence frameworks for several different decision contexts including clinical recommendations, individual perspective; clinical recommendations, population perspective; coverage decisions; and health system and public health recommendations or decisions. According to Norris and colleagues (2021), all applications of GRADE include roughly the same considerations:

- Is the problem a priority?
- How substantial are the desirable anticipated effects?
- How substantial are the undesirable anticipated effects?
- What is the overall certainty of the evidence of effects?
- Does the balance between desirable and undesirable effects favor the intervention or the comparison?
- Is there important uncertainty about, or variability in, how much people value the main outcomes?
- How large are the resource requirements (costs)?
- What is the certainty of the evidence of resource requirements (costs)?
- Does the cost-effectiveness of the intervention (the out-of-pocket cost relative to the net desirable effect) favor the intervention or the comparison?

It is important to recognize that GRADE emphasizes financial considerations, equity, acceptability, and feasibility more for health systems and public health decisions than for individual decisions (Norris et al., 2021).

U.S. Preventive Services Task Force

The USPSTF is an independent panel of primary care, prevention, and evidence-based medicine experts. It uses an analytic framework (see Figure 2-1) to evaluate systematically the benefits and harms of a particular preventive service. In evaluating evidence concerning the benefits and harms of widespread implementation, both the evidence's certainty and the magnitude of the benefits and harms are assessed. As shown in Table 2-2, the USPSTF assigns a letter grade to each preventive service signifying its recommendation about the service's provision (Sawaya et al., 2007). It is important to note that the USPSTF does not consider cost when determining a letter grade.[2] If evidence is not available, the USPSTF considers the "chain of indirect evidence, including evidence about the accuracy of screening tests, the efficiency and harms of early treatment, and the association between changes in intermediate endpoints due to treatment and changes in health endpoints" (Mabry-Hernandez et al., 2018). As Table 2-2 indicates, when the certainty, based on the evidence, for the net benefit of screening is rated low, the USPSTF considers the evidence insufficient to make a recommendation. The USPSTF views its evaluation of the benefits and harms of a preventive service as a process, and continually revisits and reevaluates based on new advances in research, testing, and treatment capabilities (Sawaya et al., 2007).

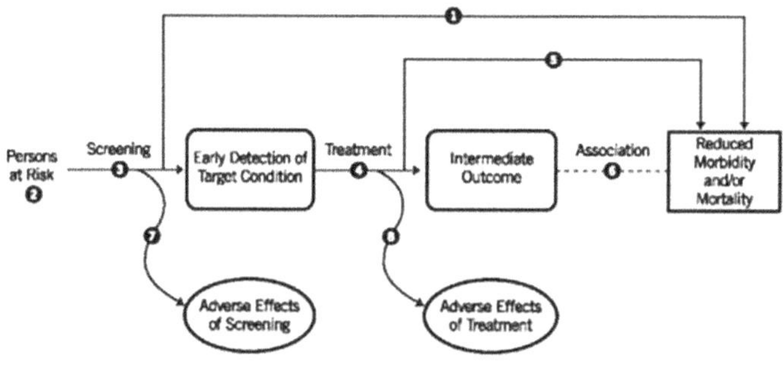

Figure 1. Generic analytic framework for a screening preventive service.
Note: Numbers correspond to the following key questions:
1. Is there evidence that randomizing a cohort of persons to screening or not screening results in greater reduced morbidity and mortality than the observed harms from screening and treatment?
2. Who are the persons at risk for the condition?
3. What are the performance characteristics of the screening test for the condition?
4. Does treatment of the screen detected condition result in improved intermediate outcomes?
5. Does treatment of the screen detected condition result in reduced morbidity and mortality?
6. Does an improvement in intermediate outcomes lead to improved health outcomes?
7. What are the harms associated with screening and diagnostic testing for the condition?
8. What are the harms associated with treatment of the condition?

FIGURE 2-1 U.S. Preventive Services Task Force's generic analytic framework for a screening preventive service.
SOURCE: Mabry-Hernandez et al., 2018. CC BY-NC-ND 4.0.

---

[2] See https://www.uspreventiveservicestaskforce.org/uspstf/about-uspstf/task-force-resources/uspstf-and-cost-considerations (accessed September 16, 2021).

TABLE 2-2 U.S. Preventive Services Task Force's (USPSTF's) Recommendation Grid[a]

Certainty of Net Benefit	Magnitude of Net Benefit			
	Substantial	Moderate	Small	Zero/Negative
High	A	B	C	D
Moderate	B	B	C	D
Low	Insufficient			

[a] *A, B, C, D,* and *Insufficient* represent the letter grades of recommendation or statement of insufficient evidence assigned by the U.S. Preventive Services Task Force after assessing certainty and magnitude of the net benefit of the service.
SOURCE: Sawaya et al., 2007.

Evidence and Value: Impact on DEcision-Making (EVIDEM) Framework

Goetghebeur and colleagues (2008) developed this framework to facilitate health care decision making. They conducted an extensive review and analysis of the literature and determined that a framework for addressing the value judgments related to assessing health care interventions should be able to do the following: "1) disentangle intrinsic and extrinsic value components; 2) develop a simple and rigorous system that applies multiple criteria decision analysis (MCDA) from a pragmatic standpoint based on actual thought processes; 3) provide practical access to the evidence on which value judgments are based; and 4) provide a practical method for decision-makers to provide feedback to data producers and all other stakeholders" (Goetghebeur et al., 2008, p. 273). Based on these needs, the authors built the EVIDEM framework, tailored to provide a "comprehensive, transparent structure grounded in global standards and local needs" (Goetghebeur et al., 2008, p. 283). The framework aims to consider all perspectives, values, and rationales related to the decision at hand.

National Academies Evidence Framework for Genetic Testing

In 2017, a National Academies committee developed an evidence framework for genetic testing that calls for a clear definition of the genetic scenario being considered and a triage process for evaluating whether the purpose of a test is worthwhile and an expedited provisional decision can be made (see Figure 2-2). In addition, given the rapidly advancing nature of the field of genetic testing, the framework includes the opportunity for review and revisions (NASEM, 2017).

C-8 Medical Panel

In February 2005, the West Virginia Circuit Court approved a class action settlement agreement in a lawsuit resulting from contamination of drinking water with perfluorooctanoic acid (PFOA, also known as C-8) from DuPont's Washington Works facility in Wood County, West Virginia (see Box 1-2 in Chapter 1). The agreement included establishing a science panel that conducted epidemiologic evaluations to determine probable link conditions associated with exposure to PFOA, and a separate medical panel that defined medical monitoring for members of the class (C-8 Medical Panel, 2013). The settlement agreement stated that the medical panel needed to consider the following factors in developing the medical monitoring protocol:

1. Increased risk … that the Class Member has a significantly increased risk of contracting the particular diseases relative to the risk in the absence of exposure;
2. Necessity of Diagnostic Testing … that the Class Member should undergo specific periodic diagnostic testing that would not be required in the absence of exposure to

C-8. The Settlement further specified that the desires of Class Members for reassurance that they did not have a probable link condition was a sufficient rationale for testing, and factors such as financial cost and the frequency of testing need not be given significant weight in assessing the need for testing; and

3. Existence of Monitoring Procedures … that testing procedures must exist that it is not necessary to show that detection and treatment in a pre-symptomatic state reduces the burden of the probable link condition. (C-8 Medical Panel, 2013, p. 1)

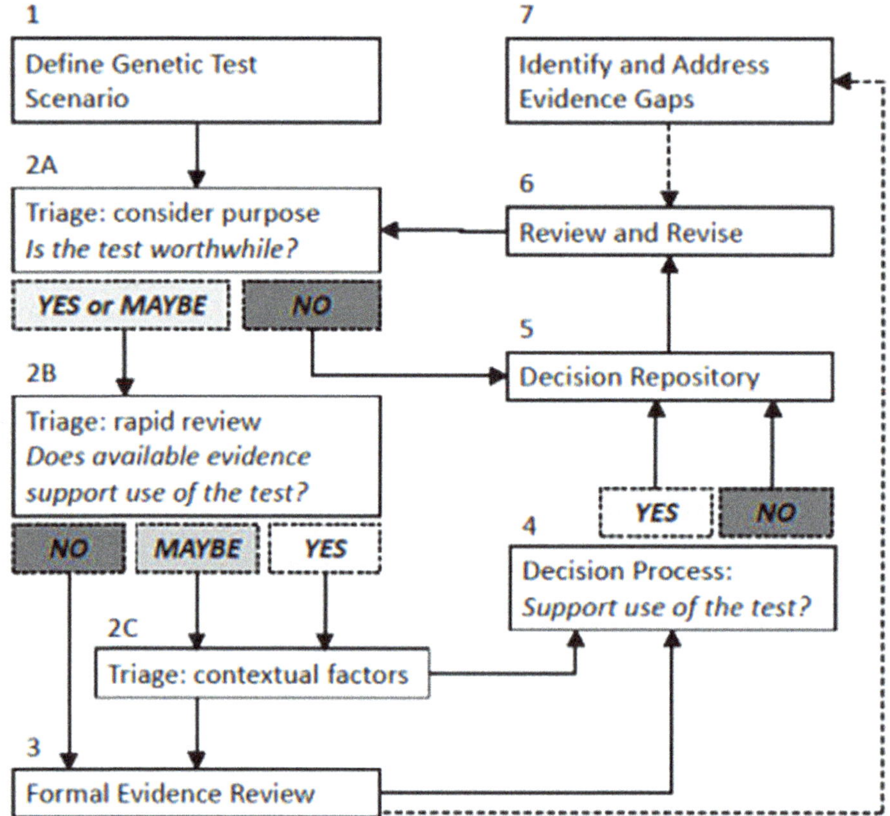

FIGURE 2-2 Visual representation of the seven components of the evidence framework for genetic testing developed by the National Academies of Sciences, Engineering, and Medicine.
SOURCE: NASEM, 2017.

Belmont Report

The Belmont Report (HHS, 1979) was developed as a set of guidelines for protecting research participants. This seminal report on ethics and health care research outlines basic ethical principles and guidelines meant to assist in resolving ethical problems surrounding the conduct of research with human subjects. The three main principles are respect for persons, beneficence, and justice.

-   *Respect for persons* requires that people be considered autonomous. Autonomous individuals can consider their personal goals and make their own decisions. To respect persons is to give weight to their opinions and choices while refraining from obstructing their actions or choices unless detrimental to others. Conversely, a lack of respect for persons or autonomous agents is to refuse to accept or validate their judgments, deny them the freedom to act on their judgments, or withhold information necessary to make a judgment.

- *Beneficence* is an obligation to respect individuals and their decisions and protect them from harm. Two general rules guide the committee's understanding of beneficence: (1) do no harm, and (2) maximize possible benefits and minimize harms. The Hippocratic Oath requires physicians to make medical recommendations and decisions that benefit their patients "according to their best judgment." The committee recognizes that zero harm or risk may not be possible, but medical decisions need to consider when the benefits outweigh the risks.
- *Justice* refers to "fairness in distribution." In the present context, it means that PFAS research should have a fair distribution such that a certain group of people is not selected for the research to the exclusion of others. An injustice occurs when some individuals benefit from the research or intervention or are denied that benefit without good, ethical reason.

Beauchamp and Childress (2001) have developed these principles further in the established textbook *Principles of Biomedical Ethics* to guide biomedical decision making beyond the context of research ethics. For example, these authors refer to respect for autonomy instead of respect for persons and divide the principle of beneficence into nonmaleficence (avoiding harm) and beneficence (providing benefits).

Precautionary Reasoning

An extensive body of scholarship in medical and environmental decision making is devoted to precautionary reasoning and the precautionary principle. This work can inform decision making under substantial scientific uncertainty. Resnik (2021) proposes four criteria for assessing the reasonableness of precautionary measures, as shown in Table 2-3.

CONCLUSION

Building on the work of other experts and evidence-to-decision frameworks, the committee developed five principles to guide decision making under uncertainty for use throughout this report: proportionality, justice, autonomy, feasibility, and adaptability. The principles of proportionality, justice, and autonomy build on the ethical principles outlined in the Belmont Report (HHS, 1979) and developed further by Beauchamp and Childress (2001). These principles affirm that decision making in response to PFAS should weigh plausible benefits and harms proportionally while treating all individuals fairly, promoting health equity, respecting human rights, and supporting informed decision making by individuals in accordance with their values. The principles of feasibility and adaptability affirm that decisions should be responsive to emerging information about harms, benefits, and other considerations.

TABLE 2-3 Criteria for Reasonableness of Precautionary Measures

Proportionality	Reasonable measures balance plausible risks and possible benefits proportionally
Fairness	Reasonable measures are based on a fair balancing of risks and benefits; fairness includes distributive and procedural fairness
Epistemic Responsibility	Reasonable measures comply with norms for the responsible acquisition and utilization of evidence, knowledge, and expertise
Consistency	Reasonable measures are based on a consistent rationale for decision making

SOURCE: Resnik, 2021.

In the clinical setting, the committee's five principles converge under the principle of autonomy and shared decision making[3] between patient and clinician. For this reason, shared decision making is likely to be the practical way a clinician can incorporate the principles into the clinical encounter.

The committee's principles may provide a framework that ATSDR can use when writing clinical guidance for environmental exposures. Despite findings of associations between adverse health effects and chemical exposures, substantial scientific uncertainty exists about the causal role of environmental chemicals in many health outcomes. Regulatory agencies, public health agencies, and clinicians need to make thoughtful decisions about how to act in response to this uncertainty. The principles of proportionality, justice (particularly environmental justice), autonomy, feasibility, and adaptability can play a central role in these decisions.

REFERENCES

Barry, M. J., and S. Edgman-Levitan. 2012. Shared decision making—pinnacle of patient-centered care. *New England Journal of Medicine* 366(9):780–781. https://doi.org/10.1056/NEJMp1109283.
Beauchamp, T. L., and J. F. Childress. 2001. *Principles of biomedical ethics.* New York: Oxford University Press.
Blair, A., R. Saracci, P. Vineis, P. Cocco, F. Forastiere, P. Grandjean, M. Kogevinas, D. Kriebel, A. McMichael, N. Pearce, M. Porta, J. Samet, D. P. Sandler, A. S. Costantini, and H. Vainio. 2009. Epidemiology, public health, and the rhetoric of false positives. *Environmental Health Perspectives* 117(12):1809–1813. https://doi.org/10.1289/ehp.0901194.
Boffetta, P., J. K. McLaughlin, C. La Vecchia, R. E. Tarone, L. Lipworth, and W. J. Blot. 2008. False-positive results in cancer epidemiology: A plea for epistemological modesty. *Journal of the National Cancer Institute* 100(14):988–995. https://doi.org/10.1093/jnci/djn191.
Boronow, K. E., H. P. Susmann, K. Z. Gajos, R. A. Rudel, K. C. Arnold, P. Brown, R. Morello-Frosch, Laurie Havas, and J. Green Brody. 2017. DERBI: A digital method to help researchers offer "right-to-know" personal exposure results. *Environmental Health Perspectives* 125(2):A27–A33.
Brody, J. G., S. C. Dunagan, R. Morello-Frosch, P. Brown, S. Patton, and R. A. Rudel. 2014. Reporting individual results for biomonitoring and environmental exposures: Lessons learned from environmental communication case studies. *Environmental Health* 13:40. https://doi.org/10.1186/1476-069x-13-40.
Bullard, R. D. 2018. *Dumping in Dixie: Race, class, and environmental quality.* New York: Routledge.
C-8 Medical Panel. 2013. *C-8 Medical Panel report.* https://www.hpcbd.com/wp-content/uploads/sites/1603732/2021/01/Medical-Panel-Report-2013-05-24.pdf (accessed June 15, 2022).
Charles, C., A. Gafni, and T. Whelan. 1997. Shared decision-making in the medical encounter: What does it mean? (or it takes at least two to tango). *Sociological Science in Medicine* 44:681–692. https://doi.org.10.1016/s0277-9536(96)00221-3.
Corburn, J. 2005. *Street science: Community knowledge and environmental health justice.* Berkeley, CA: Department of City and Regional Planning, University of California, Berkeley. https://doi.org/10.5070/BP31911296.
Douglas, H. E. 2009. *Science, policy, and the value-free ideal.* Pittsburgh, PA: University of Pittsburgh Press.

---

[3] Barry and Edgman-Levitan (2012) builds on concepts in Charles (1997) and the Institute of Medicine's (2001) *Crossing the Quality Chasm: a New Health System for the 21st Century* and describes shared decision making as the pinnacle of patient-centered care: "The process by which the optimal decision may be reached ... is called shared decision-making and involves, at minimum, a clinician and the patient.... In shared decision-making, both parties share information: the clinician offers options and describes their risks and benefits, and the patient expresses his or her preferences and values. Each participant is thus armed with a better understanding of the relevant factors and shares responsibility in the decision about how to proceed" (p. 780).

Elliott, K. C. 2017. *A tapestry of values: An introduction to values in science.* New York: Oxford University Press.

Elliott, K. C., and T. Richards. 2017. *Exploring inductive risk: Case studies of values in science.* New York: Oxford University Press.

Fischhoff, B., and A. L. Davis. 2014. Communicating scientific uncertainty. *Proceedings of the National Academy of Sciences of the United States of America* 111(Suppl 4):13664–13671. https://doi.org/10.1073/pnas.1317504111.

Ginsberg, G. L., K. Pullen Fedinick, G. M. Solomon, K. C. Elliott, J. J. Vandenberg, S. Barone, Jr., and J. R. Bucher. 2019. New toxicology tools and the emerging paradigm shift in environmental health decision-making. *Environmental Health Perspectives* 127(12):125002.

Goetghebeur, M. M., M. Wagner, H. Khoury, R. J. Levitt, L. J. Erickson, and D. Rindress. 2008. Evidence and value: Impact on DEcisionMaking—the EVIDEM framework and potential applications. *BMC Health Services Research* 8(1):1–16.

Guyatt, G. H., A. D. Oxman, G. E. Vist, R. Kunz, Y. Falck-Ytter, P. Alonso-Coello, and H. J. Schünemann. 2008. GRADE: An emerging consensus on rating quality of evidence and strength of recommendations. *British Medicial Journal* 336(7650):924–926. https://doi.org/10.1136/bmj.39489.470347.AD.

Harris, R. P., S. L. Sheridan, C. L. Lewis, C. Barclay, M. B. Vu, C. E. Kistler, C. E. Golin, J. T. DeFrank, and N. T. Brewer. 2014. The harms of screening: A proposed taxonomy and application to lung cancer screening. *JAMA Internal Medicine* 174(2):281–286.

HHS (U.S. Department of Health and Human Services). 1979. *The Belmont report: Ethical principles and guidelines for the protection of human subjects of research.* Washington, DC: U.S. Department of Health and Human Services.

IOM (Institute of Medicine). 2001. *Crossing the quality chasm: A new health system for the 21st century.* Washington, DC: National Academy Press. https://doi.org/10.17226/10027.

IOM. 2013. *Environmental decisions in the face of uncertainty.* Washington, DC: The National Academies Press. https://doi.org/10.17226/12568.

IOM, NAS, and NAE (Institute of Medicine, National Academy of Sciences, and National Academy of Engineering). 1995. *On being a scientist: Responsible conduct in research* (2nd ed.). Washington, DC: National Academy Press.

Krist, A. H., T. A. Wolff, D. E. Jonas, C. M. Mangione, C-W. Tseng, and D. Grossman. 2018. Update on the methods of the U.S. Preventive Services Task Force: Methods for understanding certainty and net benefit when making recommendations. *American Journal of Preventive Medicine* 54(Suppl 1): S11–S18.

London, A. J. 2022. *For the common good: Philosophical foundations of research ethics.* New York: Oxford University Press.

Mabry-Hernandez, I. R., S. J. Curry, W. R. Phillips, F. A. García, K. W. Davidson, J. W. Epling, Jr., Q. Ngo-Metzger, and A. S. Bierman. 2018. U.S. Preventive Services Task Force priorities for prevention research. *American Journal of Preventive Medicine* 54(Suppl 1):S95–S103. https://doi.org/10.1016/j.amepre.2017.08.014.

Moberg, J., A. D. Oxman, S. Rosenbaum, H. J. Schünemann, G. Guyatt, S. Flottorp, C. Glenton, S. Lewin, A. Morelli, and G. Rada. 2018. The GRADE Evidence to Decision (EtD) framework for health system and public health decisions. *Health Research Policy and Systems* 16(1):1–15.

Morello-Frosch, R., J. G. Brody, P. Brown, R. G. Altman, R. A. Rudel, and C. Perez. 2009. Toxic ignorance and right-to-know in biomonitoring results communication: A survey of scientists and study participants. *Environmental Health* 8:6. https://doi.org/10.1186/1476-069X-8-6.

NASEM (National Academies of Sciences, Engineering, and Medicine). 2017. *An evidence framework for genetic testing.* Washington, DC: The National Academies Press.

Norris, S. L., M. T. Aung, N. Chartres, and T. J. Woodruff. 2021. Evidence-to-decision frameworks: A review and analysis to inform decision-making for environmental health interventions. *Environmental Health* 20(1):124. https://doi.org/10.1186/s12940-021-00794-z.

NRC (National Research Council). 2004. *Adaptive management for water resources project planning.* Washington, DC: The National Academies Press. https://doi.org/10.17226/10972.

Rehfuess, E. A., J. M. Stratil, I. B. Scheel, A. Portela, S. L. Norris, and R. Baltussen. 2019. The WHO-INTEGRATE evidence-to-decision framework version 1.0: Integrating WHO norms and values and a complexity perspective. *BMJ Global Health* 4(Suppl 1):e000844.

Resnik, D. B. 2021. Precautionary reasoning and the precautionary principle. In *Precautionary reasoning in environmental and public health policy.* Cham, Switzerland: Springer. Pp. 111–128.

Sawaya, G. F., J. Guirguis-Blake, M. LeFevre, R. Harris, and D. Petitti. 2007. Update on the methods of the U.S. Preventive Services Task Force: Estimating certainty and magnitude of net benefit. *Annals of Internal Medicine* 147(12):871–875. https://doi.org/10.7326/0003-4819-147-12-200712180-00007.

Shrader-Frechette, K. 2002. *Environmental justice: Creating equality, reclaiming democracy.* New York: Oxford University Press.

Wandersman, A. 2003. Community science: Bridging the gap between science and practice with community-centered models. *American Journal of Community Psychology* 31(3–4):227–242. https://doi.org/10.1023/A:1023954503247.

Workman, M., K. Dooley, G. Lomax, J. Maltby, and G. Darch. 2020. Decision-making in contexts of deep uncertainty—An alternative approach for long-term climate policy. *Environmental Science & Policy* 103:77–84.

3

Potential Health Effects of PFAS

This chapter summarizes results of the committee's review of the potential health effects of per- and polyfluoroalkyl substances (PFAS) in response to the following element of the committee's Statement of Task: "to establish a basis for prioritized clinical surveillance or monitoring of PFAS health effects." The aim of this review was to identify a set of health effects that may be associated with PFAS to support preventive medicine recommendations and decisions. Based on the review, the committee developed strength-of-evidence conclusions for the various health effects associated with PFAS.

The committee's Statement of Task limited this review to those PFAS included in the Center for Disease Control and Prevention's (CDC's) *National Report on Human Exposure to Environmental Chemicals* (see Table 3-1). The PFAS in Table 3-1 are those most commonly studied in epidemiological research, although other PFAS may also cause harm given some similarities with those in Table 3-1 with respect to biological persistence and toxicities (Kwiatkowski et al., 2020). It is important to note as well that while different PFAS have distinct physical, chemical, and toxicological properties, people are exposed to more than a single PFAS. As a result, exposures are often to mixtures of PFAS such that specific effects are difficult to disentangle. Considering these issues, and recognizing that some PFAS are less frequently measured, the committee ultimately decided to provide one strength-of-evidence determination for all PFAS for each health effect.

The Statement of Task did not limit the health effects included in the committee's review. Speakers at the committee's town halls described a variety of health effects of concern that they had observed in their communities and that may be associated with PFAS exposure. Cancers, endocrine effects, immune function, and fertility were the most frequently mentioned health effects (see Table 3-2). The committee considered this input to be valuable and paid special attention to the health effects of concern observed in communities when describing the evidence.

TABLE 3-1 PFAS Species Currently Included in the Centers for Disease Control and Prevention's (CDC's) *National Report on Human Exposure to Environmental Chemicals*

Abbreviated Name	Full Name	CAS Registry No.
MeFOSAA	Methylperfluorooctane sulfonamidoacetic acid	2355-31-9
PFHxS	Perfluorohexanesulfonic acid perfluorohexane sulfonic acid	355-46-4
n-PFOA (linear isomer), Sb-PFOA (branched isomers)	Perfluorooctanoic acid	335-67-1*
PFDA	Perfluorodecanoic acid	335-76-2
PFUnDA	Perfluoroundecanoic acid	2058-94-8
n-PFOS (linear isomer), Sm-PFOS (branched isomers)	Perfluorooctanesulfonic acid	1763-23-1*
PFNA	Perfluorononanoic acid	375-95-1

NOTES: CAS = Chemical Abstracts Service. * = CAS number refers to linear isomer only. Previous survey years have also included perfluorobutane sulfonic acid (PFBS), perfluoropentanoic acid (PFpA), perfluorododecanoic acid (PFDoDA), perfluorooctane sulfonamide (FOSA), and 2-(N-ethyl-perfluorooctane sulfonamido)acetate (EtFOSAA), according to Patrick N. Breysse's presentation to the committee on February 4, 2021.

TABLE 3-2 Categories of Health Effects Mentioned by Speakers at the Committee's Town Halls

Health Effect Category	No. of Speakers Mentioned
Cancers, including bladder, urinary tract, liver, breast, testicular, thyroid, bone, kidney, pancreatic, and ovarian, as well as melanoma, leukemia, multiple myeloma, and lymphoma	16
Disruption of the endocrine system, including impaired thyroid and disease	10
Impaired immune function	9
Fertility and reproductive issues, including menstruation and lactation concerns	9
Diabetes, other metabolic concerns, and obesity	8
Liver disease and impairment, including nonalcoholic fatty liver disease	7
Disease and impairment of the digestive system, including ulcerative colitis, gallbladder dysfunction, irritable bowel syndrome, and other colon impairment	6
High cholesterol	6
Autoimmune disease	5
Birth defects	5
Developmental and neurological impacts, including learning disorders and autism	5
High blood pressure, gestational hypertension, and preeclampsia	5
Asthma, pulmonary disease, and cardiovascular disease	5
Premature and underweight births	4
Skin rashes, hair loss, and other skin concerns	3
Chronic inflammation and allergic reactions	3
Arthritis, osteoporosis, and other impacts to the skeletal system	2
Multiple sclerosis	1

OVERVIEW OF EVIDENCE REVIEW APPROACH

Appendix D provides full details of the methods used for the committee's review. Briefly, the committee aimed to build upon existing decisions from other authoritative bodies. The focus was on more recent studies, both high-quality systematic reviews and published epidemiologic research articles, which could inform updates to authoritative conclusions regarding PFAS exposure and any human health effects. The review approach improved efficiency while minimizing the risk of excluding scientific findings that could inform the committee's recommendations.

Consideration of Authoritative Reviews

The committee focused on authoritative reviews produced by government agencies or other bodies that publish strength-of-evidence determinations through a process that includes peer review. As detailed in Appendix D, the committee focused on the findings from the following organizations, presented in chronological order by year the literature search was completed:

- C-8 Science Panel reports (last report published 2012)
- European Food Safety Authority (EFSA) (search complete: 2013)
- Organisation for Economic Co-operation and Development (OECD) (published 2013)

- International Agency for Research on Cancer (IARC) (search complete: 2014)
- U.S. Environmental Protection Agency (EPA) (search complete: 2015)
- National Toxicology Program (NTP) (search complete: 2016)
- Agency for Toxic Substances and Disease Registry (ATSDR) (search complete: 2018)

Among the authoritative reviews, ATSDR's *Toxicological Profile for Perfluoroalkyls* included the greatest number of PFAS that were included in the committee's Statement of Task (Methylperfluorooctane sulfonamidoacetic acid [MeFOSAA] was not included), was the most recent (literature search completed in September 2018), and was not limited in terms of health effects covered (ATSDR, 2021). Therefore, the committee used this source as the primary basis for the next stages of the review process.

The committee did not formally critique the quality of the any authoritative reviews, as each authoritative body has its own procedures for public comment and peer review to ensure that biases are limited in its reviews. Nonetheless, the committee noted several areas in which ATSDR's *Toxicological Profile for Perfluoroalkyls* could be strengthened. First, the toxicological profile does not provide a detailed description of the evidence identification methods or document its reasons for excluding specific studies. Second, the study quality assessment does not appear to follow a standard approach, and in some cases, it is difficult to identify the study designs that were included in the review. Third, the process for assessing the strength of the evidence is not well described. Despite these limitations, however, the committee concluded that the ATSDR review provided a basis for assessing evidence of association between PFAS and health effects.

Review of Systematic Reviews

As detailed in Appendix D, the committee's review of systematic reviews consisted of the following steps: a literature search, screening of abstracts, a full-text review of studies identified in the abstract screening, and evaluation of a final set of relevant studies. The literature search identified 639 potentially relevant articles, of which 26 systematic reviews met the committee's inclusion criteria and were evaluated using AMSTAR-2 (A MeaSurement Tool to Assess systematic Reviews) (Shea et al., 2017). The committee conducted a critical appraisal of the systematic reviews because such reviews can be subject to a range of biases. All high-quality systematic reviews included studies that were also reviewed in ATSDR's *Toxicological Profile for Perfluoroalkyls*. Thus, the systematic reviews were used as sources for reference in the committee's assessment, but they were not formally included as part of the final strength-of-evidence determination.

Review of Recent Epidemiologic Studies

As detailed in Appendix D, the committee's review of original literature consisted of the following steps: a literature search, screening of abstracts, a full-text review of studies identified in the abstract screening, evaluation of a final set of studies identified as relevant after the full-text review, data abstraction, and an evidence synthesis step.

The literature search identified 5,172 potentially relevant studies. After removal of duplicates (112 articles), 5,060 articles were subject to title and abstract screening by two independent reviewers. After 4,434 articles had been excluded because the titles and abstracts did not meet the committee's inclusion criteria, 626 articles were subjected to full-text review, during which additional articles were excluded if they had been published before 2018 or were listed among the references in ATSDR's *Toxicological Profile for Perfluoroalkyls* (n = 320); were cross-sectional in design (n = 160); were not published in English (n = 1); did not provide risk estimates associated with PFAS exposure (n = 3); or documented studies not conducted in humans (n = 3). Cross-sectional studies were excluded largely because this study design measures exposure and disease at the same time, and so cannot determine cause and effect. The remaining 139 articles were categorized according to the human health outcomes studied, as shown in Figure 3-1.

The committee's review of the recent literature focused mainly on those health effect categories for which additional evidence might have changed the committee's understanding of the association between PFAS exposure and health outcomes. The committee formally evaluated the individual studies for internal validity or "risk of bias," using a tool adapted from the Navigation Guide of Woodruff and Sutton (2014), assigning to each an overall assessment of its risk of bias (low, probably low, probably high, or high risk of bias). Effect estimates from the individual studies included in the review were extracted into a database and uploaded to a public website (Tableau Public) to allow for visualizations such as evidence maps and forest plots.[1] The effect estimates in the Tableau represent those from the model most adjusted for confounders. Appendix D provides the critical domains used by the committee to assess risk of bias, as well as the data abstraction procedure.

Evidence Synthesis

To assess the strength of evidence regarding the potential for PFAS to cause a particular health effect, the committee integrated the evidence reviewed in ATSDR's *Toxicological Profile for Perfluoroalkyls* with the evidence from its review of the original epidemiologic studies. A framework based on the Hill considerations, which help determine whether associations are causal, guided the synthesis of available evidence (Fedak et al., 2015; Hill, 1965; NASEM, 2018a). The committee considered the animal studies discussed in ATSDR's *Toxicological Profile for Perfluoroalkyls* and in the systematic reviews examined by the committee in making its determinations, as an aid to interpretation of the human studies. Toxicologic evidence, whether it supports or conflicts with evidence from epidemiologic studies, provides insights about biologic processes and informs how an observed association might be interpreted (NASEM, 2018a).

The committee's strength-of-evidence conclusions reflect one of the four categories described below (see Figure 3-2).

Sufficient Evidence of an Association

For effects in this category, a positive association between PFAS and the outcome must be observed in studies in which chance, bias, and confounding can be ruled out with reasonable confidence. For example, the committee might regard as sufficient evidence of an association evidence from several small studies that is unlikely to be due to confounding or to otherwise be biased and that shows an association that is consistent in magnitude and direction. Experimental data supporting biologic plausibility strengthen the evidence of an association but are not a prerequisite, nor are they sufficient to establish an association without corresponding epidemiologic findings.

Limited or Suggestive Evidence of an Association

In this category, the evidence must suggest an association between exposure to PFAS and the outcome in studies of humans, but the evidence can be limited by an inability to rule out chance, bias, or confounding with confidence. One high-quality study may indicate a positive association, but the results of other studies of lower quality may be inconsistent.

---

[1] The committee's public Tableau can be accessed at the following link: https://public.tableau.com/app/profile/nationalacademies/viz/NASEMPFASEvidenceMaps/PFASEvidenceMap. The information may be viewed as an evidence map or as a forest plot. Within forest plots, filters can be accessed using the "toggle filters" function in order to restrict the view to data on specific health effect categories and other factors (reference, chemical, study design, study population, etc.).

Health Effect Category	Chemical								Grand Total
	MeFOSAA	PFDA	PFHxS	PFNA	PFOA	PFOS	PFUnDA	sumPFAS	
Bodyweight	4	4	10	10	13	12	2	0	13
Cancer	2	3	5	4	7	7	4	0	7
Cardiometabolic	7	9	21	20	19	20	6	3	21
Developmental	6	25	36	38	44	46	16	4	47
Endocrine	3	8	12	12	11	12	6	2	13
Hepatic	1	2	3	2	4	3	1	0	4
Immunological	0	8	13	13	14	14	10	2	14
Musculoskeletal	0	1	2	2	2	2	1	1	2
Neurological	1	5	9	10	10	11	3	2	11
Other	0	3	3	3	4	4	2	0	4
Renal	3	1	3	3	4	3	0	1	4
Reproductive	3	16	25	24	25	27	11	1	29
Respiratory	0	6	9	10	10	10	8	0	10
Grand Total	20	72	118	118	130	133	56	13	139

FIGURE 3-1 Evidence map describing the number of studies found, by PFAS, for each health outcome category.
NOTES: MeFOSAA = methylperfluorooctane sulfonamidoacetic acid; PFDA = perfluorodecanoic acid; PFHxS = perfluorohexanesulfonic acid; PFNA = perfluorononanoic acid; PFOA = perfluorooctanoic acid; PFOS = perfluorooctanesulfonic acid; PFUnDA = perfluoroundecanoic acid; sumPFAS = the additive sum of PFAS.

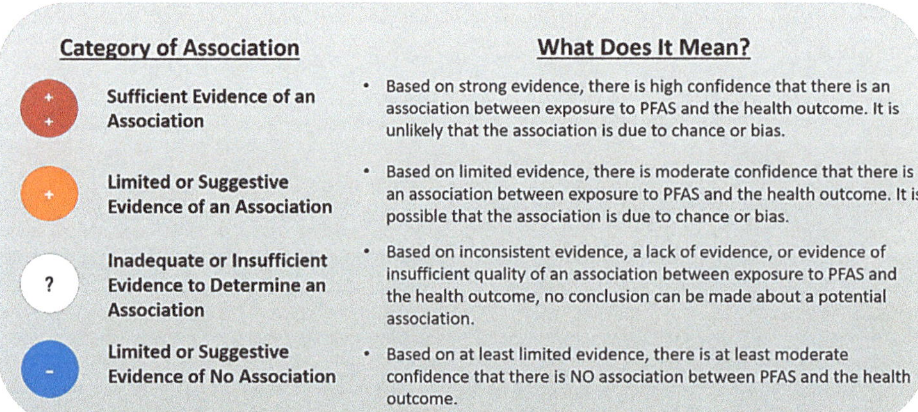

FIGURE 3-2 Categories of association used in this report.
NOTES: The categories of association only describe how strong the evidence is between PFAS and the health outcome. The risk of developing an outcome from exposure to PFAS for things in the same category can vastly differ and are dependent on whether an individual has other risk factors for developing the outcome.

Inadequate or Insufficient Evidence to Determine an Association

If there was not enough reliable scientific data to categorize the potential association with a health effect as "sufficient evidence of an association," "limited or suggestive evidence of an association," or on the other end of the spectrum, "limited or suggestive evidence of no association," the health outcome was placed in the category of "inadequate or insufficient evidence to determine an association" by default. In this category, the available human studies may have inconsistent findings or be of insufficient quality, validity, consistency, or statistical power to support a conclusion regarding the presence of an association. Such studies may have failed to control for confounding factors or may have had inadequate assessment of exposure.

Limited or Suggestive Evidence of No Association

A conclusion of "no association" is inevitably limited to the conditions, exposures, and observation periods covered by the available studies, and the possibility of a small increase in risk related to the magnitude of exposure studied can never be excluded. However, a change in classification from inadequate or insufficient evidence of an association to limited or suggestive evidence of *no* association would require new studies that corrected for the methodologic problems of previous studies and that had samples large enough to limit the possible study results attributable to chance.

COMMITTEE'S CONCLUSIONS

Annex Table 3-1 at the end of the chapter summarizes the health outcomes considered by the committee, the relevant conclusions from authoritative reviews, and the committee's overall conclusions for endpoints relevant to each outcome. The committee's conclusions reflect its integration of evidence reviewed in ATSDR's *Toxicological Profile for Perfluoroalkyls* and other authoritative reviews with the evidence garnered from the review of recent epidemiologic studies.

The committee found *sufficient* evidence of an association for the following diseases and health outcomes:

- decreased antibody response (in adults and children),
- dyslipidemia (in adults and children),

- decreased infant and fetal growth, and
- increased risk of kidney cancer (in adults).

The committee found *limited* or *suggestive* evidence of an association for the following diseases and health outcomes:

- increased risk of breast cancer (in adults),
- liver enzyme alterations (in adults and children),
- increased risk of pregnancy-induced hypertension (gestational hypertension and preeclampsia),
- increased risk of testicular cancer (in adults),
- increased risk of thyroid disease and dysfunction (in adults), and
- increased risk of ulcerative colitis (in adults).

For a range of other health effects, the evidence was *inadequate* or *insufficient*. These include type 1 and gestational diabetes; cardiovascular disease; metabolic syndrome; obesity; infertility; male and female reproductive effects; reproductive hormone levels; and cancers other than kidney, breast, and testicular.

The committee's rationale for these conclusions is provided in the sections that follow, organized by human health outcomes. For effects with *limited* or *sufficient* evidence, the range of effect size estimates considered by the committee is indicated. Additional information, including an evidence map of recent studies, information about the quality of individual studies, and forest plots showing effect size estimates in a searchable format, can also be accessed from the committee's public Tableau.

With one exception (decreased infant and fetal growth), the committee provided forest plots for all outcomes with *sufficient* evidence of an association to display effect estimates from the studies with low or probably low risk of bias. Specifically, forest plots based on the data in the Tableau are displayed in this chapter for the following outcomes: changes in antibody response (a measure of immune function), total cholesterol (a marker of dyslipidemia), and hypertensive disorders of pregnancy. For these outcomes, the committee chose to display the effect estimate that was the most common across studies if more than one was available in the Tableau. For cancer, forest plots were created based on studies included in ATSDR's *Toxicological Profile and Polyfluoroalkyls* and the more recent epidemiologic studies. The committee handled cancer in this way because it was upgrading the previously observed association between PFAS and kidney cancer and drawing a new conclusion on breast cancer. Each cancer figure displays the PFAS with the strongest effect. For birthweight, there were many studies with probably or definitely low risk of bias, and those with definitely low risk of bias are displayed in a table rather than a forest plot. In addition to the outcomes with *sufficient* evidence of an association, one outcome with *limited* or *suggestive* evidence, hypertensive disorders of pregnancy, is displayed in a forest plot. This was an important outcome category to speakers at the committee's town halls.

SUMMARY AND RATIONALE FOR THE COMMITTEE'S CONCLUSIONS BY HUMAN HEALTH OUTCOMES

Immune System Effects

The committee's evaluation of the impact of PFAS on immune function considered evidence relevant to the three basic functions of the immune system: response to infection, response to foreign substances (allergy), and response to self (autoimmunity). The committee found *sufficient* evidence for an association of PFAS exposure with decreased antibody response to vaccination or infection, and *limited* or *suggestive* evidence of an association with ulcerative colitis (relevant to autoimmunity). There was *inadequate* evidence for other immune system endpoints, including infectious disease (response to infection) and response to allergens.

Response to Infection

ATSDR concluded that there is suggestive evidence for an association between serum levels of perfluorooctanoic acid (PFOA), perfluorooctanesulfonic acid (PFOS), perfluorohexanesulfonic acid (PFHxS), and perfluorodecanoic acid (PFDA) and decreased antibody response to vaccines, and limited evidence of an association for perfluorononanoic acid (PFNA), perfluoroundecanoic acid (PFUnDA), and perfluorododecanoic acid (PFDoDA). Other authoritative reviews (including those of EFSA, EPA, NTP, and OECD) also found associations between PFAS and decreased antibody response to vaccines. Three more recent papers focus on antibody response early in life. Huang and colleagues (2020) conducted a study, with probably low risk of bias, which found that immunoglobulin G (IgG) levels were slightly lower among children with higher PFAS serum levels. The regression coefficient per 1-log10 nanograms per milliliter (ng /mL) increase in PFAS ranged from −0.04 (95% confidence interval [CI]: −0.11–0.04) for PFHxS to 0.00 for PFUnDA (95% CI: −0.06–0.05) (see Figure 3-3). In another study with probably low risk of bias, Timmermann and colleagues (2020) evaluated measles antibodies at three time points: before first vaccination (measuring antibodies passed from the mother) and after first and second vaccination; for most PFAS, no strong associations with antibody levels were found. For PFOS, there was an inverse association between PFAS levels in blood and antibody levels before first vaccination in boys (see Figure 3-4). In a study of bias of antibodies to hand, foot, and mouth disease (CA 16 and EV71 antibody), with probably high risk of bias, Zeng and colleagues (2019) observed that 3-month-old infants with higher PFAS cord blood levels were two to four times more likely to have levels of antibodies to hand, foot, and mouth disease that were below the clinically protective level (see Figure 3-5). This finding suggests that higher PFAS blood levels may contribute to lower antibody levels over time. Taken together, these three recent studies add support for the conclusion of *sufficient* evidence of an association between PFAS exposure and decreased antibody response.

The committee reviewed four papers focused on specific infectious diseases (chicken pox, common cold, otitis media, pneumonia, and respiratory tract infection) in children (Ait Bamai et al., 2020; Huang et al., 2020; Kvalem et al., 2020; Manzano-Salgado et al., 2019). All four studies were rated as having a probably low risk of bias; however, three of the studies used parental reports of infection to ascertain outcomes, which could result in information bias, leading in turn to null findings. The fourth study, by Huang and colleagues (2020), collected data on infections from medical records. These four studies did not provide strong evidence of an association of PFAS with these common illnesses, although there was some suggestion that for children without siblings, PFAS may be associated with respiratory syncytial virus (Ait Bamai et al., 2020). There was no evidence for fever among children with higher PFAS blood levels (Timmermann et al., 2020). The committee concluded there is *inadequate* or *insufficient* evidence of an association between PFAS exposure and risk of infection, although this is an area worthy of future research, including the relationship of PFAS to novel infections such as SARS-CoV-2 (see Box 3-1).

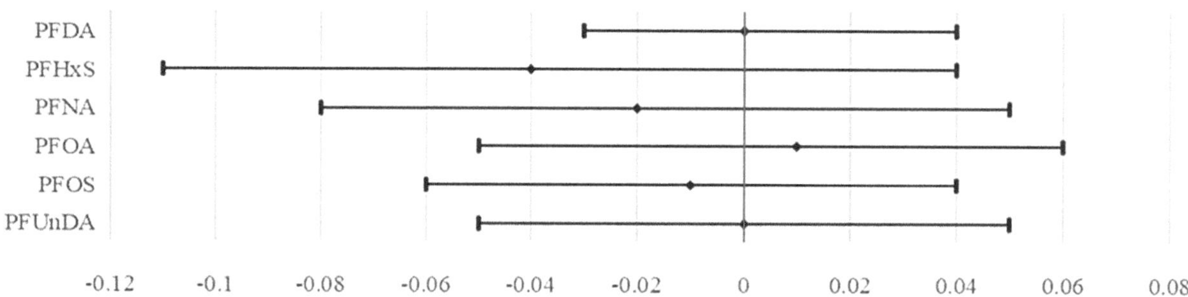

FIGURE 3-3 Regression coefficients for changes in immunoglobulin (IgG) concentrations per 1-log10 nanograms per milliliter (ng /mL) increase in PFAS serum level.
DATA SOURCE: Huang et al., 2020.

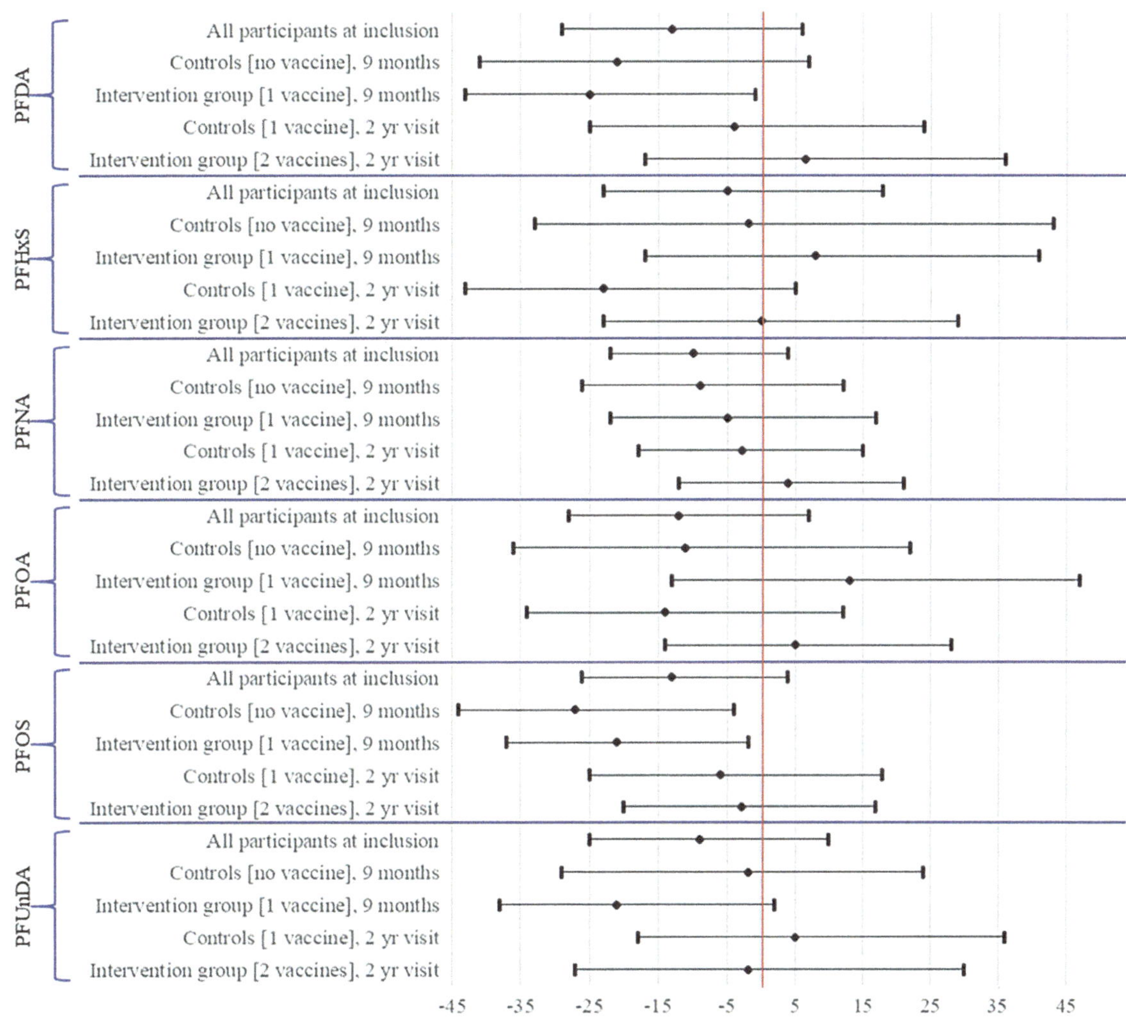

FIGURE 3-4 Regression coefficients for percent difference in measles antibody response per doubling of log10 nanograms per milliliter (ng /mL) serum PFOS.
DATA SOURCE: Timmerman et al., 2020.

Response to Foreign Substances (Allergy)

The committee reviewed several studies evaluating the impact of PFAS exposure on allergic symptoms and disease, all with a probably low risk of bias. The specific outcomes studied included allergies to food and inhaled substances, atopic dermatitis, dermatitis, changes in serum immunoglobulin E (IgE) levels, rhinitis, rhinoconjunctivitis, and results of skin prick tests. Six studies (Ait Bamai et al., 2020; Chen et al., 2018; Huang et al., 2020; Impinen et al., 2019; Kvalem et al., 2020; Wen et al., 2019) evaluated some aspect of allergic response, with most not providing strong evidence of an association with PFAS. The one exception is the study by Wen and colleagues (2019), which showed increased atopic dermatitis among those in the highest tertile of PFOA levels, but not for other PFAS. One study included in ATSDR's *Toxicological Profile for Perfluoroalkyls* found that PFAS exposure was associated with increased odds of asthma diagnosis among children at ages 5 and 13, but only those children who had not received measles, mumps, and rubella vaccination (Timmerman et al., 2020). Overall, the evidence for an association between PFAS and allergy response is *inadequate* or *insufficient*, a finding consistent with the authoritative review by EFSA.

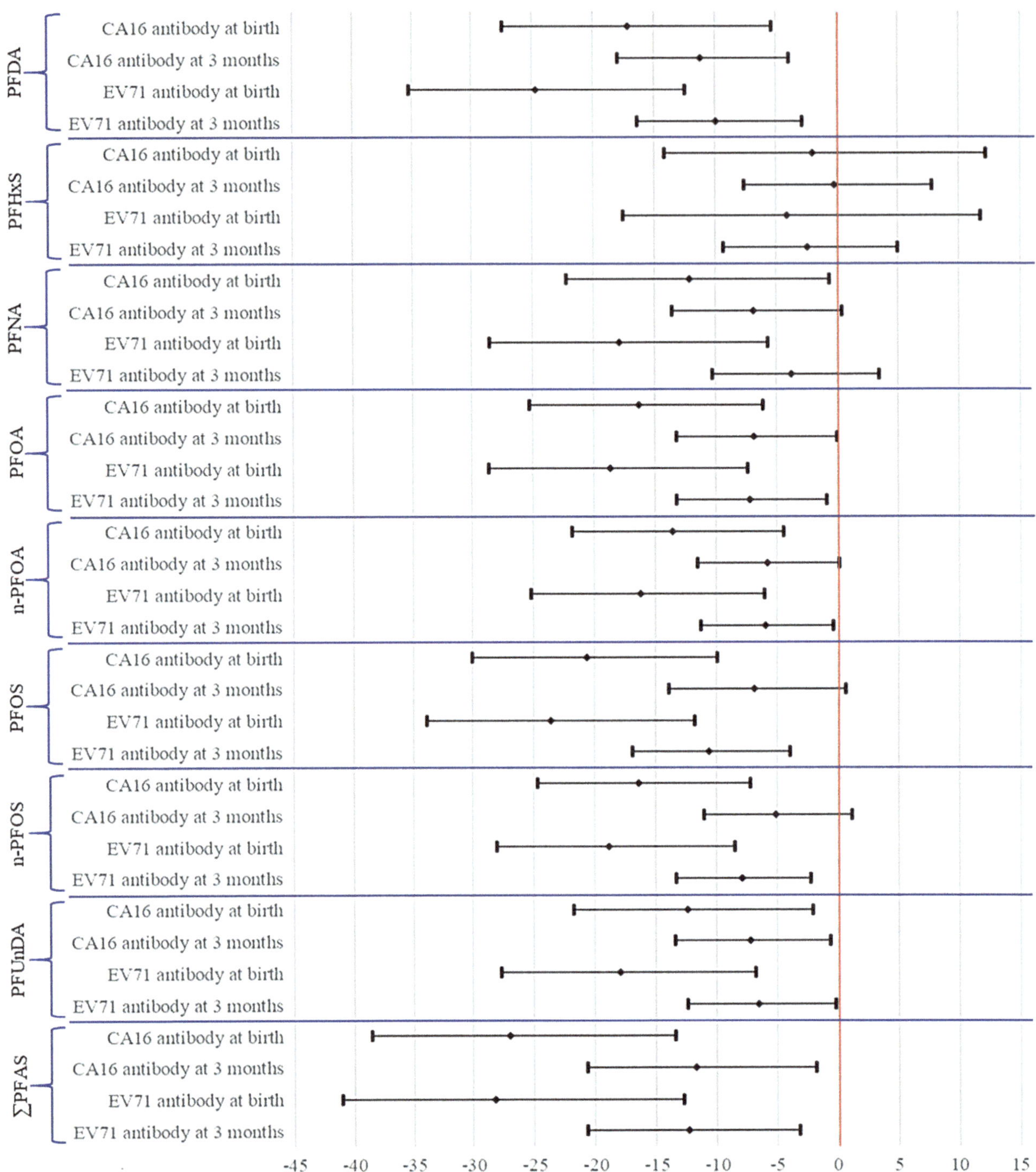

FIGURE 3-5 Regression coefficients for percent change in hand, foot, and mouth disease antibody response per doubling of natural logarithm (ln)-nanograms per milliliter (ng /mL) sum of PFAS.
DATA SOURCE: Zeng et al., 2019.

BOX 3-1
PFAS Exposure and Risk of SARS-CoV-2 Infection

SARS-CoV-2 is a novel virus with low immunity worldwide prior to the pandemic, which provides an opportunity to evaluate the impact of PFAS exposure on response to infection.

What types of studies have been conducted? Available studies evaluate SARS-CoV-2 infection, severity, and mortality.

What have the results shown?
- *Studies on SARS-CoV-2 infection.* The results of two studies show a slightly increased risk of SARS-CoV-2 infection associated with PFAS exposure, but the designs of both preclude drawing causal conclusions. One study, conducted in China, used a case-control format to compare PFAS levels in urine in cases of SARS-CoV-2 infection versus noninfected controls. PFOS, PFOA, and total PFAS levels in urine were higher in those with infection than those without. The observed odds ratios, adjusted for age, sex, body mass index (BMI), comorbidities, and urine albumin-to-creatinine ratio, were 1.94 (95% confidence interval [CI]: 1.39–2.96) for PFOS, 2.73 (95% CI: 1.71–4.55) for PFOA, and 2.82 (95% CI: 1.97–3.51) for the sum of 12 PFAS (Ji et al., 2021). Another study, of ecologic design, calculated sex- and age-standardized incidence ratios (SIRs) for SARS-CoV-2 infection among adults in the Ronneby, Sweden, area, where drinking water was highly contaminated with PFAS, and in a neighboring reference town with similar demographic characteristics but background levels of PFAS exposure (Nielsen and Jöud, 2021). The authors found a slight, but significant, elevated risk of SARS-CoV-2 infection in the former group (SIR = 1.19; 95% CI: 1.12–1.27).
- *SARS-CoV-2 severity.* The results of one study indicate that PFAS may influence the severity of COVID-19 disease, but the study design does not allow for causal determinations. Researchers in Denmark analyzed plasma samples from 323 people with COVID-19 (Grandjean et al., 2021). The authors observed that perfluorobutanoic acid (PFBA) showed an unadjusted odds ratio of 2.19 (95% CI; 1.39–3.46) for increasing the severity of the disease.
- *Studies on SARS-CoV-2 mortality.* One study showed a slight but not statistically significant increased risk of mortality following SARS-CoV-2 infection among individuals exposed to high levels of PFAS contamination; however, its design does not allow for causal determinations. Investigators compared SARS-CoV-2 infection in a community in Venetto, Italy, where drinking water was highly contaminated with PFAS, and in a similar Italian community without contaminated water, using a Bayesian ecological regression model adjusted for education level and baseline all-cause mortality. They found that the rate ratio for COVID-19 mortality for the area of high exposure versus the area of low exposure was 1.60 (90% CI: 0.94–2.51) (Catelan et al., 2021).

Conclusion: Taken as a whole, the extant research does not allow conclusions to be drawn on whether PFAS exposures may influence COVID-19 infection, severity, and mortality. Given that most of the available studies have been ecological, more evidence is necessary to address this question.

Response to Self (Autoimmunity)

As noted in authoritative reviews by the EPA and OECD, the C-8 Science Panel identified an association between PFAS and ulcerative colitis, a rare autoimmune condition of the gastrointestinal (GI) tract. In a follow-up to that study, Steenland and colleagues (2018) further evaluated the exposure–response relationship between PFAS and ulcerative colitis, examining Crohn's disease as well. They found that ulcerative colitis was positively associated with PFOA but not with other PFAS. The odds ratio for ulcerative colitis per 1 unit of log PFOA was 1.60 (95% CI: 1.14–2.24), but the trend by quintiles was not monotonic (1, 0.84, 40.98, 33.36, 2.86) (Steenland et al., 2018). The only other analysis of the association of PFAS with ulcerative colitis is a 2022 case-control study from the Nurses' Health Study using blood specimens collected between 1989 and 1999 (Lochead et al., 2022), identified after the committee had completed its review. This study did not find an association between ulcerative colitis and any PFAS measured; the median concentration of PFOA among the 80 cases was 3.97 ng/mL, higher than the median value of 2.93 ng/mL for the 114 ulcerative colitis cases in the Steenland study. Given the low incidence of ulcerative colitis, it will be difficult to replicate the findings from either of these studies in other populations. The Nurses' Health Study also observed a statistically significant decreased risk of Crohn's disease with PFAS; the Steenland study found no association with Crohn's disease. To assess the relationship between PFAS and inflammatory bowel disease, Xu and colleagues (2020b) measured two proteins (calprotectin and Zonulin) in feces and saw no association with PFAS.

Given the difficulty of evaluating these diseases, future studies are needed to characterize the impact of PFAS on autoimmunity. Although there is more recent inconsistent evidence, it is not strong enough to override the previous conclusion from the C-8 study. Overall, the committee concluded that there is *limited* or *suggestive* evidence of an association of PFAS with ulcerative colitis. The committee did not review studies that considered other autoimmune endpoints.

Cardiometabolic Outcomes

The committee's evaluation of the impact of PFAS on cardiometabolic outcomes considered evidence relevant to four disorders of the three basic functions of the cardiometabolic system: dyslipidemia, high blood pressure or hypertension, metabolic syndrome, and elevated body mass index (BMI) or obesity. The committee did not identify any studies that evaluated the association between PFAS and cardiovascular disease, a group of disorders involving the heart and blood vessels (coronary [ischemic] heart disease, cerebrovascular disease [stroke], peripheral arterial disease, rheumatic heart disease, congenital heart disease, and deep vein thrombosis and pulmonary embolism).[2] The committee concluded that there is *sufficient* evidence of an association of PFAS exposure with dyslipidemia in adults and children. This conclusion builds on those of the authoritative reviews considered by the committee, all of which (including those of ATSDR, EFSA, EPA, OECD, and the C-8 Science Panel) found associations between PFAS and dyslipidemia. The committee concluded that the evidence is *insufficient* for other findings related to cardiovascular risk factors and nonpregnancy clinical cardiovascular diseases. This conclusion is consistent with those of the authoritative reviews, which concluded that there was mixed to limited evidence supporting associations between PFAS and cardiovascular risk factors and diseases other than the four discussed above because of inconsistencies in measurement, differences in study designs and populations, and differences in adjustment for potential confounding factors.

Dyslipidemia

The studies identified by the committee that evaluated dyslipidemia used several types of study designs, including cohort and nested case-control approaches, and the studies encompassed both children and adults. Some of the major challenges to interpretation of their findings were that outcome definitions were inconsistent across studies (total triglycerides, total cholesterol, and low-density lipoprotein [LDL], and high-density lipoprotein [HDL], sometimes measured in a variety of different units). In addition, the study populations included were very broad with respect to age range, sex, and race or ethnicity representation, making it difficult to interpret and generalize the results across studies. In addition, the timing of exposure to PFAS was often unclear, particularly in the studies of adults, in which confounding could be an issue. The study designs and sample sizes varied; there were six cohort studies and one nested case-control study, most of which were rated as having probably low risk of bias. Figures 3-6 and 3-7 display effect estimates from those studies of low or probably low risk of bias that evaluated the impact of PFAS exposure on total cholesterol (the most consistent effect measured across

---

[2] See https://www.heart.org/en/health-topics/consumer-healthcare/what-is-cardiovascular-disease (accessed June 28, 2022).

studies) in adults (Donat-Vargas et al., 2019b; Lin et al., 2019; Tian et al., 2021) and children (Mora et al., 2018). The effects documented across studies were heterogenous, possibly because of the timing of the exposures and outcome measurements. Overall, the committee concluded that there is *sufficient* evidence of an association between PFAS and dyslipidemia, as the recent epidemiologic literature provides additional confidence in the conclusions of authoritative reviews regarding this association.

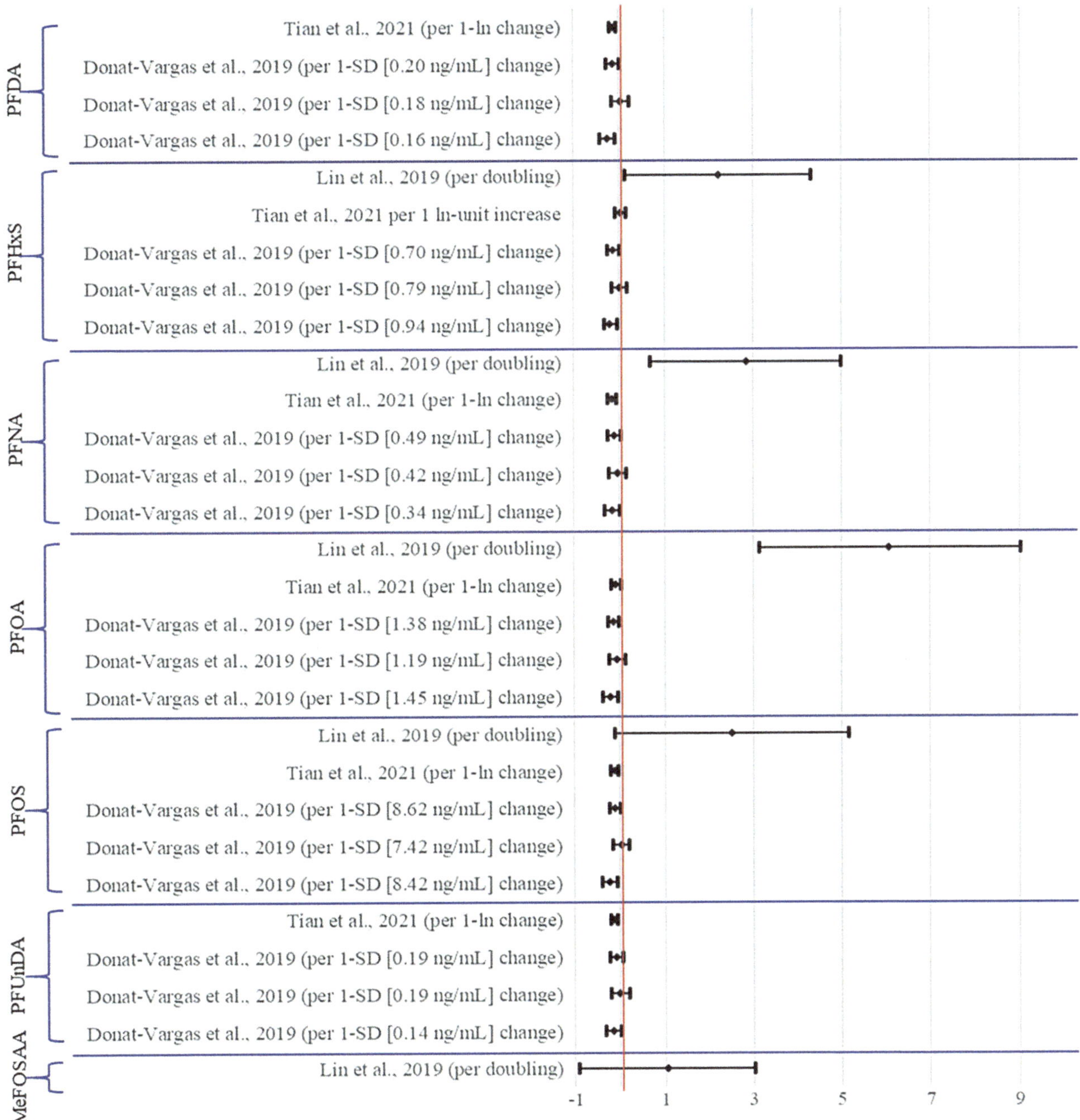

FIGURE 3-6 Regression coefficients for changes in total cholesterol in adults.
DATA SOURCES: Donat-Vargas et al., 2019b; Lin et al., 2019; Tian et al., 2021.

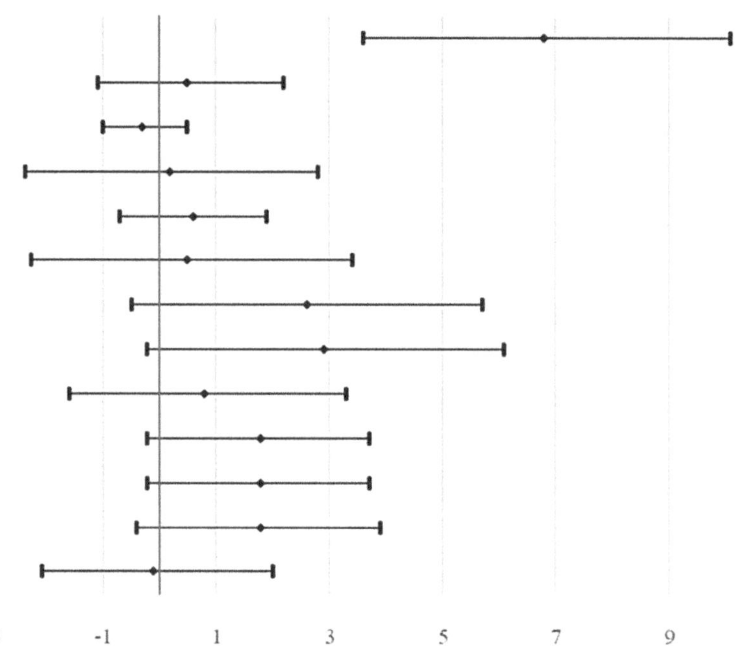

FIGURE 3-7 Regression coefficients for total cholesterol per interquartile range (IQR) increase in PFAS exposure in children.
DATA SOURCE: Mora et al., 2018.

High Blood Pressure or Hypertension

The authoritative reviews do not identify an association between PFAS and high blood pressure or hypertension. The committee identified four studies rated as having probably low or definitely low risk of bias that evaluated the impact of PFAS exposure on hypertension, blood pressure, systolic blood pressure, and diastolic blood pressure (Donat-Vargas et al., 2019b; Lin et al., 2020b; Mitro et al., 2020a). The populations and designs varied greatly across these studies: Donat-Vargus and colleagues (2019b) was a nested case-control study of middle-aged women and men; Lin and colleagues (2020b) was a randomized, controlled clinical trial conducted at 27 clinical centers around the United States from 1996 to 2001; and Mitro and colleagues (2020a) was a cohort study among postpartum females. Donat-Vargus and colleagues (2019b observed that the effect estimates for the impact of exposure to PFAS and hypertension were inconsistent. Lin and colleagues (2020b) observed modest and mostly null associations of plasma PFAS concentrations with high blood pressure and hypertension. Mitro and colleagues (2020a) observed higher systolic blood pressure (e.g., 1.2 mm Hg [95% CI: 0.3, 2.2] per doubling of PFOS) at 3 years postpartum. Given the inconsistency of the evidence, the committee concluded that the evidence is *inadequate* or *insufficient* to determine an association between PFAS and high blood pressure or hypertension.

Metabolic Syndrome

The authoritative reviews do not identify an association between PFAS and metabolic syndrome. Metabolic syndrome is a group of risk factors that increases the risk of heart disease and stroke. Diagnosis requires that an individual have three of the following risk factors: (1) a large waist circumference (males: >102 cm, females: >88 cm); (2) high triglyceride levels (≥1.7 mmol/L); (3) low HDL cholesterol (males <1.04 mmol/L, females <1.30 mmol/L); (4) high blood pressure (≥130 over ≥85 mm Hg); and (5) high fasting glucose levels (≥6.1 mmol/L) (Beilby, 2004). The committee did not identify any new epidemiologic studies reporting associations between exposure to PFAS and a diagnosis of metabolic syndrome. The committee concluded that the evidence is *inadequate* or *insufficient* to determine such an association, although the relationship is plausible given the association between PFAS and dyslipidemia.

Elevated Body Mass Index or Obesity

The authoritative reviews do not identify an association between PFAS and elevated BMI or obesity. Epidemiologic studies have assessed associations between exposure to PFAS and anthropometric outcomes because some PFAS may activate peroxisome proliferator-activated receptor (PPAR) gamma, which promotes adipogenesis (Liu et al., 2018; Takacs and Abbott, 2007). Sex differences in obesity, coupled with differences in exposures within certain subpopulations, may place certain groups at increased risk of overweight and obesity (Fenton et al., 2021; Mitro et al., 2020b; Starling et al., 2019). In addition, exposure during certain periods of growth and development may have short- and long-term consequences for overweight and obesity among children and adolescents, with later-life health consequences (Araújo and Ramos, 2017; Fenton et al., 2021; Gross et al., 2020; Hruby and Hu, 2015; Yeung et al., 2019). Co-exposures to psychosocial factors and other chemicals that also promote fat cell development may work synergistically with PFAS chemicals to impact fat growth and development, with impacts on body weight and growth measures (Araújo and Ramos, 2017; Braun et al., 2021; Chen et al., 2019; Fenton et al., 2021; Jensen et al., 2020a; Liu et al., 2018, 2020; Mitro et al., 2020b; Rahman et al., 2019; Romano et al., 2021; Shoaff et al., 2018; Starling et al., 2019, 2020).

The committee identified several studies that evaluated the impact of exposure to PFAS on obesity, four being rated as having a definitely low risk of bias (Braun et al., 2021; Chen et al., 2019; Mitro et al., 2020a,b; Shoaff et al., 2018). These four studies varied greatly in the ages of the populations assessed. Two of the studies used data from the Health Outcomes and Measures of the Environment (HOME) study cohort (Braun et al., 2021; Shoaff et al., 2018), and one was a study of children in Shanghai, China (Chen et al., 2019). The fourth investigated the impact of PFAS exposure on adiposity among postpartum women (Mitro et al., 2020b). The study by Braun and colleagues (2020) (which largely updates Shoaff et al., 2018)—assessed exposure to PFAS (PFOA, PFOS, and PFHxS) at 16 weeks' gestation and delivery, and measured weight and length or height and calculated child BMI at several time points across the life course (4 weeks to 12 years). The authors observed some suggestive evidence of an impact on BMI trajectory (age × PFOA interaction p value = 0.03) for PFOA, but not for PFOS and PFHxS. The study in Shanghai, China, was a prospective birth cohort study that measured 10 PFAS in cord blood plasma and assessed child adiposity measures at 5 years of age. The authors observed no association for the PFAS considered in this review, but did observe that, among girls, perfluorobutane sulfonic acid (PFBS) exposure had a significant positive association with waist circumference and waist-to-height ratio (p values <0.05). Mitro and colleagues (2020b) observed that PFOS and PFOA were associated with greater adiposity at 3 years' postpartum. The heterogeneity of the effects found across studies is the reason the committee concluded that the evidence for an association between PFAS exposure and elevated BMI or obesity is *inadequate* or *insufficient* in adults and children, although this is an area worthy of future study.

Developmental Outcomes

The committee's evaluation of the impact of PFAS on developmental outcomes considered evidence relevant to fetal growth, development of genitalia, and neurodevelopment. The committee concluded that there is *sufficient* evidence of an association between PFAS exposure and reductions in birthweight. This conclusion builds on those of the authoritative reviews (including ATSDR, EFSA, and OECD). ATSDR concluded that the evidence is suggestive of association between serum PFOA and PFOS and small decreases in birthweight. The committee concluded that there is *insufficient* evidence of an association between PFAS exposure and either development of the external genitalia or neurodevelopmental outcomes.

Birthweight

The committee identified numerous studies with low or probably low risk of bias that examined the relationship between exposure to PFAS and birthweight (Buck Louis et al., 2018; Chu et al., 2020; Gao et al., 2019; Kashino et al., 2020; Marks et al., 2019; Wikstrom et al., 2020; Workman et al., 2019). The magnitude and precision of the estimates of the impacts of PFAS exposure on birthweight varied across and within studies, but the direction of the effect was consistent. None of the studies found a statistically significant effect of PFAS exposure on an increase in birthweight. Table 3-3 presents the estimated impact of PFAS on birthweight from the two studies reviewed by the committee that were rated as having low risk of bias. Overall, these studies strengthened conclusions and supported the committee's assessment that there is *sufficient* evidence of an association of PFAS exposure with small reductions in birthweight.

TABLE 3-3 Effect Estimates Change in Birthweight Per Change in PFAS, from Studies Rated as Having Low Risk of Bias[3]

PFAS	Short Citation	Population	Estimated Change in Birthweight (g) (95% CI)	Units
PFOA	Buck Louis et al., 2018	infants	−5.9 (−28.75 to 16.94)[a]	standard deviation (SD) increase in log-PFOA
	Chu et al., 2020	infants	−73.64 (−126.39 to − 20.88)[b]	1 natural logarithm (ln) change in PFOA
		female infants	−56.04 (−129.32 to 17.24)[c]	1 ln change in PFOA
		male infants	−71.8 (−148.61 to 5)[c]	1 ln change in PFOA
PFOS	Chu et al., 2020	infants	−83.28 (−133.2 to − 33.36)[b]	1 ln change in PFOS
		female infants	−71.91 (−143.86 to 0.05)[c]	1 ln change in PFOS
		male infants	−71.52 (−142.44 to − 0.61)[c]	1 ln change in PFOS

[a] Adjusted for maternal age, education, prepregnant body mass index, serum cotinine, infant sex, and a chemical–maternal race/ethnicity interaction term.
[b] Adjusted for gestational age, maternal age, maternal occupation, maternal education, family income, parity, and infant sex.
[c] Adjusted for gestational age, maternal age, maternal occupation, maternal education, family income, and parity.

Development of Genitalia

Authoritative reviews have yet to associate PFAS exposure with the development of genitalia, a possible indicator of reproductive disorders (Bonde et al., 2016). The committee identified one study that evaluated PFAS exposure and hypospadias and cryptorchidism, potential manifestations of testicular dysgenesis syndrome at birth (Anand-Ivell et al., 2018). Two studies (Arbuckle et al., 2020; Tian et al., 2019) evaluated PFAS and measures of anogenital distance (distance from the anus to the penis or scrotum in males or to the clitoris in females).

Anand-Ivell and colleagues (2018) conducted a case-control study within a large national biobank of amniotic fluid samples, which was rated as having a probably high risk of bias due to the high potential for confounding. The study found no influence of PFOS on cryptorchid or hypospadias (comparison of mean PFOS in amniotic fluid: control versus cryptorchid versus hypospadias). Arbuckle and colleagues (2020) conducted a cohort study with probably low risk of bias and reported inconsistent findings regarding PFAS and anogenital distance. Although the authors observed an association between PFOA (measured in first-trimester maternal plasma) and increased anoscrotal distance (adjusted for active smoking status during pregnancy and gestational age), when they examined the data by quartiles, they found no consistent patterns of association, and the effect estimates were imprecise with wide confidence intervals. Tian and colleagues (2019) conducted a cohort study with probably low risk of

---

[3] This table was altered after release of the pre-publication version of the report to correct data entry errors.

bias to evaluate the impacts of a PFAS on anogenital distance measures. They observed that maternal plasma concentrations (ln-transformed) of PFOS, PFDA, and PFUnDA were inversely associated with anoscrotal distance and anopenile distance measures at birth. For anoscrotal distance, they found per ln unit increase in PFAS concentrations −0.65 (−1.27 to −0.02) mm for PFOS; −0.58 (−1.11 to −0.06) mm for PFDA; and −0.57 (−1.09 to −0.06) mm for PFUnDA. For anopenile distance, they found per ln unit increase in PFAS concentrations −0.63 (−1.24 to −0.01) mm for PFDA and −0.76 (−1.36 to −0.16) mm for PFUnDA. The committee determined that, taken together, the evidence is *inadequate* or *insufficient* to determine an association between PFAS exposure and the development of external genitalia, largely because effects were inconsistent across studies.

Neurodevelopmental Effects

The committee divided the literature on neurodevelopmental effects of PFAS into studies of learning and behavior and of autism spectrum disorder.

Learning and behavior The committee identified 12 studies with low or possibly low risk of bias that examined learning and behavior using psychometrically valid tools to evaluate the impact of PFAS on neurodevelopment in children. The ages of the children varied greatly across studies, as did the timing of the exposure measurement used in the analysis, making it difficult to generalize the findings. For example, Hoyer and colleagues (2018) observed weak effects on child behavior of prenatal exposure to some PFAS. In analysis that combined results from birth cohorts in Greenland and Ukraine, the odds ratio (OR) for hyperactivity was 1.8 (CI: 1.0–3.2) for one nanoliter-unit (nL-unit) increase in prenatal PFNA and 1.7 (CI: 1.0–3.1) for one nL-unit increase in prenatal PFDA exposure. Using the Ages and Stages Questionnaire, Niu and colleagues (2019) found that prenatal plasma concentrations of most PFAS, including PFHxS, PFOS, PFOA, PFNA, and PFDA tended to be associated with an increased risk of developmental problems in personal/social skills, and the associations for PFNA and PFDA were significant (per nL-unit increase). The committee concluded that the evidence is *inadequate* or *insufficient* to determine an association of PFAS exposure with neurodevelopmental effects, largely because of the heterogeneity of both the effects measured and the results observed.

Autism spectrum disorder The committee identified four studies with low or probably low risk of bias evaluating the impacts of PFAS exposure on autism spectrum disorder (Long et al., 2019; Lyall et al., 2018; Oh et al., 2021; Shin et al., 2020). Lyall and colleagues (2018) conducted a population-based, nested case-control study of children born from 2000 to 2003 in southern California and did not observe an association between exposure to PFAS and autism. Long and colleagues (2019) conducted a case-control study that compared exposure to PFAS in amniotic fluid between cases and controls, and observed a negative association between PFAS in amniotic fluid and autism spectrum disorder diagnosis (OR: 0.410; 95% CI: 0.174–0.967). Shin and colleagues (2020) conducted a case-control study of autism spectrum disorder and observed that PFHxS and PFOS were borderline associated with increased odds of child diagnosis of autism spectrum disorder (per ng/mL increase: OR = 1.46; 95% CI: 0.98–2.18 for PFHxS, OR = 1.03; 95% CI: 0.99–1.08 for PFOS). Oh and colleagues (2021) conducted an analysis of the impacts of PFAS exposure on the risk of developing autism spectrum disorder and found that increased PFOA exposures were associated with negative trend Early Learning Composite scores and all four subscales. When they compared trajectories of the scores between low- and high-scoring groups, PFOA was associated with having lower or decreasing Early Learning Composite scores (risk ratio [RR] = 1.49; 95% CI: 1.09–2.03). The committee determined that the evidence is *inadequate* or *insufficient* to determine an association between exposure to PFAS and neurodevelopment, largely because effects were inconsistent across studies.

Cancers

Authoritative reviews have considered the carcinogenic potential of PFAS, and IARC has classified PFOA as possibly carcinogenic to humans (Benbrahim-Tallaa et al., 2014). The committee concluded that there is *sufficient* evidence for an association between PFAS and kidney cancer. This conclusion builds on those of the authoritative reviews (C-8 Science Panel, EPA, ATSDR), which concluded that the evidence for an association between PFAS and cancer in humans is limited, and takes into account robust findings from more recent epidemiological studies. The committee concluded that there is *limited* or *suggestive* evidence for cancers of the testis and breast. The conclusion on testicular cancer is consistent with the authoritative reviews, and the finding on breast cancer is based on the more recent epidemiological studies considered by the committee. The committee found that the existing body of literature on other cancers constituted *inadequate* or *insufficient* evidence to determine an association with PFAS.

Kidney Cancer

The committee's assessment that there is *sufficient* evidence of an association between PFAS exposure and kidney cancer was motivated in large part by the study with low risk of bias conducted by Shearer and colleagues (2021). These investigators conducted a nested case-control study within the Prostate, Lung, Colorectal and Ovarian Cancer Screening Trial with a large sample size, appropriate controls, and validated endpoints (renal cell carcinoma diagnosis [C64.9 in the *International Classification of Diseases for Oncology*, Second Edition]). The statistical analyses conducted by the authors were robust and adjusted for relevant confounders, and sensitivity analysis was performed to assess whether effects were still observed regardless of kidney function. The study clearly showed that ORs for kidney cancer were significantly elevated among individuals in the highest PFOA exposure category, with a strong exposure-response trend. This study and earlier studies (Barry et al., 2013; Vieira et al., 2013) demonstrating a consistency in the direction and magnitude of this effect among those with the highest exposure form a body of literature that the committee concluded constitutes *sufficient* evidence of an association. Effect estimates for associations between PFOA exposure and kidney cancer are summarized by study in Figure 3-8.

Testicular Cancer

The committee's conclusion of *limited* or *suggestive* evidence of an association between PFAS exposure and testicular cancer is consistent with the conclusions of the authoritative reviews. ATSDR does not draw a clear conclusion with respect to PFAS and testicular cancer, but the C-8 Science Panel identified a probable link with incident testicular cancer based on evidence from studies by Barry and colleagues (2013) and Vieira and colleagues (2013). These two studies were also included in the EPA review (2016), which noted the positive associations they found with PFOA and their overlap in cases. IARC (2016) stated that the evidence for an association with testicular cancer was credible and unlikely to be explained by bias and confounding, but limited by small sample numbers. EFSA (2020) found that there was insufficient support for the carcinogenicity of PFOA and PFOS in humans. Given a lack of new studies on this association, the committee found the existing evidence to be supportive of a conclusion of *limited* or *suggestive* evidence of an association between PFAS exposure and testicular cancer (see Figure 3-9).

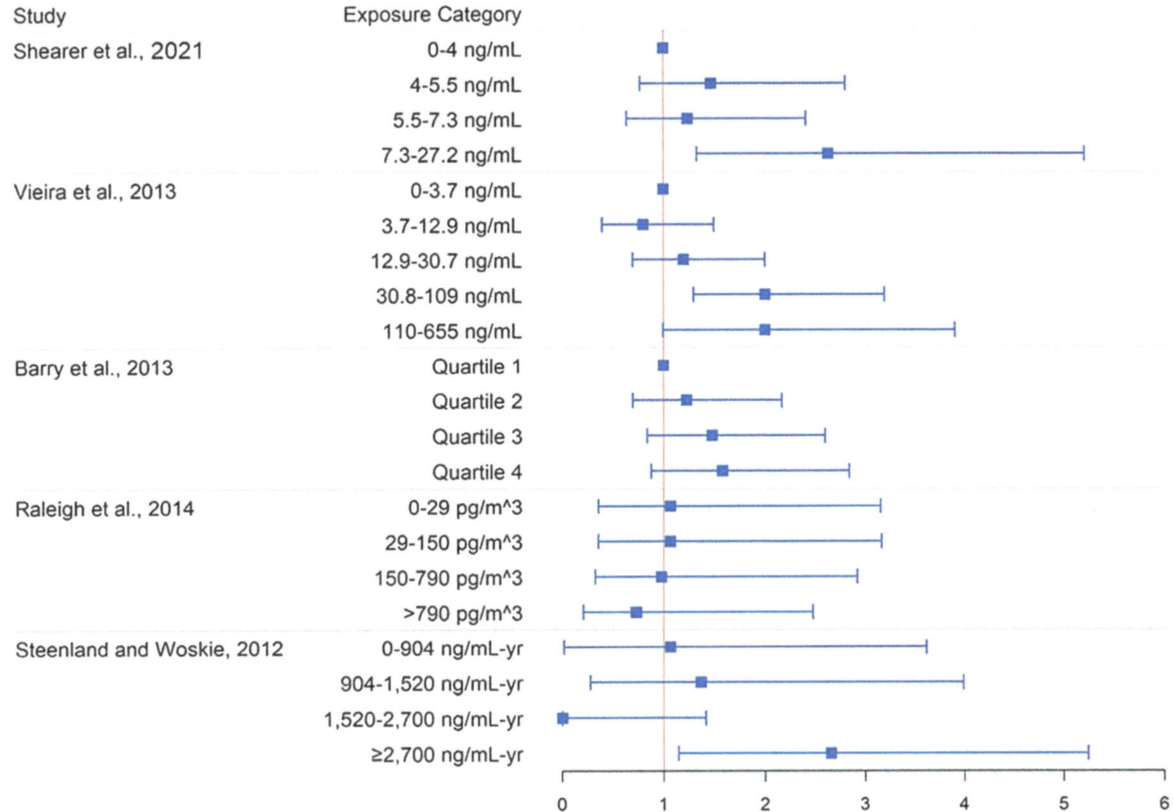

FIGURE 3-8 Kidney cancer adjusted rate ratios and 95% confidence intervals by study and PFOA exposure category.
DATA SOURCES: Barry et al., 2013; Raleigh et al., 2014; Shearer et al., 2021; Steenland and Woskie, 2012; Vieira et al., 2013.

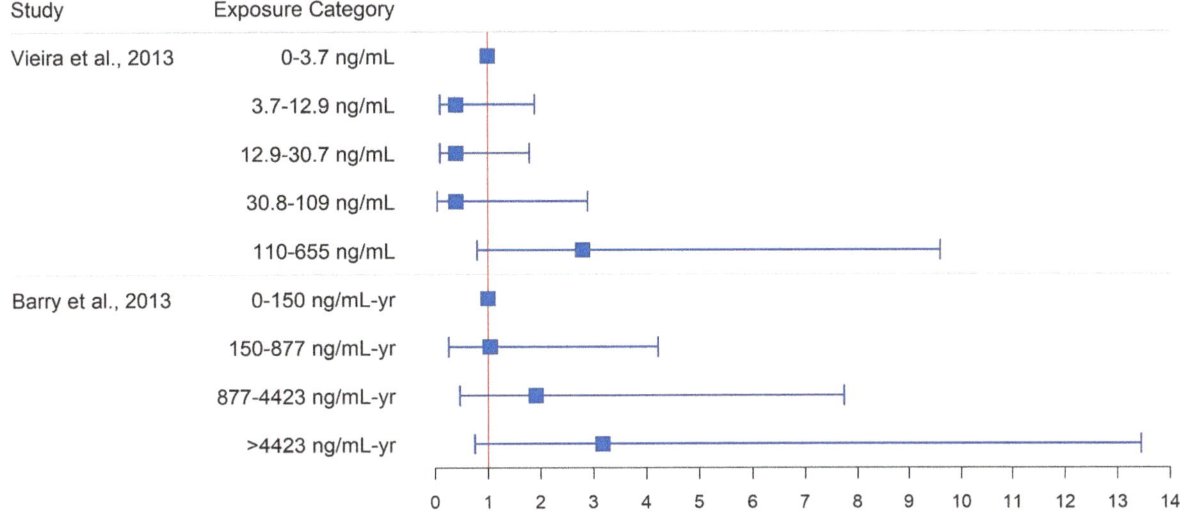

FIGURE 3-9 Testicular cancer adjusted rate ratios and 95% confidence intervals by study and PFOA exposure category.
DATA SOURCES: Barry et al., 2013; Vieira et al., 2013.

Breast Cancer

ATSDR did not draw a clear conclusion with respect to PFAS and breast cancer. Studies included in the ATSDR review found inconsistent associations, with Bonefeld-Jørgensen and colleagues (2014) finding an inverse association with PFHxS and null association for PFOS and PFNA, while Wielsøe and colleagues (2017) found positive associations with all these chemicals. Wielsøe and colleagues (2017) also found positive associations between PFDA, PFUnDA, and PFHpA (not statistically significant), and no association for PFDoDA. Bonefeld-Jorgensen and colleagues (2014) found a positive association with FOSA. IARC (2016) also reviewed the null study of Bonefeld-Jorgensen and colleagues (2014), as did EPA (2016), which stated that no associations were found in the general community. Recent studies, however, found more associations suggestive of a relationship between PFAS and breast cancer. A nested case-control study with low risk of bias found an association between estrogen receptor–positive breast cancer and PFOS (Mancini et al., 2020). An additional study with probably low risk of bias found no evidence of associations between PFAS and breast cancer (Cohn et al., 2020). Two additional recent studies had high risk of bias because the exposure was measured after the cancer diagnosis had been made (Hurley et al., 2018; Tsai et al., 2020). Hurley and colleagues (2018) found no evidence of associations between PFAS exposure and breast cancer, whereas Tsai and colleagues (2020) found evidence of associations among women aged ≤50. Those associations were stronger among women with estrogen receptor–positive tumors. Figure 3-10 summarizes the effect estimates for associations between PFOS exposure and breast cancer by study. The committee found that this body of literature constitutes *limited* or *suggestive* evidence of an association of PFAS exposure with breast cancer.

Reproductive Outcomes

In its review of reproductive outcomes, the committee considered evidence on hypertensive disorders of pregnancy, female reproductive effects, reproductive hormone levels, infertility and subfecundity, and gestational diabetes. The committee concluded that there is *limited* or *suggestive* evidence of an association between PFAS exposure and hypertensive disorders of pregnancy, preeclampsia, and gestational hypertension without preeclampsia. This conclusion is consistent with that of the C-8 Science Panel and subsequent authoritative reviews by ATSDR, EPA, and OECD. Consistent with ATSDR, the committee concluded that there is *insufficient* evidence of an association between PFAS exposure and other reproductive outcomes.

Hypertensive Disorders of Pregnancy (Gestational Hypertension and Preeclampsia)

The committee identified five recent studies examining PFAS exposure and preeclampsia, all having probably low risk of bias (see Figures 3-11–3-15). The four cohort studies (Birukov et al., 2021; Borghese et al., 2020; Huang et al., 2019; Wikstrom et al., 2019) supported the conclusions from the authoritative reviews. Wikstrom and colleagues (2019) observed modestly elevated risk for preeclampsia, but not necessarily a consistent exposure-response trend, and Borghese and colleagues (2020) observed no association with gestational hypertension without preeclampsia. Birukov and colleagues (2021) evaluated exposure to PFAS in early pregnancy and maternal blood pressure trajectories in pregnancy, gestational hypertension, and preeclampsia. No clear associations were observed with gestational hypertension (de novo blood pressure >140/90 mm Hg after 20 weeks' gestation on two or more episodes with at least 4 h in between or significant aggravation of preexisting hypertension) or preeclampsia (gestational hypertension with proteinuria [>0.3 g/24 h or at least +1 on sterile urine dipstick]). Birukov and colleagues (2021) did observe modest but not statistically significant increases in blood pressure for PFOS and PFOA. Another cohort study measuring PFAS in cord blood found an association with preeclampsia but not gestational hypertension (Huang et al., 2019).

A case-control study of preeclampsia and PFAS measured in maternal serum categorized PFAS into quartiles; the women in the highest quartiles had no significant increased risks of developing preeclampsia compared with the women in the lowest quartile in adjusted analyses (Rylander et al., 2020). Given that the studies showed a tendency toward an association between PFAS and hypertensive disorders of pregnancy, the committee concluded that there is *limited* or *suggestive* evidence of an association.

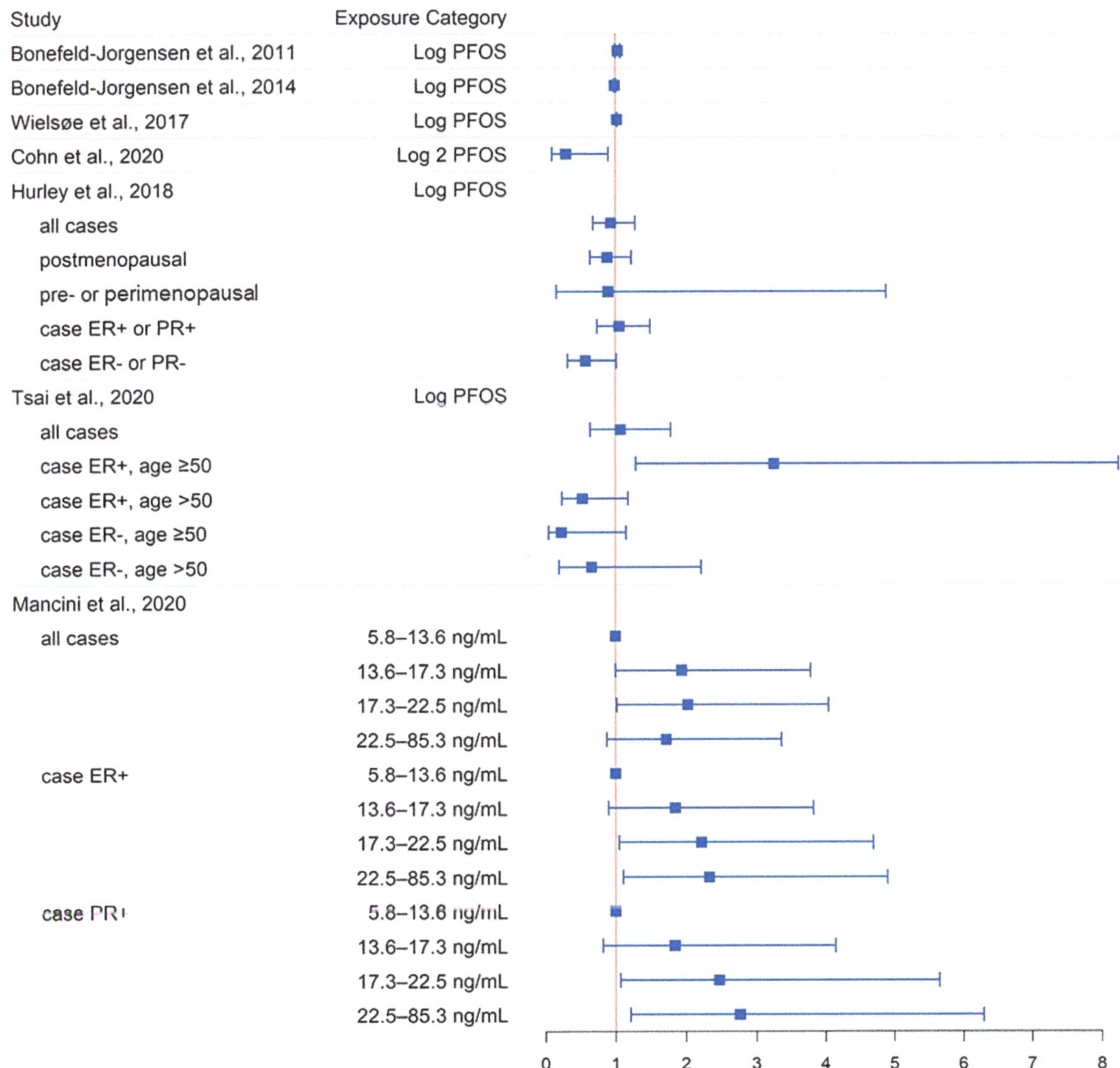

FIGURE 3-10 Breast cancer adjusted rate ratios and 95% confidence intervals by study and PFOS exposure category.
NOTE: ER = estrogen receptor; PR = progesterone receptor.
DATA SOURCES: Bonefeld-Jorgensen et al., 2011; Cohn et al., 2020; Hurley et al., 2018; Mancini et al., 2020; Tsai et al., 2020; Wielsøe et al., 2017.

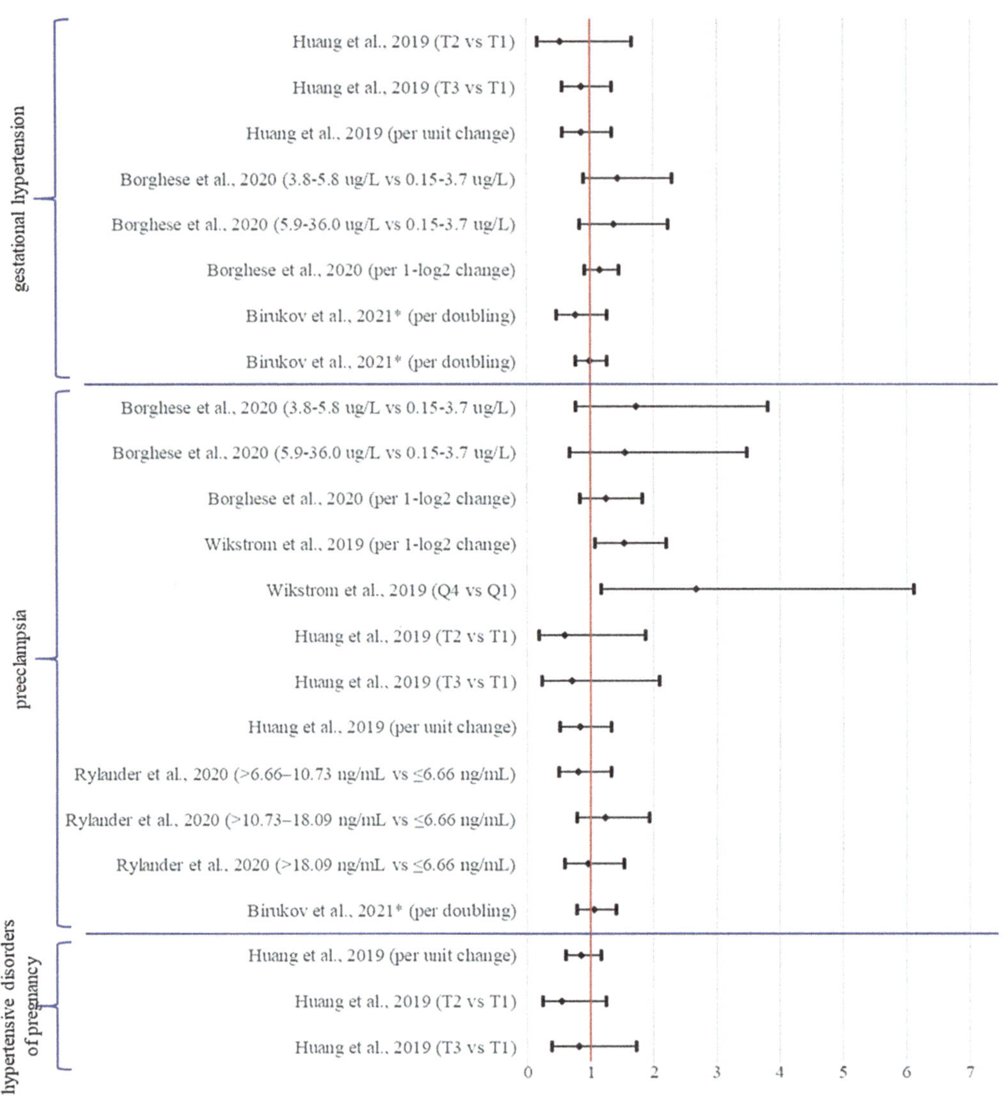

FIGURE 3-11 Adjusted risk estimates for preeclampsia, gestational hypertension (hypertension without preeclampsia), and hypertensive disorders of pregnancy and 95% confidence intervals by study and PFOS exposure category.
NOTE: All studies present odds ratios except Birukov et al., 2021, which presents a hazard ratio.
DATA SOURCES: Birukov et al., 2021; Borghese et al., 2020; Huang et al., 2019; Rylander et al., 2020; Wikstrom et al., 2019.

Fertility and Fecundity

A conclusion in ATSDR's *Toxicological Profile for Perfluoroalkyls* is that "epidemiological studies provided mixed evidence of impaired fertility (increased risks of longer time to pregnancy and infertility) for PFOA, PFOS, PFHxS PFNA, PFHpA, and PFBS: the results are not consistent across studies or were only based on a single study. The small number of studies evaluating fertility for PFDA, PFUnDA, PFDoDA, and FOSA did not find associations and no study has evaluated reproductive outcomes and PFBA" (ATSDR, 2021, p. 359). The committee identified a few more recent studies to update that authoritative review.

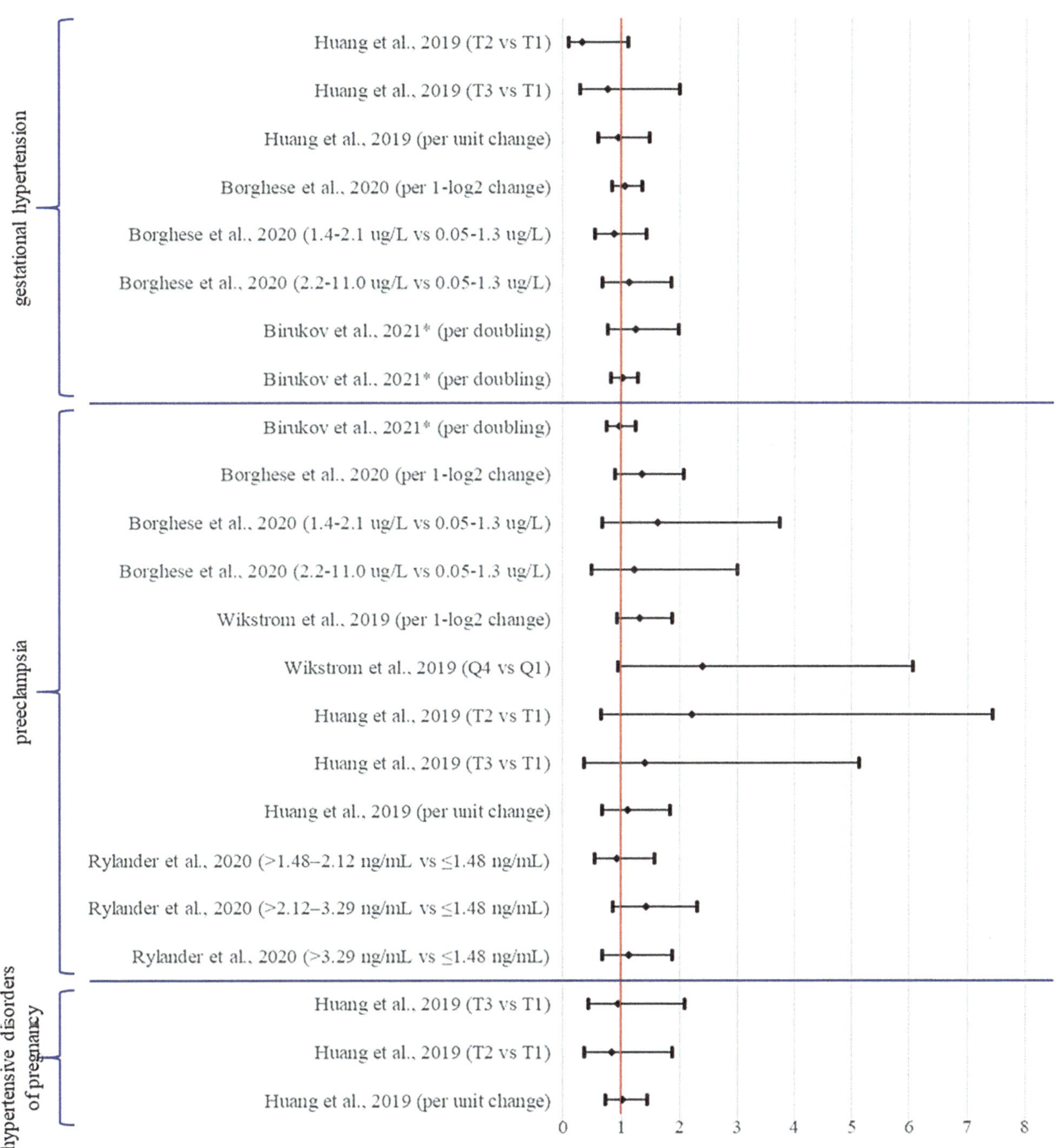

FIGURE 3-12 Adjusted risk estimates for preeclampsia, gestational hypertension (hypertension without preeclampsia), and hypertensive disorders of pregnancy and 95% confidence intervals by study and PFOA exposure category.
NOTE: All studies present odds ratios except Birukov et al., 2021, which presents a hazard ratio.
DATA SOURCES: Birukov et al., 2021; Borghese et al., 2020; Huang et al., 2019; Rylander et al., 2020; Wikstrom et al., 2019.

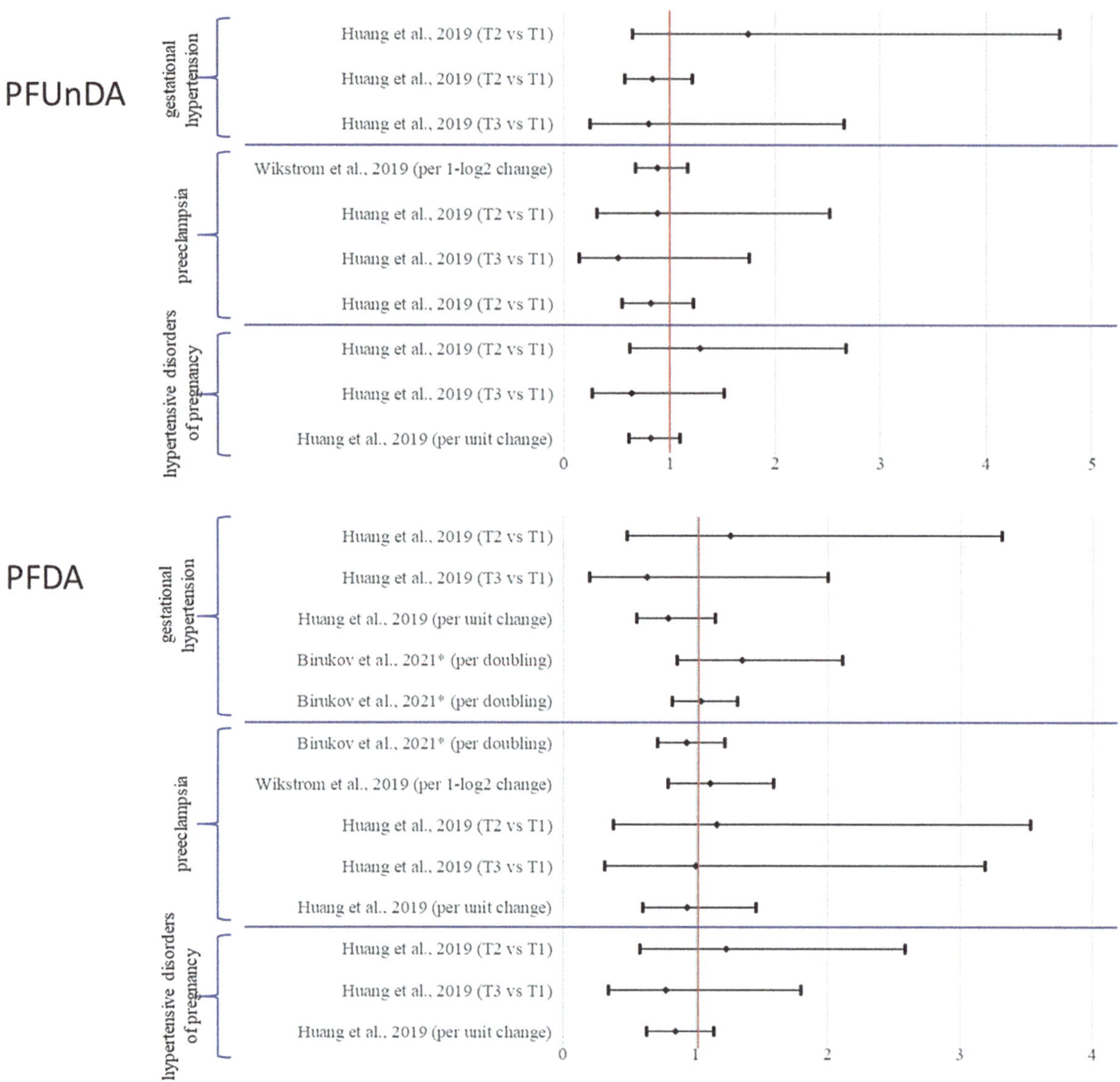

FIGURE 3-13 Adjusted risk estimates for preeclampsia, gestational hypertension (hypertension without preeclampsia), and hypertensive disorders of pregnancy and 95% confidence intervals by study and PFuDA and PFDA exposure category.
NOTE: All studies present odds ratios except Birukov et al., 2021, which presents a hazard ratio.
DATA SOURCES: Birukov et al., 2021; Borghese et al., 2020; Huang et al., 2019; Rylander et al., 2020; Wikstrom et al., 2019.

Zhang and colleagues (2018) conducted a case-control study to evaluate the impact of PFAS on risks of premature ovarian insufficiency, and observed positive associations with PFOA, PFOS, and PFHxS (highest versus lowest tertile, PFOA: OR, 3.80; 95% CI: 1.92–7.49; PFOS: OR, 2.81; 95% CI: 1.46–5.41; PFHxS: OR, 6.63; 95% CI: 3.22–13.65). The study was rated as having high risk of bias for a potential for reverse causality (Zhang et al., 2018). Ma and colleagues (2021) conducted a small cohort study in a fertility clinic in Zhejiang, China, that evaluated the association between PFAS exposure and fertility measures (numbers of retrieved oocytes, mature oocytes, two-pronuclei (2 PN) zygotes, good-quality embryos, and semen parameters). The authors found that maternal plasma concentrations of PFOA were negatively associated with the numbers of retrieved oocytes (p-trend 0.023), mature oocytes (p-trend 0.015), 2 PN zygotes (p-trend 0.014), and good-quality embryos (p-trend 0.012). Higher paternal plasma PFOA concentrations were found to be significantly associated with reduced numbers of 2 PN zygotes (p-trend 0.047), but no associations were found between maternal or paternal PFAS levels and the probability of implantation, clinical pregnancy, or live birth. Given the mixed evidence, the committee concluded that there is *insufficient* evidence of an association between PFAS exposure and fertility or fecundity.

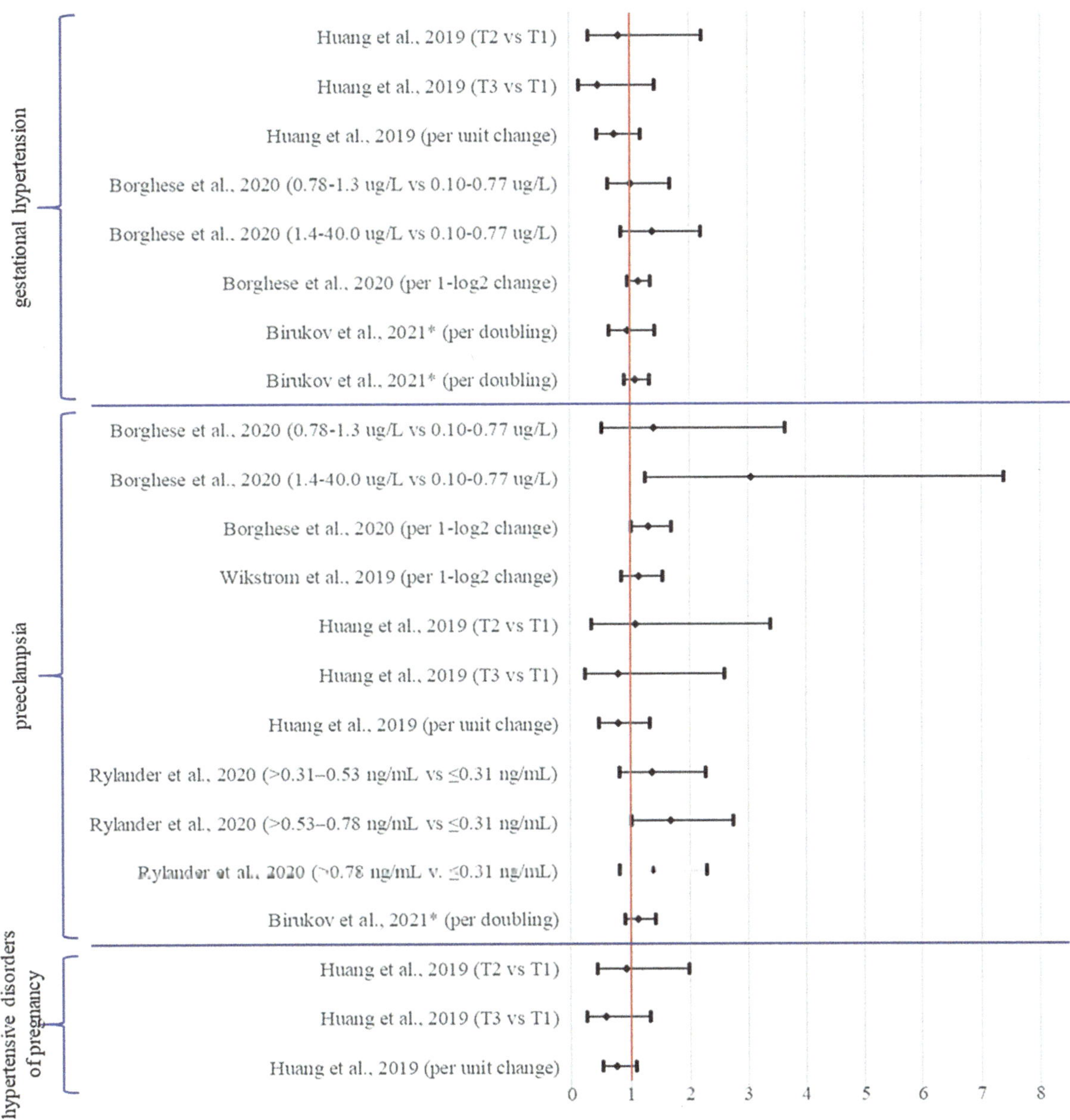

FIGURE 3-14 Adjusted risk estimates for preeclampsia, gestational hypertension (hypertension without preeclampsia), and hypertensive disorders of pregnancy and 95% confidence intervals by study and PFHxS exposure category.
NOTE: All studies present odds ratios except Birukov et al., 2021, which presents a hazard ratio.
DATA SOURCES: Birukov et al., 2021; Borghese et al., 2020; Huang et al., 2019; Rylander et al., 2020; Wikstrom et al., 2019.

Male Reproductive Effects

For its authoritative review, ATSDR looked at articles examining the relationship between PFAS and sperm quality and concluded that while some associations with serum perfluoroalkyl levels were observed for some markers of sperm quality, the markers measured were not consistent across studies. The committee identified only one new study evaluating the effect on PFAS on male reproduction (Ma et al., 2021). This study found no significant association between PFAS and sperm progressive motility rate, but did find associations of some PFAS with decreased sperm concentration. Given that this study has some potential for bias, the committee concluded that the evidence is *inadequate* or *insufficient* to determine an association.

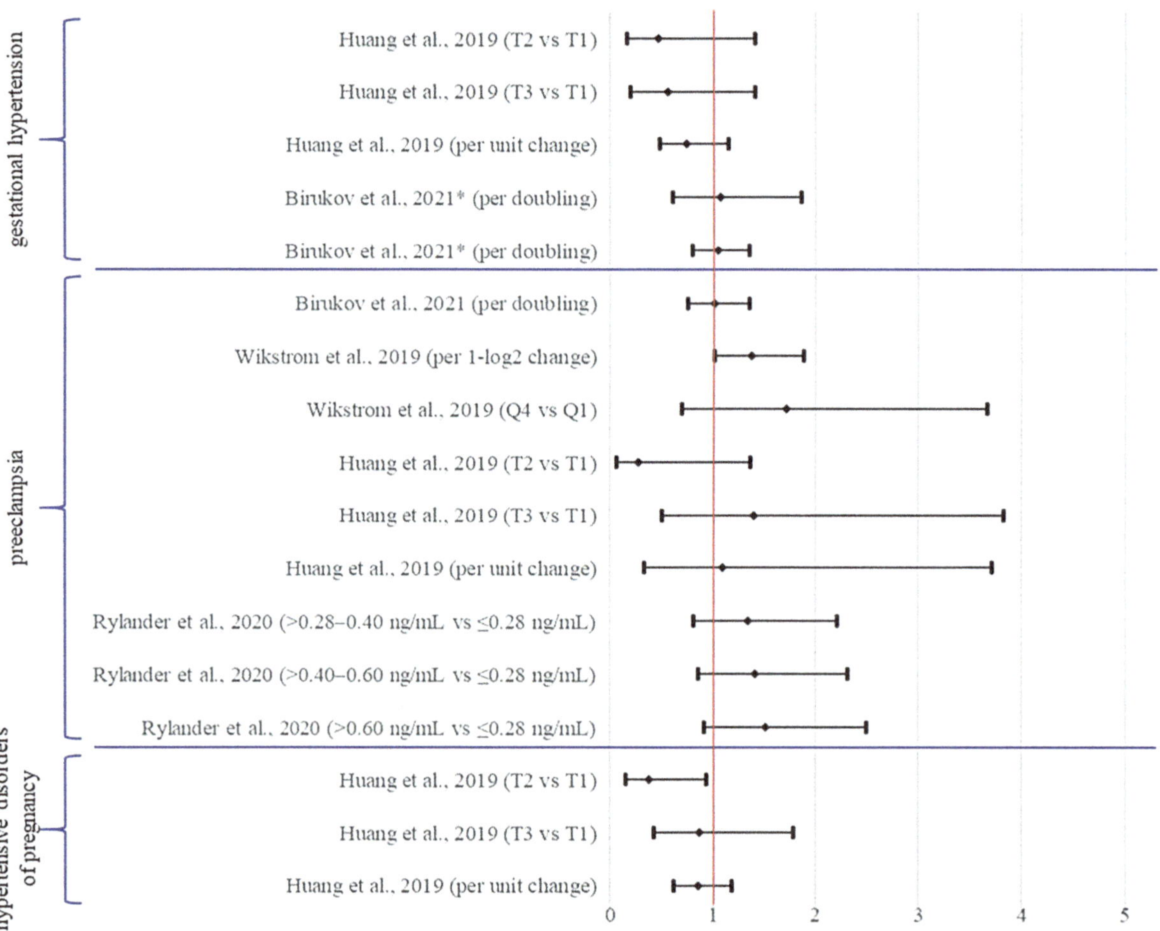

FIGURE 3-15 Adjusted risk estimates for preeclampsia, gestational hypertension (hypertension without preeclampsia), and hypertensive disorders of pregnancy and 95% confidence intervals by study and PFNA exposure category.
NOTE: All studies present odds ratios except Birukov et al., 2021, which presents a hazard ratio.
DATA SOURCES: Birukov et al., 2021; Borghese et al., 2020; Huang et al., 2019; Rylander et al., 2020; Wikstrom et al., 2019.

Female Reproductive Effects

Female reproductive effects discussed here include menopause, age at menarche, and duration of breastfeeding. The authoritative reviews did not find associations for other female reproductive outcomes (polycystic ovary syndrome, endometriosis), and the committee identified no new studies of these effects.

In its *Toxicological Profile for Perfluoroalkyls*, ATSDR concludes that there is some suggestive evidence of an association between serum PFAS levels and an increased risk of early menopause; however, this finding may be due to reverse causation since an earlier onset of menopause would result in a decrease in the removal of perfluoroalkyls in menstrual blood. One more recent cohort study (Ding et al., 2020), with low risk of bias, found an association of PFAS with earlier onset of menopause. Mixed results were observed in a study of age at menarche (Ernst et al., 2019) and a study of cycle irregularity (Singer et al., 2018) (rated as probably having and having low risk of bias, respectively).

ATDSR's *Toxicological Profile* does not offer conclusions on the impact of PFAS exposure on duration of breastfeeding. The committee identified one more recent study, with probably low risk of bias, that observed mostly null associations but also a decreased hazard of breastfeeding cessation by 3 and 6 months with increasing maternal serum concentrations of PFNA, PFDA, and PFUnDA during pregnancy (Rosen et al., 2018). The committee concluded that there is *insufficient* human evidence of an association of PFAS with female reproductive effects, including breastfeeding duration.

Reproductive Hormone Levels

ATSDR's *Toxicology Profile* reviews the literature on associations between PFAS concentrations and reproductive hormones. The conclusion of this review is that while some studies examining reproductive hormone levels have observed associations with PFAS, the findings are inconsistent across studies, and there are too few studies to enable interpretation of the results.

The committee found several more recent studies evaluating the relationship between PFAS and reproductive hormone levels, but these studies varied in the populations they included and the hormones measured, among other factors, making it difficult to synthesize the evidence. For example, three studies evaluated the impact of PFAS on estradiol levels (Ma et al., 2021; Yao et al., 2019; Zhang et al., 2018). The study by Ma and colleagues (2021) was conducted among couples visiting a fertility clinic in China; the study by Zhang and colleagues (2018) was a case-control study of adult women in China; and Yao and colleagues (2019) analyzed PFAS and hormone levels in infant cord blood. The studies that measured testosterone were also distinct; two were analyses of data from birth cohorts (Jensen et al., 2020b; Nian et al., 2020); two were studies in pregnant women (Anand-Ivell et al., 2018; Yao et al., 2019); and two were studies in Chinese adults (Ma et al., 2021; Zhang et al., 2018). The committee concluded that the evidence is too heterogeneous to support drawing conclusions and therefore *inadequate* or *insufficient* to determine an association.

Gestational Diabetes

ATSDR's *Toxicology Profile* reviews the evidence for an association between PFAS and gestational diabetes and concludes that the results of the studies reviewed do not suggest an association. The committee identified four more recent studies evaluating the impact of PFAS on gestational diabetes. These studies had varying designs, including case-control, cohort, and nested case-control (Preston et al., 2020a; Rahman et al., 2019; Wang et al., 2018; Xu et al., 2020a) and had somewhat inconsistent results. Xu and colleagues (2020a) and Rahman and colleagues (2019) observed an effect on gestational diabetes, whereas Preston and colleagues (2020a) and Wang and colleagues (2018) observed an effect on glucose homeostasis. Given the inconsistent effects reported, the committee concluded the evidence is *inadequate* or *insufficient* to determine an association.

Endocrine Outcomes

The committee's evaluation of the impact of PFAS on endocrine outcomes considered evidence in the following subcategories: thyroid disease, hyper- and hypothyroidism and thyroid hormones, and diabetes. The committee found *limited* or *suggestive* evidence of an association with thyroid hormones and disease, and *inadequate* evidence for diabetes.

Thyroid Hormones and Disease

The authoritative review of the C-8 Science Panel and subsequent authoritative reviews completed by ATSDR, EPA, and OECD found associations between PFAS exposure and thyroid hormones and disease. Among the more recent cohort studies with probably low or low risk of bias, the majority observed weak to no association with thyroid hormone levels or subclinical hypothyroidism in children and adults (Blake et al., 2018; Itoh et al., 2019; Jansen et al., 2020; Kim et al., 2020; Lebeaux et al., 2020; Liang et al., 2020; Preston et al., 2020b; Reardon et al., 2019; Xiao et al., 2020). Timing of exposure, life stage, and dietary factors all are likely to modify the relationship between PFAS and thyroid hormones, which could account for the weak observations observed in recent studies. For example, in a cohort study with low risk of bias that measured maternal and cord sera for PFAS and thyroid hormones, Lebeaux and colleagues (2020) found that individual PFAS or mixtures of PFAS were generally not associated with any thyroid hormones, although they did observe a slight association between PFAS and cord serum thyroid-stimulating hormone, with PFOS being the major contributor of the mixture (Bayesian kernel machine regression model estimate per doubling of PFOS [$\beta = 0.09$; 95% credible interval: -0.08–0.27]). They also observed some indication of effect measure modification by maternal thyroid peroxidase antibody status for the associations of PFAS with cord free thyroxine. The committee concluded that there is *limited* or *suggestive* evidence of an association between PFAS and thyroid hormones and thyroid disease.

Diabetes

Authoritative reviews have not found evidence of an association between PFAS and diabetes. ATSDR concluded that, while one prospective cohort study (Sun et al., 2018) suggested an association of PFOA and PFOS with risk of diabetes, overall the epidemiological studies did not provide support for an association between serum PFAS levels and increased risk of diabetes or related outcomes (e.g., increases in blood glucose, glucose tolerance). The committee identified three more recent studies with a probably low risk of bias examining type 2 diabetes. Their results were mixed, with elevated associations observed by Charles and colleagues (2020), an inverse association by Donat-Vargas and colleagues (2019a), and weak or null associations observed by Cardenas and colleagues (2019). One study identified after the committee had completed its literature review (Valvi et al., 2021) observed a modest association with decreased insulin sensitivity and increased pancreatic beta cell function in a cohort study of young adults, but associations with type 2 diabetes were not examined. No new studies examined type 1 diabetes. Given the mixed effects observed, the committee concluded that evidence is *inadequate* or *insufficient* to determine an association between PFAS exposure and diabetes.

Hepatic Outcomes

Authoritative reviews, including those by ATSDR, EFSA, and EPA, have consistently found associations between PFAS and liver effects. ATSDR noted that decreases in serum bilirubin were observed in studies of PFOA, PFOS, and PFHxS, suggestive of liver alterations. The committee identified four more recent studies on PFAS and liver effects with probably low or definitely low risk of bias, including a prebirth cohort study in Boston, a cohort study in Sweden, a study of pooled data from longitudinal birth cohorts across Europe, and a study based on a liver registry in Atlanta (Jin et al., 2020;

Mora et al., 2018; Salihovic et al., 2018; Stratakis et al., 2020). All studies observed some association between PFAS and the liver, but the effects observed were slightly heterogeneous. Mora and colleagues (2018) observed an inverse association between PFAS exposure and alanine transaminase in the prenatal period and in childhood. Salihovic and colleagues (2018) observed that changes in levels of many measured PFAS were positively associated with alanine transaminase and alkaline phosphatase levels and negatively associated with bilirubin. Stratakis and colleagues (2020) observed that higher prenatal exposure to a PFAS mixture was associated with increased risk of liver injury during childhood, as indicated by enzyme levels exceeding the 90th percentile for the study population. And Jin and colleagues (2020) observed that PFAS exposure was associated with more severe disease in children with nonalcoholic fatty liver disease. Taken together, the committee concludes that the available studies provide *limited* or *suggestive* evidence of an association between PFAS exposure and liver enzyme levels.

Respiratory Outcomes

Authoritative reviews, including those of ATSDR, EFSA, and EPA, have not yet drawn conclusions about PFAS exposure and respiratory effects. Respiratory outcomes considered by the committee include pulmonary function tests (objective measures of how well the respiratory system is working); respiratory diseases, including obstructive airway diseases, such as asthma and chronic obstructive pulmonary disease; restrictive diseases, such as pulmonary fibrosis; and respiratory symptoms, such as wheeze, cough, phlegm, and dyspnea (shortness of breath). The committee identified 10 recent studies evaluating the association between PFAS exposure and respiratory outcomes. Three studies with probably low risk of bias evaluated PFAS and pulmonary function in cohorts of children (Agier et al., 2019; Kung et al., 2021; Manzano-Salgado et al., 2019); results were mixed within each study. Several studies evaluated the impact of PFAS exposure on asthma and respiratory symptoms, such as cough and wheeze; results were mixed both between and within studies (Ait Bamai et al., 2020; Beck et al., 2019; Impinen et al., 2019; Kvalem et al., 2020; Manzano-Salgado et al., 2019; Timmermann et al., 2020; Zeng et al., 2019). Taken together, the committee concludes that the available studies provide *inadequate* or *insufficient* evidence of an association between PFAS exposure and liver enzyme levels.

Hematological Outcomes

ATSDR evaluated the impact of PFAS on hematological parameters and concluded that PFAS are associated with no consistent alteration in hematological parameters. The committee did not identify any more recent studies on hematological effects. Thus, the committee concludes that the evidence is *inadequate* or *insufficient* to determine an association between PFAS exposure and hematological effects.

Musculoskeletal Outcomes

The authoritative reviews identify several studies evaluating possible PFAS-associated risk of osteoarthritis, osteoporosis, and reduced bone mineral density (BMD). The committee identified two more recent studies evaluating the relationship between PFAS exposure and bone health (Banjabi et al., 2020; Hu et al., 2019). A case-control study with high risk of bias found that serum PFAS concentrations increased the odds of diagnosis of osteoporosis among adults in Saudi Arabia (Banjabi et al., 2020). A study with probably low risk of bias evaluated BMD within a weight-loss trial of U.S. adults aged 30–70 and found associations between higher plasma PFAS concentrations and lower BMD at baseline, as well as a faster decline in BMD (Hu et al., 2019). The study by Hu and colleagues (2019) provides longitudinal evidence. However, the study was designed to measure weight loss, and its conclusions may not be generalizable to the broader population. In addition, the study is at risk of selection bias because only a small percentage of the study participants had both measures of BMD. There is also a risk of information bias from residual confounding because the analysis did not account for nutritional and menopausal status. Because of the small sample size, moreover, the authors did not adjust for multiple comparisons. Taken together with the evidence presented in ATSDR's *Toxicological Profile*, these findings are intriguing and merit further study. The committee concluded that the evidence is *inadequate* or *insufficient* to determine an association between PFAS and bone health. Nevertheless, the available evidence does raise concerns about the potential adverse effects of PFAS on bone health in both children and adults that warrant further investigation.

Renal Outcomes

ATSDR evaluated the impact of PFAS on kidney disease and biomarkers of renal function, but did not draw any conclusions because results were mixed across studies for most outcomes, and most studies were cross-sectional, so causal determinations could not be made. The committee identified three more recent studies evaluating impacts of PFAS on glomular filtration rate. A cohort study of participants in a medical surveillance program for residents near a former U.S. Department of Energy uranium-processing site assessed serum PFAS and measures of glomular filtration rate at repeated time points from 1990 to 2008 and found decreased glomular filtration rate to be associated with serum PFAS (Blake et al., 2018). The study had a probably high risk of bias because of a small potential for selection bias into the cohort and a slight risk of reverse causality. Two studies analyzing a diabetes prevention trial (Lin et al., 2021, which updates Cardenas et al., 2019) found that plasma PFAS concentrations during the diabetes prevention program were inversely associated with glomular filtration. Each quartile increase in baseline plasma measures of six PFAS was associated with 2.26 ml/min/1.73 m^2 lower glomular filtration (95% CI: −4.12, −0.39) at years 5 and 9. The study's strengths included tests of reverse causation and a lengthy follow-up period, although there is potential for residual confounding. The study's principal limitation is related to the generalizability of the findings to the general population given the inclusion criteria of overweight or obesity and prediabetes. Overall, the committee concluded that the evidence is *inadequate* or *insufficient* to draw a conclusion regarding an association between PFAS exposure and renal function. Nevertheless, the available evidence does raise concerns about the potential adverse effects of PFAS on renal function that warrant further investigation.

Neurological Outcomes

The authoritative reviews do not draw conclusions about the impact of PFAS on neurological outcomes, such as changes in motor function; behavioral changes; mood disorders; sensory disorders; cognitive disorders; and changes in neurochemistry, neurophysiology, or neuropathology. One recent case-control study nested within the diabetes prevention trial mentioned above found no association between any PFAS measured in serum and neuropathy in either the diabetes cases or the controls (Cardenas et al., 2019). The committee concluded that the evidence is *inadequate* or *insufficient* to draw a conclusion regarding an association between PFAS exposure and neurological outcomes.

EVIDENCE GAPS

The committee found several conditions to be associated with exposure to PFAS. The effects of PFAS span many different organ systems and disease states. The human populations most at risk of these health effects include those with a family history of or other risk factors for associated health effects and those who are in vulnerable life stages, including pregnancy, fetal development or early childhood, and the elderly. The committee did not complete a meta-analysis of the impact of each individual PFAS on each health outcome or provide an overall estimate of risk because the data from the studies are highly heterogeneous, limiting the applicability of meta-analytic techniques. The committee also observed gaps in the evidence for many health effects, whereby the evidence was *inadequate* or *insufficient* to determine associations. These gaps include

- immune effects other than reduced antibody response, and ulcerative colitis;
- cardiovascular outcomes other than dyslipidemia;
- developmental outcomes other than small reductions in birthweight;
- cancers other than kidney, breast, and testicular;
- reproductive effects other than hypertensive disorders of pregnancy;
- endocrine disorders other than thyroid hormone levels;
- hepatic effects other than liver enzyme levels;
- respiratory effects;
- hematological effects;
- musculoskeletal effects, such as effects on bone mineral density;
- renal effects, such as renal disease; and
- neurological effects.

It is critical to recognize that an assessment of *inadequate* or *insufficient* evidence does not mean there is no significant and important association between PFAS exposure and the outcome under consideration. It is quite possible that further research into the association of PFAS exposure with these outcomes would provide the evidence necessary to change the assessment to the category of either *limited* or *suggestive* or *sufficient* evidence. The committee believes that ongoing research on and review of these associations will be important in updating its clinical recommendations. A gap also remains in determining which developmental effects are the most clinically meaningful. For some outcome categories, the available research spans many different tests, all of which assessed slightly different effects, making the evidence difficult to synthesize and support strong conclusions. An authoritative organization needs to determine which endpoints are the most critical to evaluate to support clinical follow-up recommendations.

Additionally, most studies reviewed by the committee were not conducted among people known to have high exposures to PFAS. As a result, there is a gap in understanding of the effects of PFAS among those highly exposed, and the evidence presented in this report may therefore underestimate the effects of PFAS.

Although the committee aimed to assess the available scientific evidence as carefully and systematically as possible, it was sensitive to the fact that value judgments are unavoidable when performing an assessment of this kind (Elliott, 2017; Elliott and Richards, 2017; Jasanoff, 1998). These judgments include decisions about what forms of evidence to include, how to weigh and categorize different pieces of evidence, and what standards of evidence to demand before drawing conclusions. Literature from the sociology of science and medicine emphasizes not only that different expert communities may disagree about how to make these decisions (Cetina, 1999) but also that lay communities may make these decisions in ways that differ from those of expert communities (Epstein, 1996; Ottinger, 2010; Suryanarayanan and Kleinman, 2016). For example, because lay communities are often particularly concerned about addressing urgent health issues or informing time-sensitive policy decisions, they may accept lower standards of evidence than experts typically do (Brown, 1992). These differing evidential approaches across different communities raise the potential for differing rates of positive and negative errors (Douglas, 2009; Elliott, 2017; Elliott and Richards, 2017). Different communities also have varying background disease risk, which may lead to differing associations with PFAS and differing needs for risk assessment as it relates to these associations.

With these observations in mind, the committee acknowledges that other expert and lay communities might draw different conclusions about PFAS health risks, either by including different lines of evidence or by making alternative judgments when assessing the available evidence. This is one of the reasons that the committee emphasizes the importance of patient autonomy and shared decision making in subsequent chapters. Although the committee's evaluation of the evidence can provide an important starting point for decision making by clinicians and their patients, some individuals and groups could employ different evidential standards. Therefore, the committee encourages ongoing efforts to make scientific information about PFAS publicly available and understandable so that patients and clinicians can make informed decisions that respect individual patient values.

REFERENCES

Agier, L., X. Basagana, L. Maitre, B. Granum, P. K. Bird, M. Casas, B. Oftedal, J. Wright, S. Andrusaityte, M. de Castro, E. Cequier, L. Chatzi, D. Donaire-Gonzalez, R. Grazuleviciene, L. S. Haug, A. K. Sakhi, V. Leventakou, R. McEachan, M. Nieuwenhuijsen, I. Petraviciene, O. Robinson, T. Roumeliotaki, J. Sunyer, I. Tamayo-Uria, C. Thomsen, J. Urquiza, A. Valentin, R. Slama, M. Vrijheid, and V. Siroux. 2019. Early-life exposome and lung function in children in Europe: An analysis of data from the longitudinal, population-based HELIX cohort. *The Lancet Planetary Health* 3(2):e81–e92.

Ait Bamai, Y., H. Goudarzi, A. Araki, E. Okada, I. Kashino, C. Miyashita, and R. Kishi. 2020. Effect of prenatal exposure to per- and polyfluoroalkyl substances on childhood allergies and common infectious diseases in children up to age 7 years: The Hokkaido study on environment and children's health. *Environment International* 143:105979.

Anand-Ivell, R., A. Cohen, B. Nørgaard-Pedersen, B. A. G. Jönsson, J. P. Bonde, D. M. Hougaard, C. H. Lindh, G. Toft, M. S. Lindhard, and R. Ivell. 2018. Amniotic fluid INSL3 measured during the critical time window in human pregnancy relates to cryptorchidism, hypospadias, and phthalate load: A large case-control study. *Frontiers in Physiology* 9:406.

Araújo, J., and E. Ramos. 2017. Paediatric obesity and cardiovascular risk factors—A life course approach. *Porto Biomedical Journal* 2(4):102–110.

Arbuckle, T. E., S. MacPherson, W. G. Foster, S. Sathyanarayana, M. Fisher, P. Monnier, B. Lanphear, G. Muckle, and W. D. Fraser. 2020. Prenatal perfluoroalkyl substances and newborn anogenital distance in a Canadian cohort. *Reproductive Toxicology (Elmsford, NY)* 94:31–39.

ATSDR (Agency for Toxic Substances and Disease Registry). 2021. *Toxicological profile for perfluoroalkyls*. Atlanta, GA: Centers for Disease Control and Prevention, U.S. Department of Health and Human Services, Public Health Service.

Banjabi, A. A., A. J. Li, T. A. Kumosani, J. M. Yousef, and K. Kannan. 2020. Serum concentrations of perfluoroalkyl substances and their association with osteoporosis in a population in Jeddah, Saudi Arabia. *Environmental Research* 187:109676.

Barry, V., A. Winquist, and K. Steenland. 2013. Perfluorooctanoic acid (PFOA) exposures and incident cancers among adults living near a chemical plant. *Environmental Health Perspectives* 121(11–12):1313–1318.

Beck, I. H., C. A. G. Timmermann, F. Nielsen, G. Schoeters, C. Johnk, H. B. Kyhl, A. Host, and T. K. Jensen. 2019. Association between prenatal exposure to perfluoroalkyl substances and asthma in 5-year-old children in the Odense Child Cohort. *Environmental Health: A Global Access Science Source* 18(1):97.

Beilby, J. 2004. Definition of metabolic syndrome: Report of the National Heart, Lung, and Blood Institute/American Heart Association Conference on Scientific Issues Related to Definition. *The Clinical Biochemist Reviews* 25(3):195–198.

Benbrahim-Tallaa, L., B. Lauby-Secretan, D. Loomis, K. Z. Guyton, Y. Grosse, F. El Ghissassi, V. Bouvard, N. Guha, H. Mattock, and K. Straif. 2014. Carcinogenicity of perfluorooctanoic acid, tetrafluoroethylene, dichloromethane, 1, 2-dichloropropane, and 1, 3-propane sultone. *Lancet Oncology* 15(9):924.

Birukov, A., L. B. Andersen, M. S. Andersen, J. H. Nielsen, F. Nielsen, H. B. Kyhl, J. S. Jorgensen, P. Grandjean, R. Dechend, and T. K. Jensen. 2021. Exposure to perfluoroalkyl substances and blood pressure in pregnancy among 1436 women from the Odense Child Cohort. *Environment International* 151:106442.

Blake, B. E., S. M. Pinney, E. P. Hines, S. E. Fenton, and K. K. Ferguson. 2018. Associations between longitudinal serum perfluoroalkyl substance (PFAS) levels and measures of thyroid hormone, kidney function, and body mass index in the Fernald Community Cohort. *Environmental Pollution (Barking, Essex 1987)* 242(P):894–904.

Bonde, J. P., E. M. Flachs, S. Rimborg, C. H. Glazer, A. Giwercman, C. H. Ramlau-Hansen, K. S. Hougaard, B. B. Hoyer, K. K. Haervig, S. B. Petersen, L. Rylander, I. O. Specht, G. Toft, and E. V. Brauner. 2016. The epidemiologic evidence linking prenatal and postnatal exposure to endocrine disrupting chemicals with male reproductive disorders: A systematic review and meta-analysis. *Human Reproductive Update* 23(1):104–125.

Bonefeld-Jorgensen, E. C., M. Long, R. Bossi, P. Ayotte, G. Asmund, T. Krüger, M. Ghisari, G. Mulvad, P. Kern, P. Nzulumiki, and E. Dewailly. 2011. Perfluorinated compounds are related to breast cancer risk in Greenlandic Inuit: A case control study. *Environmental Health* 10(88). https://doi.org/10.1186/1476-069X-10-88.

Bonefeld-Jørgensen, E. C., M. Long, S. O. Fredslund, R. Bossi, and J. Olsen. 2014. Breast cancer risk after exposure to perfluorinated compounds in Danish women: A case–control study nested in the Danish National Birth Cohort. *Cancer Causes & Control* 25(11):1439–1448.

Borghese, M. M., M. Walker, M. E. Helewa, W. D. Fraser, and T. E. Arbuckle. 2020. Association of perfluoroalkyl substances with gestational hypertension and preeclampsia in the MIREC study. *Environment International* 141:105789.

Braun, J. M., M. Eliot, G. D. Papandonatos, J. P. Buckley, K. M. Cecil, H. J. Kalkwarf, A. Chen, C. B. Eaton, K. Kelsey, B. P. Lanphear, and K. Yolton. 2021. Gestational perfluoroalkyl substance exposure and body mass index trajectories over the first 12 years of life. *International Journal of Obesity (2005)* 45(1):25–35.

Brown, P. 1992. Popular epidemiology and toxic waste contamination: Lay and professional ways of knowing. *Journal of Health and Social Behavior* 33(September):267–281.

Buck Louis, G. M., S. Zhai, M. M. Smarr, J. Grewal, C. Zhang, K. L. Grantz, S. N. Hinkle, R. Sundaram, S. Lee, M. Honda, J. Oh, and K. Kannan. 2018. Endocrine disruptors and neonatal anthropometry, NICHD Fetal Growth Studies—Singletons. *Environment International* 119:515–526.

C-8 Medical Panel. 2013. *C-8 Medical Panel report.* https://www.hpcbd.com/wp-content/uploads/sites/1603732/2021/01/Medical-Panel-Report-2013-05-24.pdf (accessed June 15, 2022).

Cardenas, A., M.-F. Hivert, D. R. Gold, R. Hauser, K. P. Kleinman, P.-I. D. Lin, A. F. Fleisch, A. M. Calafat, X. Ye, T. F. Webster, E. S. Horton, and E. Oken. 2019. Associations of perfluoroalkyl and polyfluoroalkyl substances with incident diabetes and microvascular disease. *Diabetes Care* 42(9):1824–1832.

Catelan, D., A. Biggeri, F. Russo, D. Gregori, G. Pitter, F. Da Re, T. Fletcher, and C. Canova. 2021. Exposure to perfluoroalkyl substances and mortality for COVID-19: A spatial ecological analysis in the Veneto region (Italy). *International Journal of Environmental Research and Public Health* 18(5):2734.

Cetina, K. K. 1999. *Epistemic cultures: How the sciences make knowledge.* Cambridge, MA: Harvard University Press.

Charles, D., V. Berg, T. H. Nøst, S. Huber, T. M. Sandanger, and C. Rylander. 2020. Pre- and post-diagnostic blood profiles of perfluoroalkyl acids in type 2 diabetes mellitus cases and controls. *Environment International* 145:106095. https://doi.org/10.1016/j.envint.2020.106095.

Chen, Q., R. Huang, L. Hua, Y. Guo, L. Huang, Y. Zhao, X. Wang, and J. Zhang. 2018. Prenatal exposure to perfluoroalkyl and polyfluoroalkyl substances and childhood atopic dermatitis: A prospective birth cohort study. *Environmental Health* 17(1):8. https://doi.org//10.1186/s12940-018-0352-7.

Chen, Q., X. Zhang, Y. Zhao, W. Lu, J. Wu, S. Zhao, J. Zhang, and L. Huang. 2019. Prenatal exposure to perfluorobutanesulfonic acid and childhood adiposity: A prospective birth cohort study in Shanghai, China. *Chemosphere* 226:17–23.

Chu, C., Y. Zhou, Q.-Q. Li, M. S. Bloom, S. Lin, Y.-J. Yu, D. Chen, H.-Y. Yu, L.-W. Hu, B.-Y. Yang, X.-W. Zeng, and G.-H. Dong. 2020. Are perfluorooctane sulfonate alternatives safer? New insights from a birth cohort study. *Environment International* 135:105365.

Cohn, B. A., M. A. La Merrill, N. Y. Krigbaum, M. Wang, J.-S. Park, M. Petreas, G. Yeh, R. C. Hovey, L. Zimmermann, and P. M. Cirillo. 2020. In utero exposure to poly- and perfluoroalkyl substances (PFASs) and subsequent breast cancer. *Reproductive Toxicology (Elmsford, NY)* 92:112–119.

Ding, N., S. D. Harlow, J. F. Randolph, A. M. Calafat, B. Mukherjee, S. Batterman, E. B. Gold, and S. K. Park. 2020. Associations of perfluoroalkyl substances with incident natural menopause: The Study of Women's Health Across the Nation. *The Journal of Clinical Endocrinology and Metabolism* 105(9):e3169–e3182.

Donat-Vargas, C., I. A. Bergdahl, A. Tornevi, M. Wennberg, J. Sommar, H. Kiviranta, J. Koponen, O. Rolandsson, and A. Akesson. 2019a. Perfluoroalkyl substances and risk of type II diabetes: A prospective nested case-control study. *Environment International* 123:390–398.

Donat-Vargas, C., I. A. Bergdahl, A. Tornevi, M. Wennberg, J. Sommar, J. Koponen, H. Kiviranta, and A. Akesson. 2019b. Associations between repeated measure of plasma perfluoroalkyl substances and cardiometabolic risk factors. *Environment International* 124:58–65.

Douglas, H. E. 2009. Science, policy, and the value-free ideal. Pittsburgh, PA: University of Pittsburgh Press.

EFSA (European Food Safety Authority). 2020. Risk to human health related to the presence of perfluoroalkyl substances in food. *EFSA Journal* 18(9):e06223.

Elliott, K. C. 2017. *A tapestry of values: An introduction to values in science.* New York: Oxford University Press.

Elliott, K. C., and T. Richards. 2017. *Exploring inductive risk: Case studies of values in science.* New York: Oxford University Press.

EPA (U.S. Environmental Protection Agency). 2005. *Guidelines for carcinogen risk assessment.* https://www.epa.gov/sites/default/files/2013-09/documents/cancer_guidelines_final_3-25-05.pdf (accessed June 15, 2022).

EPA. 2016. *Health effects support document for perfluorooctanoic acid (PFOA).* Washington, DC: U.S. Environmental Protection Agency.

Epstein, S. 1996. *Impure science: AIDS, activism, and the politics of knowledge.* Oakland, CA: University of California Press.

Ernst, A., N. Brix, L. L. B. Lauridsen, J. Olsen, E. T. Parner, Z. Liew, L. H. Olsen, and C. H. Ramlau-Hansen. 2019. Exposure to perfluoroalkyl substances during fetal life and pubertal development in boys and girls from the Danish National Birth Cohort. *Environmental Health Perspectives* 127(1):17004.

Fedak, K. M., A. Bernal, Z. A. Capshaw, and S. Gross. 2015. Applying the Bradford Hill criteria in the 21st century: How data integration has changed causal inference in molecular epidemiology. *Emerging Themes in Epidemiology* 12:14.

Fenton, S. E., A. Ducatman, A. Boobis, J. C. DeWitt, C. Lau, C. Ng, J. S. Smith, and S. M. Roberts. 2021. Per- and polyfluoroalkyl substance toxicity and human health review: Current state of knowledge and strategies for informing future research. *Environmental Toxicology and Chemistry* 40(3):606–630.

Gao, K., T. Zhuang, X. Liu, J. Fu, J. Zhang, J. Fu, L. Wang, A. Zhang, Y. Liang, M. Song, and G. Jiang. 2019. Prenatal exposure to per- and polyfluoroalkyl substances (PFASs) and association between the placental transfer efficiencies and dissociation constant of serum proteins-PFAS complexes. *Environmental Science & Technology* 53(1):6529–6538.

Grandjean, P., C. A. G. Timmermann, M. Kruse, F. Nielsen, P. J. Vinholt, L. Boding, C. Heilmann, and K. Mølbak. 2021. Severity of COVID-19 at elevated exposure to perfluorinated alkylates. *PLOS ONE* 15(12):e0244815.

Gross, R. S., A. Ghassabian, S. Vandyousefi, M. J. Messito, C. Gao, K. Kannan, and L. Trasande. 2020. Persistent organic pollutants exposure in newborn dried blood spots and infant weight status: A case-control study of low-income Hispanic mother-infant pairs. *Environmental Pollution* 267:115427.

Hill, A. B. 1965. The environment and disease: Association or causation? *Proceedings of the Royal Society of Medicine* 58(5):295–300.

Høyer, B. B., J. P. Bonde, S. S. Tøttenborg, C. H. Ramlau-Hansen, C. Lindh, H. S. Pedersen, G. Toft. 2018. Exposure to perfluoroalkyl substances during pregnancy and child behaviour at 5 to 9 years of age. *Hormones and Behavior* 101:105–112. https://doi.org/10.1016/j.yhbeh.2017.11.007.

Hruby, A., and F. B. Hu. 2015. The epidemiology of obesity: A big picture. *PharmacoEconomics* 33(7):673–689.

Hu, Y., G. Liu, J. Rood, L. Liang, G. A. Bray, L. de Jonge, B. Coull, J. D. Furtado, L. Qi, P. Grandjean, and Q. Sun. 2019. Perfluoroalkyl substances and changes in bone mineral density: A prospective analysis in the POUNDS-LOST study. *Environmental Research* 179(P):108775.

Huang, H., K. Yu, X. Zeng, Q. Chen, Q. Liu, Y. Zhao, J. Zhang, X. Zhang, and L. Huang. 2020. Association between prenatal exposure to perfluoroalkyl substances and respiratory tract infections in preschool children. *Environmental Research* 191:110156.

Huang, R., Q. Chen, L. Zhang, K. Luo, L. Chen, S. Zhao, L. Feng, and J. Zhang. 2019. Prenatal exposure to perfluoroalkyl and polyfluoroalkyl substances and the risk of hypertensive disorders of pregnancy. *Environmental Health: A Global Access Science Source* 18(1):5.

Hurley, S., D. Goldberg, M. Wang, J.-S. Park, M. Petreas, L. Bernstein, H. Anton-Culver, D. O. Nelson, and P. Reynolds. 2018. Breast cancer risk and serum levels of per- and poly-fluoroalkyl substances: A case-control study nested in the California Teachers Study. *Environmental Health: A Global Access Science Source* 17(1):83.

IARC (International Agency for Research on Cancer) Working Group. 2016. *IARC Monographs on the Evaluation of Carcinogenic Risks to Humans.* Some Chemicals Used as Solvents and in Polymer Manufacture. Vol. 110. Lyon, France.

Impinen, A., M. P. Longnecker, U. C. Nygaard, S. J. London, K. K. Ferguson, L. S. Haug, and B. Granum. 2019. Maternal levels of perfluoroalkyl substances (PFASs) during pregnancy and childhood allergy and asthma related outcomes and infections in the Norwegian Mother and Child (MoBa) cohort. *Environment International* 124:462–472.

Itoh, S., A. Araki, C. Miyashita, K. Yamazaki, H. Goudarzi, M. Minatoya, Y. Ait Bamai, S. Kobayashi, E. Okada, I. Kashino, M. Yuasa, T. Baba, and R. Kishi. 2019. Association between perfluoroalkyl substance exposure and thyroid hormone/thyroid antibody levels in maternal and cord blood: The Hokkaido Study. *Environment International* 133(P):105139.

Jansen, A., J. P. Berg, O. Klungsoyr, M. H. B. Muller, J. L. Lyche, and J. O. Aaseth. 2020. The influence of persistent organic pollutants on thyroidal, reproductive and adrenal hormones after bariatric surgery. *Obesity Surgery* 30(4):1368–1378.

Jasanoff, S. 1998. The political science of risk perception. *Reliability Engineering & System Safety* 59(1):91–99.

Jensen, R. C., M. S. Andersen, P. V. Larsen, D. Glintborg, C. Dalgard, C. A. G. Timmermann, F. Nielsen, M. B. Sandberg, H. R. Andersen, H. T. Christesen, P. Grandjean, and T. K. Jensen. 2020a. Prenatal exposures to perfluoroalkyl acids and associations with markers of adiposity and plasma lipids in infancy: An Odense Child Cohort Study. *Environmental Health Perspectives* 128(7):77001.

Jensen, R. C., D. Glintborg, C. A. Gade Timmermann, F. Nielsen, H. B. Kyhl, H. Frederiksen, A. M. Andersson, A. Juul, J. J. Sidelmann, H. R. Andersen, P. Grandjean, M. S. Andersen, and T. K. Jensen. 2020b. Prenatal exposure to perfluorodecanoic acid is associated with lower circulating concentration of adrenal steroid metabolites during mini puberty in human female infants: The Odense Child Cohort. *Environmental Research* 182:109101.

Ji, J., L. Song, J. Wang, Z. Yang, H. Yan, T. Li, L. Yu, L. Jian, F. Jiang, J. Li, J. Zheng, and K. Li. 2021. Association between urinary per- and poly-fluoroalkyl substances and COVID-19 susceptibility. *Environment International* 153:106524.

Jin, R., R. McConnell, C. Catherine, S. Xu, D. I. Walker, N. Stratakis, D. P. Jones, G. W. Miller, C. Peng, D. V. Conti, M. B. Vos, and L. Chatzi. 2020. Perfluoroalkyl substances and severity of nonalcoholic fatty liver in children: An untargeted metabolomics approach. *Environment International* 134:105220.

Kashino, I., S. Sasaki, E. Okada, H. Matsuura, H. Goudarzi, C. Miyashita, E. Okada, Y. M. Ito, A. Araki, and R. Kishi. 2020. Prenatal exposure to 11 perfluoroalkyl substances and fetal growth: A large-scale, prospective birth cohort study. *Environment International* 136:105355.

Kim, Y. R., N. White, J. Braunig, S. Vijayasarathy, J. F. Mueller, C. L. Knox, F. A. Harden, R. Pacella, and L.-M. L. Toms. 2020. Per- and poly-fluoroalkyl substances (PFASs) in follicular fluid from women experiencing infertility in Australia. *Environmental Research* 190:109963.

Kung, Y.-P., C.-C. Lin, M.-H. Chen, M.-S. Tsai, W.-S. Hsieh, and P.-C. Chen. 2021. Intrauterine exposure to per- and polyfluoroalkyl substances may harm children's lung function development. *Environmental Research* 192:110178.

Kvalem, H. E., U. C. Nygaard, K. C. Lodrup Carlsen, K. H. Carlsen, L. S. Haug, and B. Granum. 2020. Perfluoroalkyl substances, airways infections, allergy and asthma related health outcomes—Implications of gender, exposure period and study design. *Environment International* 134:105259.

Kwiatkowski, C. F., D. Q. Andrews, L. S. Birnbaum, T. A. Bruton, J. C. DeWitt, D. R. U. Knappe, M. V. Maffini, M. F. Miller, K. E. Pelch, A. Reade, A. Soehl, X. Trier, M. Venier, C. C. Wagner, Z. Wang, and A. Blum. 2020. Scientific basis for managing PFAS as a chemical class. *Environmental Science & Technology Letters* 7(8):532–543.

Lebeaux, R. M., B. T. Doherty, L. G. Gallagher, R. T. Zoeller, A. N. Hoofnagle, A. M. Calafat, M. R. Karagas, K. Yolton, A. Chen, B. P. Lanphear, J. M. Braun, and M. E. Romano. 2020. Maternal serum perfluoroalkyl substance mixtures and thyroid hormone concentrations in maternal and cord sera: The HOME Study. *Environmental Research* 185:109395.

Liang, H., Z. Wang, M. Miao, Y. Tian, Y. Zhou, S. Wen, Y. Chen, X. Sun, and W. Yuan. 2020. Prenatal exposure to perfluoroalkyl substances and thyroid hormone concentrations in cord plasma in a Chinese birth cohort. *Environmental Health: A Global Access Science Source* 19(1):127.

Lin, P.-I. D., A. Cardenas, R. Hauser, D. R. Gold, K. P. Kleinmman, M. F. Hivert, A. F. Fleisch, A. M. Calafat, T. F. Webster, E. S. Horton, and E. Oken. 2019. Per- and polyfluoroalkyl substances and blood lipid levels in pre-diabetic adults-longitudinal analysis of the diabetes prevention program outcomes study. *Environment International* 129:343–353.

Lin, H.-W., H.-X. Feng, L. Chen, X.-J. Yuan, and Z. Tan. 2020a. Maternal exposure to environmental endocrine disruptors during pregnancy is associated with pediatric germ cell tumors. *Nagoya Journal of Medical Science* 82(2):323–333.

Lin, P.-I. D., A. Cardenas, R. Hauser, D. R. Gold, K. P. Kleinman, M.-F. Hivert, A. M. Calafat, T. F. Webster, E. S. Horton, and E. Oken. 2020b. Per- and polyfluoroalkyl substances and blood pressure in pre-diabetic adults-cross-sectional and longitudinal analyses of the diabetes prevention program outcomes study. *Environment International* 137:105573.

Lin, P.-I. D., A. Cardenas, R. Hauser, D. R. Gold, K. P. Kleinman, M.-F. Hivert, A. M. Calafat, T. F. Webster, E. S. Horton, and E. Oken. 2021. Per- and polyfluoroalkyl substances and kidney function: Follow-up results from the Diabetes Prevention Program trial. *Environment International* 148:106375.

Liu, P., F. Yang, Y. Wang and Z. Yuan. 2018. Perfluorooctanoic acid (PFOA) exposure in early life increases risk of childhood adiposity: A meta-analysis of prospective cohort studies. *International Journal of Environmental Research and Public Health* 15(10):2070.

Liu, Y., N. Li, G. D. Papandonatos, A. M. Calafat, C. B. Eaton, K. T. Kelsey, A. Chen, B. P. Lanphear, K. M. Cecil, H. J. Kalkwarf, K. Yolton, and J. M. Braun. 2020. Exposure to per- and polyfluoroalkyl substances and adiposity at age 12 years: Evaluating periods of susceptibility. *Environmental Science & Technology* 54(2):16039–16049.

Lochhead, P, H. Khalili, A. N. Ananthakrishnan, K. E. Burke, J. M. Richter, Q. Sun, P. Grandjean, and A. T. Chan. 2022. Plasma concentrations of perfluoroalkyl substances and risk of inflammatory bowel diseases in women: A nested case control analysis in the Nurses' Health Study cohorts. *Environmental Research.* 207:112222. https://doi.org/10.1016/j.envres.2021.112222.

Long, M., M. Ghisari, L. Kjeldsen, M. Wielsoe, B. Norgaard-Pedersen, E. L. Mortensen, M. W. Abdallah, and E. C. Bonefeld-Jorgensen. 2019. Autism spectrum disorders, endocrine disrupting compounds, and heavy metals in amniotic fluid: A case-control study. *Molecular Autism* 10:1.

Lyall, K., V. M. Yau, R. Hansen, M. Kharrazi, C. K. Yoshida, A. M. Calafat, G. Windham, and L. A. Croen. 2018. Prenatal maternal serum concentrations of per- and polyfluoroalkyl substances in association with autism spectrum disorder and intellectual disability. *Environmental Health Perspectives* 126(1):017001.

Ma, X., L. Cui, L. Chen, J. Zhang, X. Zhang, Q. Kang, F. Jin, and Y. Ye. 2021. Parental plasma concentrations of perfluoroalkyl substances and in vitro fertilization outcomes. *Environmental Pollution (Barking, Essex 1987)* 269:116159.

Mancini, F. R., G. Cano-Sancho, J. Gambaretti, P. Marchand, M.-C. Boutron-Ruault, G. Severi, P. Arveux, J.-P. Antignac, and M. Kvaskoff. 2020. Perfluorinated alkylated substances serum concentration and breast cancer risk: Evidence from a nested case-control study in the French E3N cohort. *International Journal of Cancer* 146(4):917–928.

Manzano-Salgado, C. B., B. Granum, M.-J. Lopez-Espinosa, F. Ballester, C. Iniguez, M. Gascon, D. Martinez, M. Guxens, M. Basterretxea, C. Zabaleta, T. Schettgen, J. Sunyer, M. Vrijheid, and M. Casas. 2019. Prenatal exposure to perfluoroalkyl substances, immune-related outcomes, and lung function in children from a Spanish birth cohort study. *International Journal of Hygiene and Environmental Health* 222(6):945–954.

Marks, K. J., A. J. Cutler, Z. Jeddy, K. Northstone, K. Kato, and T. J. Hartman. 2019. Maternal serum concentrations of perfluoroalkyl substances and birth size in British boys. *International Journal of Hygiene and Environmental Health* 222(5):889–895.

Mitro, S. D., S. K. Sagiv, A. F. Fleisch, L. M. Jaacks, P. L. Williams, S. L. Rifas-Shiman, A. M. Calafat, M.-F. Hivert, E. Oken, and T. M. James-Todd. 2020a. Pregnancy per- and polyfluoroalkyl substance concentrations and postpartum health in Project Viva: A prospective cohort. *The Journal of Clinical Endocrinology and Metabolism* 105(9):e3415–e3426.

Mitro, S. D., S. K. Sagiv, S. L. Rifas-Shiman, A. M. Calafat, A. F. Fleisch, L. M. Jaacks, P. L. Williams, E. Oken, and T. M. James-Todd. 2020b. Per- and polyfluoroalkyl substance exposure, gestational weight gain, and postpartum weight changes in Project Viva. *Obesity (Silver Spring, MD)* 28(1):1984–1992.

Mora, A. M., A. F. Fleisch, S. L. Rifas-Shiman, J. A. Woo Baidal, L. Pardo, T. F. Webster, A. M. Calafat, X. Ye, E. Oke, and S. K. Sagiv. 2018. Early life exposure to per- and polyfluoroalkyl substances and mid-childhood lipid and alanine aminotransferase levels. *Environment International* 111:1–13. https://doi.org/10.1016/j.envint.2017.11.008.

NASEM (National Academies of Sciences, Engineering, and Medicine). 2018a. *Advances in causal understanding for human health risk-based decision-making: Proceedings of a Workshop—in Brief.* Washington, DC: The National Academies Press. https://doi.org/10.17226/25004.

NASEM. 2018b. *Gulf War and health. Volume 11: Generational health effects of serving in the Gulf War.* Washington, DC: The National Academies Press. https://doi.org/10.17226/25162.

NASEM. 2018c. *Veterans and Agent Orange: Update 11.* Washington, DC: The National Academies Press. https://doi.org/10.17226/25137.

Nian, M., K. Luo, F. Luo, R. Aimuzi, X. Huo, Q. Chen, Y. Tian, and J. Zhang. 2020. Association between prenatal exposure to PFAS and fetal sex hormones: Are the short-chain PFAS safer? *Environmental Science and Technology* 54(1):8291–8299.

Nielsen, C., and A. Jöud. 2021. Susceptibility to COVID-19 after high exposure to perfluoroalkyl substances from contaminated drinking water: An ecological study from Ronneby, Sweden. *International Journal of Environmental Research and Public Health* 18(20):10702.

Niu, J., H. Liang, Y. Tian, W. Yuan, H. Xiao, H. Hu, X. Sun, X. Song, S. Wen, L. Yang, Y. Ren, and M. Miao. 2019. Prenatal plasma concentrations of perfluoroalkyl and polyfluoroalkyl substances and neuropsychological development in children at four years of age. *Environmental Health* 18(1):53.

Oh, J., D. H. Bennett, A. M. Calafat, D. Tancredi, D. L. Roa, R. J. Schmidt, I. Hertz-Picciotto, and H.-M. Shin. 2021. Prenatal exposure to per- and polyfluoroalkyl substances in association with autism spectrum disorder in the MARBLES study. *Environment International* 147:106328.

Ottinger, G. 2010. Buckets of resistance: Standards and the effectiveness of citizen science. *Science, Technology, & Human Values* 35(2):244–270.

Preston, E. V., S. L. Rifas-Shiman, M.-F. Hivert, A. R. Zota, S. K. Sagiv, A. M. Calafat, E. Oken, and T. James-Todd. 2020a. Associations of per- and polyfluoroalkyl substances (PFAS) with glucose tolerance during pregnancy in Project Viva. *The Journal of Clinical Endocrinology and Metabolism* 105(8):e2864–e2876.

Preston, E. V., T. F. Webster, B. Claus Henn, M. D. McClean, C. Gennings, E. Oken, S. L. Rifas-Shiman, E. N. Pearce, A. M. Calafat, A. F. Fleisch, and S. K. Sagiv. 2020b. Prenatal exposure to per- and polyfluoroalkyl substances and maternal and neonatal thyroid function in the Project Viva Cohort: A mixtures approach. *Environment International* 139:105728.

Rahman, M. L., C. Zhang, M. M. Smarr, S. Lee, M. Honda, K. Kannan, F. Tekola-Ayele, and G. M. Buck Louis. 2019. Persistent organic pollutants and gestational diabetes: A multi-center prospective cohort study of healthy US women. *Environment International* 124:249–258.

Raleigh, K. K., B. H. Alexander, G. W. Olsen, G. Ramachandran, S. Z. Morey, T. R. Church, P. W. Logan, L. L. Scott, and E. M. Allen. 2014. Mortality and cancer incidence in ammonium perfluorooctanoate production workers. *Occupational and Environmental Medicine* 71(7):500–506. https://doi.org/10.1136/oemed-2014-102109.

Reardon, A. J. F., E. Khodayari Moez, I. Dinu, S. Goruk, C. J. Field, D. W. Kinniburgh, A. M. MacDonald, J. W. Martin, and A. P. Study. 2019. Longitudinal analysis reveals early-pregnancy associations between perfluoroalkyl sulfonates and thyroid hormone status in a Canadian prospective birth cohort. *Environment International* 129:389–399.

Romano, M. E., L. G. Gallagher, M. N. Eliot, A. M. Calafat, A. Chen, K. Yolton, B. Lanphear, and J. M. Braun. 2021. Per- and polyfluoroalkyl substance mixtures and gestational weight gain among mothers in the Health Outcomes and Measures of the Environment study. *International Journal of Hygiene and Environmental Health* 231:113660.

Rosen, E. M., A. L. Brantsæter, R. Carroll, L. Haug, A. B. Singer, S. Zhao, and K. K. Ferguson. 2018. Maternal plasma concentrations of per- and polyfluoroalkyl substances and breastfeeding duration in the Norwegian mother and child cohort. *Environmental Epidemiology* 2(3):e027. https://doi.org/10.1097/EE9.0000000000000027.

Rylander, L., C. H. Lindh, S. R. Hansson, K. Broberg, and K. Källén. 2020. Per- and polyfluoroalkyl substances in early pregnancy and risk for preeclampsia: A case-control study in southern Sweden. *Toxics* 8(2):43.

Salihovic, S., J. Stubleski, A. Karrman, A. Larsson, T. Fall, L. Lind, and P. M. Lind. 2018. Changes in markers of liver function in relation to changes in perfluoroalkyl substances—A longitudinal study. *Environment International* 117:196–203.

Shea, B. J., B. C. Reeves, G. Wells, M. Thuku, C. Hamel, J. Moran, D. Moher, P. Tugwell, V. Welch, E. Kristjansson, and D. A. Henry. 2017. AMSTAR 2: A critical appraisal tool for systematic reviews that include randomised or non-randomised studies of healthcare interventions, or both. *British Medical Journal* 358:j4008.

Shearer, J. J., C. L. Callahan, A. M. Calafat, W.-Y. Huang, R. R. Jones, V. S. Sabbisetti, N. D. Freedman, J. N. Sampson, D. T. Silverman, M. P. Purdue, and J. N. Hofmann. 2021. Serum concentrations of per- and polyfluoroalkyl substances and risk of renal cell carcinoma. *Journal of the National Cancer Institute* 113(5):580–587.

Shin, H.-M., D. H. Bennett, A. M. Calafat, D. Tancredi, and I. Hertz-Picciotto. 2020. Modeled prenatal exposure to per- and polyfluoroalkyl substances in association with child autism spectrum disorder: A case-control study. *Environmental Research* 186:109514.

Shoaff, J., G. D. Papandonatos, A. M. Calafat, A. Chen, B. P. Lanphear, S. Ehrlich, K. T. Kelsey, and J. M. Braun. 2018. Prenatal exposure to perfluoroalkyl substances: Infant birth weight and early life growth. *Environmental Epidemiology (Philadelphia, PA)* 2(2):e101.

Singer, A. B., K. W. Whitworth, L. S. Haug, A. Sabaredzovic, A. Impinen, E. Papadopoulou, and M. P. Longnecker. 2018. Menstrual cycle characteristics as determinants of plasma concentrations of perfluoroalkyl substances (PFASs) in the Norwegian mother and child cohort (MoBa study). *Environmental Research* 166:78–85.

Starling, A. P., J. L. Adgate, R. F. Hamman, K. Kechris, A. M. Calafat, and D. Dabelea. 2019. Prenatal exposure to per- and polyfluoroalkyl substances and infant growth and adiposity: The Healthy Start Study. *Environment International* 131:104983.

Starling, A. P., C. Liu, G. Shen, I. V. Yang, K. Kechris, S. J. Borengasser, K. E. Boyle, W. Zhang, H. A. Smith, A. M. Calafat, R. F. Hamman, J. L. Adgate, and D. Dabelea. 2020. Prenatal exposure to per- and polyfluoroalkyl substances, umbilical cord blood DNA methylation, and cardio-metabolic indicators in newborns: The Healthy Start Study. *Environmental Health Perspectives* 128(1):127014.

Steenland, K., and S. Woskie. 2012. Cohort mortality study of workers exposed to perfluorooctanoic acid. *American Journal of Epidemiology* 176(10):909–917. https://doi.org/10.1093/aje/kws171.

Steenland, K., S. Kugathasan, and D. B. Barr. 2018. PFOA and ulcerative colitis. *Environmental Research* 165:317–321.

Stratakis, N., V. C. D, R. Jin, K. Margetaki, D. Valvi, A. P. Siskos, L. Maitre, E. Garcia, N. Varo, Y. Zhao, T. Roumeliotaki, M. Vafeiadi, J. Urquiza, S. Fernandez-Barres, B. Heude, X. Basagana, M. Casas, S. Fossati, R. Grazuleviciene, S. Andrusaityte, K. Uppal, R. R. C. McEachan, E. Papadopoulou, O. Robinson, L. S. Haug, J. Wright, M. B. Vos, H. C. Keun, M. Vrijheid, K. T. Berhane, R. McConnell, and L. Chatzi. 2020. Prenatal exposure to perfluoroalkyl substances associated with increased susceptibility to liver injury in children. *Hepatology* 72(5):1758–1770.

Sun, Q., G. Zong, D. Valvi, F. Nielsen, B. Coull, and P. Grandjean. 2018. Plasma concentrations of perfluoroalkyl substances and risk of type 2 diabetes: A prospective investigation among U.S. women. *Environmental Health Perspectives* 126(3):037001. https://doi.org/10.1289/EHP2619.

Suryanarayanan, S., and D. L. Kleinman. 2016. *Vanishing bees.* New Brunswick, NJ: Rutgers University Press.

Takacs, M. L., and B. D. Abbott. 2007. Activation of mouse and human peroxisome proliferator-activated receptors (alpha, beta/delta, gamma) by perfluorooctanoic acid and perfluorooctane sulfonate. *Toxicological Sciences* 95(1):108–117.

Tian, Y., H. Liang, M. Miao, F. Yang, H. Ji, W. Cao, X. Liu, X. Zhang, A. Chen, H. Xiao, H. Hu, and W. Yuan. 2019. Maternal plasma concentrations of perfluoroalkyl and polyfluoroalkyl substances during pregnancy and anogenital distance in male infants. *Human Reproduction (Oxford, England)* 34(7):1356–1368.

Tian, Y., M. Miao, H. Ji, X. Zhang, A. Chen, Z. Wang, W. Yuan, and H. Liang. 2021. Prenatal exposure to perfluoroalkyl substances and cord plasma lipid concentrations. *Environmental Pollution* 268(P):115426. https://doi.org/10.1016/j.envpol.2020.115426.

Timmermann, C. A. G., K. J. Jensen, F. Nielsen, E. Budtz-Jorgensen, F. van der Klis, C. S. Benn, P. Grandjean, and A. B. Fisker. 2020. Serum perfluoroalkyl substances, vaccine responses, and morbidity in a cohort of Guinea-Bissau children. *Environmental Health Perspectives* 128(8):87002.

Tsai, M.-S., S.-H. Chang, W.-H. Kuo, C.-H. Kuo, S.-Y. Li, M.-Y. Wang, D.-Y. Chang, Y.-S. Lu, C.-S. Huang, A.-L. Cheng, C.-H. Lin, and P.-C. Chen. 2020. A case-control study of perfluoroalkyl substances and the risk of breast cancer in Taiwanese women. *Environment International* 142:105850.

Valvi, D., K. Højlund, B. A. Coull, F. Nielsen, P. Weihe, and P. Grandjean. 2021. Life-course exposure to perfluoroalkyl substances in relation to markers of glucose homeostasis in early adulthood. *Journal of Clinical Endocrinology & Metabolism* 106(8):2495–2504.

Vieira, V. M., K. Hoffman, H. M. Shin, J. M. Weinberg, T. F. Webster, and T. Fletcher. 2013. Perfluorooctanoic acid exposure and cancer outcomes in a contaminated community: A geographic analysis. *Environmental Health Perspectives* 121(3):318–323.

Wang, Y., L. Zhang, Y. Teng, J. Zhang, L. Yang, J. Li, J. Lai, Y. Zhao, and Y. Wu. 2018. Association of serum levels of perfluoroalkyl substances with gestational diabetes mellitus and postpartum blood glucose. *Journal of Environmental Sciences (China)* 69:5–11.

Wen, H. J., S. L. Wang, Y. C. Chuang, P. C. Chen, and Y. L. Guo. 2019. Prenatal perfluorooctanoic acid exposure is associated with early onset atopic dermatitis in 5-year-old children. *Chemosphere* 231:25–31. https://doi.org/10.1016/j.chemosphere.2019.05.100.

Wielsøe, M., P. Kern, and E. C. Bonefeld-Jørgensen. 2017. Serum levels of environmental pollutants is a risk factor for breast cancer in Inuit: A case control study. *Environmental Health* 16(1):56.

Wikstrom, S., C. H. Lindh, H. Shu, and C.-G. Bornehag. 2019. Early pregnancy serum levels of perfluoroalkyl substances and risk of preeclampsia in Swedish women. *Scientific Reports* 9(1):9179.

Wikstrom, S., P.-I. Lin, C. H. Lindh, H. Shu, and C.-G. Bornehag. 2020. Maternal serum levels of perfluoroalkyl substances in early pregnancy and offspring birth weight. *Pediatric Research* 87(6):1093–1099.

Woodruff, T. J., and P. Sutton. 2014. The Navigation Guide systematic review methodology: A rigorous and transparent method for translating environmental health science into better health outcomes. *Environmental Health Perspectives* 122(10):1007–1014.

Workman, C. E., A. B. Becker, M. B. Azad, T. J. Moraes, P. J. Mandhane, S. E. Turvey, P. Subbarao, J. R. Brook, M. R. Sears, and C. S. Wong. 2019. Associations between concentrations of perfluoroalkyl substances in human plasma and maternal, infant, and home characteristics in Winnipeg, Canada. *Environmental Pollution (Barking, Essex: 1987)* 249:758–766.

Xiao, C., P. Grandjean, D. Valvi, F. Nielsen, T. K. Jensen, P. Weihe, and Y. Oulhote. 2020. Associations of exposure to perfluoroalkyl substances with thyroid hormone concentrations and birth size. *The Journal of Clinical Endocrinology & Metabolism* 105(3):735–745.

Xu, H., Q. Zhou, J. Zhang, X. Chen, H. Zhao, H. Lu, B. Ma, Z. Wang, C. Wu, C. Ying, Y. Xiong, Z. Zhou, and X. Li. 2020a. Exposure to elevated per- and polyfluoroalkyl substances in early pregnancy is related to increased risk of gestational diabetes mellitus: A nested case-control study in Shanghai, China. *Environment International* 143:105952.

Xu, Y., Y. Li, K. Scott, C. H. Lindh, K. Jakobsson, T. Fletcher, B. Ohlsson, and E. M. Andersson. 2020b. Inflammatory bowel disease and biomarkers of gut inflammation and permeability in a community with high exposure to perfluoroalkyl substances through drinking water. *Environmental Research* 181:108923.

Yao, Q., R. Shi, C. Wang, W. Han, Y. Gao, Y. Zhang, Y. Zhou, G. Ding, and Y. Tian. 2019. Cord blood per- and polyfluoroalkyl substances, placental steroidogenic enzyme, and cord blood reproductive hormone. *Environment International* 129:573–582.

Yeung, E. H., E. M. Bell, R. Sundaram, A. Ghassabian, W. Ma, K. Kannan, and G. M. Louis. 2019. Examining endocrine disruptors measured in newborn dried blood spots and early childhood growth in a prospective cohort. *Obesity (Silver Spring, MD)* 27(1):145–151.

Zeng, X. W., M. S. Bloom, S. C. Dharmage, C. J. Lodge, D. Chen, S. Li, Y. Guo, M. Roponen, P. Jalava, M. R. Hirvonen, H. Ma, Y. T. Hao, W. Chen, M. Yang, C. Chu, Q. Q. Li, L. W. Hu, K. K. Liu, B. Y. Yang, S. Liu, C. Fu, and G. H. Dong. 2019. "Prenatal exposure to perfluoroalkyl substances is associated with lower hand, foot and mouth disease viruses antibody response in infancy: Findings from the Guangzhou Birth Cohort Study." Science of the Total Environment 663:60-67. doi: 10.1016/j.scitotenv.2019.01.325.

Zhang, S., R. Tan, R. Pan, J. Xiong, Y. Tian, J. Wu, and L. Chen. 2018. Association of perfluoroalkyl and polyfluoroalkyl substances with premature ovarian insufficiency in Chinese Women. *The Journal of Clinical Endocrinology and Metabolism* 103(7):2543–2551.

ANNEX TABLE 3-1 Health Effects of PFAS by Category

Health Effect Category	Specific Health Effects Included	Authoritative Strength-of-Evidence Conclusion	Committee's Strength-of-Evidence Conclusion
Immunological	Response to infection Response to allergens Response to self: autoimmunity	ATSDR: Evidence is suggestive of an association between serum levels of PFOA, PFOS, PFHxS, and PFDA and decreased antibody responses to vaccines; there is also limited evidence of this effect for PFNA, PFUnDA, and PFDoDA. EFSA: PFOS and PFOA are associated with reduced antibody response to vaccination, observed in several studies. Some of the studies suggest that serum levels of PFOS and PFOA are associated with increased propensity for infection. Epidemiological studies provide insufficient evidence to draw conclusions on associations between exposure to PFASs and asthma and allergies. EPA: Human epidemiology data show associations between PFOA exposure and decreased vaccination response. Increased risk of ulcerative colitis was found in the PFOA high-exposure community study, as well as in a study limited to workers in that population. NTP: The evidence indicating that PFOA affects multiple aspects of the immune system supports the overall conclusion that PFOA alters immune function in humans. However, the mechanism(s) of PFOA-associated immunotoxicity is not clearly understood, and effects on diverse endpoints, such as suppression of the antibody response and increased hypersensitivity, may be unrelated. The NTP concludes that PFOS is presumed to be an immune hazard to humans based on a high level of evidence from animal studies that PFOS suppressed the antibody response and a moderate level of evidence from studies in humans. Although the mechanism(s) of PFOS-associated immunotoxicity is not clearly understood, potential mechanisms by which PFOS may reduce disease resistance include suppression of the antibody response and of NK cell function. OECD: A study with 656 children demonstrated that elevated exposures to PFOA and PFOS are associated with reduced humoral immune response to routine childhood immunizations in children aged 5 and 7. A large epidemiological study of 69,000 persons conducted by the C-8 Science Panel found probable links between elevated PFOA blood levels and ulcerative colitis.	Sufficient: decreased antibody response (in adults and children) Limited or suggestive: ulcerative colitis in adults Inadequate or insufficient: response to allergens, all other immune outcomes

Cardiometabolic	Cardiovascular disease, dyslipidemia, metabolic syndrome, obesity	ATSDR: The results of epidemiological studies of PFOA, PFOS, PFNA, and PFDA suggest an association between perfluoroalkyl exposure and increases in serum lipid levels, particularly total cholesterol and low-density lipoprotein (LDL) cholesterol.	Sufficient: dyslipidemia in adults and children Inadequate or insufficient: other outcomes
		EFSA: Epidemiological studies provide clear evidence of an association between exposure to PFOS, PFOA, and PFNA and increased serum levels of cholesterol. There is insufficient evidence of associations with diabetes, obesity, and metabolic syndrome.	
		EPA: Human epidemiological data show associations between PFOA exposure and high cholesterol. These epidemiological studies have generally found positive associations between serum PFOA concentration and total cholesterol in PFOA-exposed workers and high-exposure communities (i.e., increasing lipid level with increasing PFOA); similar patterns are seen with LDL cholesterol but not with high-density lipoprotein (HDL) cholesterol.	
		OECD: The C-8 Science Panel's epidemiological study of 69,000 persons found probable links between elevated PFOA blood levels and high cholesterol (hypercholesterolemia).	
		C-8 Science Panel: There is a probable link between exposure to C-8 and high cholesterol (hypercholesterolemia).	
Developmental	Infant and fetal growth, neurodevelopment, development of the reproductive system	ATSDR: Evidence is suggestive of an association between serum levels of PFOA and PFOS and small decreases in birthweight.	Sufficient: small reductions in birthweight Inadequate or insufficient: all other outcomes, such as development of the reproductive system, neurodevelopment
		EFSA: There may well be a causal association between PFOS and PFOA and birthweight. Maternal serum levels in studies reporting results on other PFASs were generally much lower, and those studies provide no evidence of an adverse association between other PFASs and birthweight.	
		EPA: The epidemiological studies found no association of PFOA with neurodevelopmental effects or with preterm birth and other complications of pregnancy.	
		OECD: High levels of PFOS and PFOA are toxic for reproduction and development of the fetus (such as reducing birthweight and lowering semen quality).	
Cancer	Kidney cancer, testicular cancer, breast cancer	EPA: Under EPA's *Guidelines for Carcinogen Risk Assessment* (EPA, 2005), there is "suggestive evidence of carcinogenic potential" for PFOA. Epidemiology studies demonstrate an association of serum PFOA with kidney and testicular tumors among highly exposed members of the general population.	Sufficient: kidney cancer in adults Limited or suggestive: testicular cancer, breast cancer in adults Inadequate or insufficient: all other cancers
		There is suggestive evidence of carcinogenic potential for PFOS. Human epidemiology studies found no direct correlation between PFOS exposure and the incidence of carcinogenicity in worker-based populations. Although one worker cohort showed an increase in bladder cancer, smoking was a major confounding factor, and the standardized incidence ratios were not significantly different from those for the general population. Other worker and general population studies found no statistically significant trends for any cancer type.	

continued

ANNEX TABLE 3-1 *continued*

Health Effect Category	Specific Health Effects Included	Authoritative Strength-of-Evidence Conclusion	Committee's Strength-of-Evidence Conclusion
		IARC: There is "limited" evidence in humans for the carcinogenicity of PFOA. A positive association was observed for cancers of the testes and kidney. Overall, PFOA is possibly carcinogenic to humans (Group 2B). OECD: High levels of PFOS and PFOA have been found to be potentially carcinogenic in animal tests. The C-8 Science Panel's epidemiological study of 69,000 persons found probable links between elevated PFOA blood levels and testicular and kidney cancer. C-8 Science Panel: There is a probable link between exposure to C-8 and kidney and testicular cancer.	
Reproductive	Infertility, male reproductive effects, female reproductive effects, hormone levels, hypertension during pregnancy, gestational diabetes	ATSDR: There is suggestive epidemiological evidence of an association between serum PFOA and PFOS and pregnancy-induced hypertension and/or preeclampsia. EPA: Human epidemiological data show associations between PFOA exposure and pregnancy-induced hypertension and preeclampsia. OECD: High levels of PFOS and PFOA are toxic for reproduction and development of the fetus (e.g., reducing birthweight and lowering semen quality). In addition, 8:2 fluorotelomer phosphate diesters (8:2 PAPs), 8:2 FTOH, and PFOA show endocrine effects in different in vitro and in vivo tests. The C-8 Science Panel's epidemiological study of 69,000 persons found probable links between elevated PFOA blood levels and preeclampsia and elevated blood pressure during pregnancy. C-8 Science Panel: There is a probable link between exposure to C-8 and pregnancy-induced hypertension (including preeclampsia).	Limited or suggestive: pregnancy-induced hypertension (gestational hypertension and preeclampsia) Inadequate or insufficient: all other effects, such as fecundity, infertility, male reproductive effects, female reproductive effects, reproductive hormone levels, gestational diabetes
Endocrine	Thyroid disease, hypothyroidism, hyperthyroidism, thyroid hormone levels, type 1 and 2 diabetes	EPA: Human epidemiology data show associations between PFOA exposure and thyroid disorders. The epidemiological studies found no associations between PFOA and diabetes. OECD: 8:2 fluorotelomer phosphate diesters (8:2 PAPs), 8:2 FTOH, and PFOA show endocrine effects in different in vitro and in vivo tests. The C-8 Science Panel's epidemiological study of 69,000 persons found probable links between elevated PFOA blood levels and thyroid disease. C-8 Science Panel: There is a probable link between exposure to C-8 and thyroid disease.	Limited or suggestive: thyroid disease or dysfunction in adults Inadequate or insufficient: all other outcomes, including type 1 and 2 diabetes

Hepatic	Liver disease, altered serum liver enzymes and bilirubin	ATSDR: Increases in serum enzymes and decreases in serum bilirubin, observed in studies of PFOA, PFOS, and PFHxS, are suggestive of liver alterations. EFSA: Epidemiological studies provide evidence for an association between exposure to PFAS and increased serum levels of the liver enzyme alanine transferase (ALT). The magnitude of the association was small, however, and few studies found an association with ALT outside the reference range. There were no associations with liver disease. EPA: Human epidemiology data show associations between PFOA exposure and increased liver enzymes.	Limited or suggestive: liver enzyme alterations in children and adults
Respiratory	Respiratory symptoms, pulmonary function, respiratory diseases	None	Insufficient
Hematological	Blood disorders or impacts on blood-forming organs	None	Insufficient
Musculoskeletal	Loss of muscle tone or strength, muscular rigidity, muscular atrophy, arthritis, altered bone density, arthralgia (joint pain)	None	Insufficient
Renal	Kidney disease and biomarkers of renal function	None	Insufficient
Neurological	Parkinson's disease, memory loss, neuropsychological function	None	Insufficient

NOTE: ATSDR = Agency for Toxic Substances and Disease Registry; EFSA = European Food Safety Authority; EPA = U.S. Environmental Protection Agency; IARC = International Agency for Research on Cancer; NK = natural killer; NTP = National Toxicology Program; OECD = Organisation for Economic Co-operation and Development.

4

PFAS Exposure Reduction

It is difficult to provide clear advice on how to reduce exposure to per- and polyfluoroalkyl substances (PFAS) because there are many potential exposure sources. Importantly, even if the source of a person's exposure is completely removed, it will take years for the internal body burden (levels in the body) of many PFAS to be fully eliminated. Biological half-lives (i.e., the time it takes for plasma concentration to decrease by 50 percent after exposure) vary depending on the PFAS; half-life estimates for the four most studied PFAS (perfluorooctanoic acid [PFOA], perfluorooctane sulfonic acid [PFOS], perfluorohexane sulfonate [PFHxS], and perfluorononanoic acid [PFNA]) range from around 2 to 8 years (Li et al., 2018; Olsen et al., 2007). These long biological half-lives are due to reabsorption in the kidney and enterohepatic recirculation, both processes greatly reducing the capacity to eliminate PFAS (Harada et al., 2007).

In developing this chapter, the committee first created a conceptual model for exposure reduction (see Figure 4-1). The premise of this model is that if any exposure pathway is interrupted, exposure should be reduced. For example, if PFAS were no longer used in industries, they would not be in consumer products or waste streams, so exposures would be reduced. Similarly, if PFAS in the environment are cleaned up, exposures will decrease.

Support for the committee's conceptual model is found in the changes in exposure at the population level due to industry-wide changes in PFAS production. For example, serum PFAS concentrations in the United States declined over time following the 3M company's (Maplewood, Minnesota) voluntary phase-out of perfluorooctanyl chemicals and related precursors, including PFOS and PFOA, in 2000 (ITRC, 2017) (see Figure 4-2). Therefore, there is evidence that removing these chemicals from products on a large scale can result in lower levels in a population.

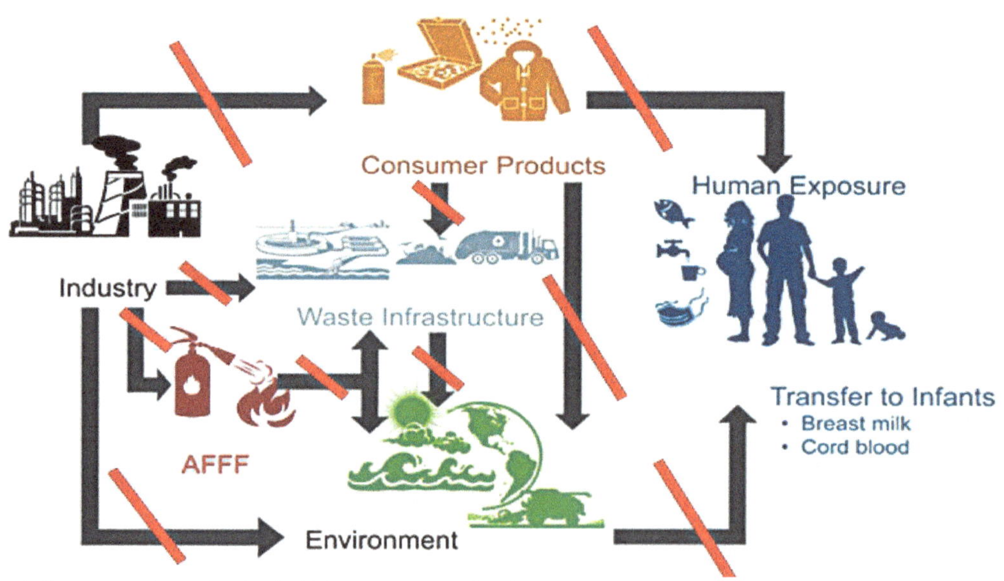

FIGURE 4-1 Conceptual model for PFAS exposure reduction.
NOTE: Red lines indicate a break in the pathway that could reduce human exposure.
SOURCE: Adapted from Sunderland et al., 2019.

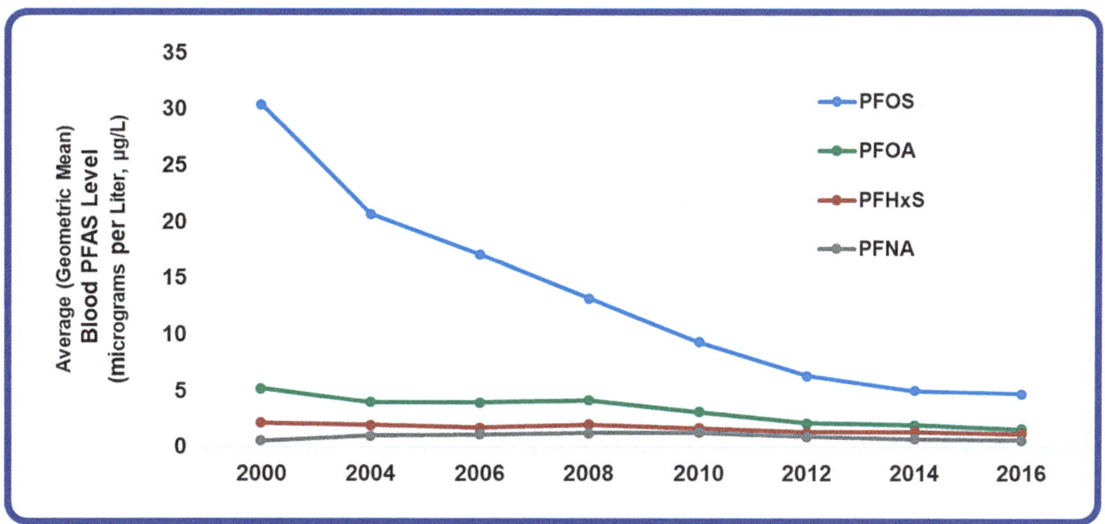

FIGURE 4-2 Blood (serum) levels of PFAS, United States, 2000–2016.
NOTE: Average = geometric mean.
SOURCE: Patrick N. Breysse's presentation to the committee on February 4, 2021. DATA SOURCE: Centers for Disease Control and Prevention. Fourth Report on Human Exposure to Environmental Chemicals, Updated Tables, (January, 2019). Atlanta, GA: U.S. Department of Health and Human Services, Centers for Disease Control and Prevention.

SOURCES AND ROUTES OF EXPOSURE TO PFAS

The sources of and routes of exposure to PFAS are an active area of investigation. What is well known is that PFAS exposure is highly complex, with pathways that include occupational exposures, environmental contamination, consumer product use, and food exposures. Specific sources of exposure include, for example, jobs in fluorochemical manufacturing facilities or where PFAS-containing products, such as textiles or food contact materials, are made. Other jobs with a known increased risk of exposure to PFAS include electroplating; painting; carpet installation and treatment; serving as a military or civilian firefighter, which entails using PFAS-containing foams in training exercises and wearing PFAS-impregnated gear; and jobs that require prolonged work with ski wax (ATSDR, 2021). In addition, food workers and others in the hospitality industry may have elevated exposure since they handle PFAS-containing food packaging as part of their job duties (Carnero et al., 2021; Curtzwiler et al., 2021; Schaider et al., 2017).

Ingestion is the most well-studied route of exposure in nonoccupational settings (Trudel et al., 2008). Ingestion of PFAS can occur through drinking contaminated water; eating contaminated seafood; or consuming other contaminated foods, such as vegetables, game, or dairy products (Bao et al., 2019, 2020; Death et al., 2021; Domingo and Nadal, 2017; Herzke et al., 2013; Li et al., 2019). PFAS are often used in cookware and in materials that came in contact with food, such as microwave popcorn bags or packaging used for fast foods or processed foods (Carnero et al., 2021; Curtzwiler et al., 2021; Schaider et al., 2017). Exposure can also occur through accidental ingestion of PFAS-containing dusts (Fraser et al., 2013). PFAS cross the placenta, and PFAS from the mother's body burden can be passed on to her developing fetus (Gao et al., 2019; Manzano-Salgado et al., 2015). Maternal transfer of PFAS can also occur through breastfeeding (Serrano et al., 2021; Zheng et al., 2021).

Inhalation and transdermal exposures are less well studied. Inhalation of PFAS is well documented in occupational settings that use aerosolized PFAS (Gilliland, 1992). Volatile PFAS have been detected indoors (Fromme et al., 2015; Morales-McDevitt et al., 2021), and inhalation near factory emissions and incinerators contributes to exposures in nearby communities (Fenton et al., 2021). There are as yet no data formally evaluating inhalation from showering in contaminated water, but this is an active area of research.

PFAS are used in thousands of products (e.g., products containing water; stain-resistant clothing; and personal care products, such as sunscreen, makeup, and dental floss). PFAS are also used in such products as paint, textiles, firefighting foam, electroplating material, ammunition, climbing ropes, guitar strings, artificial turf, and soil remediation products (Glüge et al., 2020). The extent to which use of products contributes to human exposures is unkown, however, because the relative contribution of PFAS exposures from sources other than food or water is not well characterized (DeLuca et al., 2021).

The presence of PFAS in everyday consumer products may be an important source of exposure for the general population, but this likely varies greatly by individual (Rodgers et al., 2022). Consumer products are often treated with fluoropolymers to impart water and stain resistance, and fluoropolymers can result in exposure to the nonpolymer PFAS discussed in this report in several ways. Nonpolymer PFAS can be present as impurities in fluoropolymer-containing products because these PFAS are used as processing aids in chemical production or are degradation products of the fluoropolymers (Rodgers et al., 2022; Schellenberger et al., 2019). Moreover, fluoropolymer PFAS can degrade (biodegradation) slowly when exposed to water and release fluorotelomer alcohols, which are precursors to nonpolymer PFAS. Once the precursor PFAS are in the human body, they can transform to such PFAS as PFOA, PFOS, and PFHxS (Washington and Jenkins, 2015).

APPROACH TO DETERMINING ADVICE ON PFAS EXPOSURE REDUCTION

To inform its recommendations on how clinicians can advise patients on PFAS exposure reduction, the committee looked at several sources of evidence. The committee contracted with a consultant to review the literature evaluating the effectiveness of behavioral interventions in reducing exposure to PFAS. The consultant also reviewed several studies that model estimates of PFAS intake, which could inform predicted changes in serum PFAS levels if exposure routes were modified (see Appendix E). The committee reviewed studies on PFAS exposure through breastfeeding since this route impacts a vulnerable population and is of particular interest to people in PFAS-impacted communities, as voiced by speakers at the committee's town halls (see Appendix B). The committee also evaluated studies of medical interventions for reducing internal levels of PFAS (e.g., phlebotomy or taking of prescription drugs). PFAS exposure pathways discussed in Chapter 1 were used to outline strategies clinicians can use to determine whether a patient may be at risk of PFAS exposure based on residential and work history. In addition, the committee reviewed advice from other entities on PFAS exposure reduction, such as that provided by nongovernmental organizations, state and local governments, and other countries.

CONTRIBUTION OF INDIVIDUAL EXPOSURE SOURCES TO HUMAN EXPOSURE

Behavioral Intervention Studies

Studies evaluating the impact of behavior change on reducing exposures would help determine the impact of behavior on human exposure. This section, based on a literature review conducted by LaKind Associates (see Appendix E), provides an overview of the literature available to inform what individuals can do to reduce their serum PFAS levels. The literature review used a three-step approach to identify relevant publications in PubMed, EMBASE, and Google Scholar (see Appendix E for keywords and the search strategy). Studies were selected that entailed interventions designed to reduce human exposure to PFAS, specifically interventions that could be carried out by individuals. Secondary references of retrieved articles were reviewed to identify publications not found through the electronic search. An additional literature search was then conducted to identify reviews containing estimates of human PFAS intakes. The final search date was March 5, 2021.

The reviewed intervention studies are presented here by exposure route, with the primary focus being on drinking water and diet. Literature on interventions for other exposure sources, such as dust and consumer products, is more limited. Lastly, breastfeeding is an important potential source of exposure for infants, and lactating could reduce parents' PFAS levels. After reviewing the relevant studies, this section concludes by assessing the effectiveness of behavior modifications based on the current literature.

Drinking water has been identified as a substantial source of PFAS exposure for many populations (Andrews and Naidenko, 2020; Domingo and Nadal, 2019). Studies of interventions focused on drinking water are consistent in demonstrating the effectiveness of water filtration at reducing levels of certain PFAS. Consumers have a variety of options for filtering PFAS from drinking water, including whole-house, under-sink, and filtering-pitcher devices. The committee identified seven publications and one agency report evaluating possible drinking water interventions. Of these, five (Ao et al., 2019; Herkert et al., 2020; Iwabuchi and Sato, 2021; MDH, 2008; Patterson et al., 2019) evaluate use of whole-house, under-sink, and filtering-pitcher devices, and three (Ao et al., 2019; Gellrich et al., 2013; Heo et al., 2014) evaluate differences in PFAS concentrations between tap water and bottled water.

In a study of real-world uses of water filtration options, Herkert and colleagues (2020) tested municipal, well, and filtered (n = 89) and unfiltered (n = 87) tap water in residences (N = 73) in North Carolina for 11 PFAS. The filters tested varied in both type and filtration method (reverse osmosis, granulated activated carbon, single-stage, dual-stage). Notably, reverse osmosis and dual-stage filters were found consistently to remove most measured compounds at an average of ≥90 percent efficiency; some short-chain replacement PFAS are difficult to remove with carbon filtration (Herkert et al., 2020). Use of bottled water can also be a way to reduce PFAS exposure from drinking water, although some bottled waters contain detectable levels of C-3–C-10 perfluorocarboxylic acids (PFCAs) and C-3–C-6 and C-8 perfluorosulfonic acids (PFSAs) (Chow et al., 2021). In Parkersburg, West Virginia, for example, the use of bottled water resulted in a dramatic reduction in serum PFOA levels among a contaminated community living near a Teflon-manufacturing plant (Emmett et al., 2006).

Changes in diet may potentially reduce PFAS exposure, given that PFAS can be present in a number of food products, including wild-caught fish and game, livestock, and produce, as well as prepared foods. Fish and seafood have been identified as sources of PFAS, but the levels of PFAS vary by fish type and water body (Sunderland et al., 2019). PFAS-contaminated drinking water can also impact home-grown vegetables (Brown et al., 2020; Emmett et al., 2006). The majority of ingestion-based intervention studies focused on seafood preparation, but there was no strong evidence that fish preparation methods influenced PFAS levels (Alves et al., 2017; Barbosa et al., 2018; Bhavsar et al., 2014; Del Gobbo et al., 2008; Hu et al., 2020; Kim et al., 2020; Luo et al., 2019; Taylor et al., 2019).

A study on PFAS exposure from indoor dust found significantly lower PFAS levels in vacuumed dust from rooms with PFAS-free furnishings relative to control rooms (78 percent reduction, 95% CI: 38–92). Results suggest that modifying personal behavior to be capable of identifying and purchasing PFAS-free furnishings can decrease exposure levels from indoor dust (Scher et al., 2019).

Literature on PFAS in consumer products is available only for nonstick pans and dental floss, and these studies have several weaknesses, including recall bias, small sample sizes, and lack of replication. Their results do not provide enough evidence to suggest that modifications in behavior relative to those products would decrease PFAS exposure, but in the absence of such evidence, consumers should be aware of which products contain PFAS (Scher et al., 2019; Young et al., 2021).

Another factor that complicates consumer choices aimed at avoiding PFAS is the lack of consistent labeling of products. A recent study (identified after Appendix E of this report had been completed) that screened 93 market items across three different product types (furnishings, apparel, and bedding) found that PFAS were present in many items that were labeled as green or nontoxic (Rodgers et al., 2022).

To summarize, the available literature is limited in presenting recommendations for effective behavior modifications to reduce internal levels of PFAS. In places with water contamination, individuals can reduce their exposure through use of water filtration. In places without PFAS water contamination or workplace exposure, diet is believed to be the primary exposure route, but there is limited information with which to recommend dietary interventions. No intervention study has examined exposure reduction and its impact on serum concentrations, likely in part because to fully show effectiveness for an intervention, it would have to be conducted over a long enough time to account for the long half-lives of PFAS.

Modeled Estimates of PFAS Intakes

In the absence of studies demonstrating the impact of interventions on reducing PFAS exposure, exposure models may help inform predicted changes in serum PFAS levels if exposure routes are modified. Pharmacokinetic modeling is useful for estimating the body burden from different exposure routes, but such models are often limited by the many assumptions made about intake factors (e.g., food contamination). High-quality data on the distribution of PFAS in different food types and consumer products are sparse. Nonetheless, pharmacokinetic models can be used to estimate the variability possible in exposure reduction and the impact of changes with different parameters. Although imperfect, exposure modeling can provide, at a minimum, an estimate of the change in internal PFAS body burden if PFAS levels in diet or water were decreased.

Studies that have estimated intake of PFOS and PFOA have been used to determine the dominant routes of exposure in communities without contaminated drinking water (see Appendix E). Estimates of how much PFAS exposure comes from diet in adults vary widely, from 16–99 percent for PFOA to 66–100 percent for PFOS (Egeghy and Lorber, 2011; Haug et al., 2011; Lorber and Egeghy, 2011; Sunderland et al., 2019; Vestergren and Cousins, 2009); no estimates are available for individual food products. For dust, the estimates are 1–11 percent for PFOA and 1–15 percent for PFOS (Sunderland et al., 2019). For PFOA, the dominant routes are thought to be oral exposure resulting from consumption of fish and seafood, drinking water, and ingestion of dust. For PFOS, the dominant routes are thought to be ingestion of food and water, ingestion of dust, and hand-to-mouth transfer from treated carpets (Trudel et al., 2008). Residual PFOA in food packaging (used to greaseproof food-containing paper products) is another potential route of exposure (Trudel et al., 2008); polyfluoroalkyl phosphoric acids in food packaging can also be metabolized in the body to PFOA (Begley et al., 2005; Carnero et al., 2021; Curtzwiler et al., 2021; Schaider et al., 2017).

Interpretation of the estimated intake studies is challenged by several factors. First, while diet appears to be a major pathway of exposure, there is little information on PFAS in commercial foods commonly consumed in the United States. The U.S. Food and Drug Administration (FDA) has released PFAS data for certain foods that could be used to model source contributions to PFAS intake in future studies. However, the FDA data for produce, meat, dairy, and grain products are based on a small sample size, and the results "cannot be used to draw definitive conclusions about the levels of PFAS in the general food supply."[1]

The relative importance of different PFAS sources varies by study, population, and time period of exposure (Sunderland et al., 2019). The production and use of individual PFAS have changed over time and will continue to do so. Serum PFAS levels in the United States dropped following the phase-out of production of PFOS and PFOA; however, exposures to C9–C11 PFCAs have not followed the same trend. Thus, it is important to use recent environmental, consumer product, and dietary data to develop robust estimates of current dominant pathways of PFAS exposure. In a recent review evaluating nonoccupational intakes via background PFAS exposures, De Silva and colleagues (2021) observed that the inconsistency among studies in the relative importance of different exposure sources may be due to differing concentrations of PFAS in sources, as well as the assignment of differing values for exposure intake factors (e.g., exposure frequency and duration). The authors conclude, "Without rigorously conducted exposure studies it is challenging to rank order the most important human exposure pathways and without these data, our ability to design evidence-based exposure intervention strategies will be limited."

---

[1] See https://www.fda.gov/food/chemicals/analytical-results-testing-food-pfas-environmental-contamination (accessed May 12, 2021).

Exposure to PFAS Through Breastfeeding

Breastfeeding is a route of exposure of great interest to people who spoke at the committee's town halls. Breast milk is the only food many infants receive in their first 6 months of life, and if the breast milk they receive is contaminated with PFAS, it may take years for their body burden to be reduced, given the long half-lives of some PFAS.

In her testimony at the committee's first town hall (April 7, 2021), Loreen Hackett (PFOA Project New York) stated that in her view, health care mantras such as "breast milk is best" need to be thoroughly reevaluated in exposed communities, noting that breastfeeding "may double or triple PFAS levels in an infant compared to the mothers thereby increasing risks to their developing systems." She stressed that families in exposed communities cannot make informed reproductive choices or other family decisions without improved information tailored to their situation, and she relayed concerns among community members expressing guilt for unknowingly poisoning their child over the course of pregnancy and breastfeeding.

Nonetheless, data on PFAS in breast milk are very limited. A few studies measuring PFAS in breast milk in North America (Kubwabo et al., 2013; Tao et al., 2008; Zheng et al., 2021) indicate transfer from the parent to the child during the first months of life. And although the concentrations in breast milk are generally much lower than the concentrations in maternal serum (Cariou et al., 2015; Kärrman et al., 2007; Kim et al., 2011; Liu et al., 2011), breastfeeding has been shown to contribute significantly to children's serum levels of some PFAS (Gyllenhammar et al., 2018, 2019; Koponen et al., 2018). In a cohort of 2- to 4-month-old infants in Sweden, for example, bottle-fed infants had mean serum concentrations twice as low as those of their exclusively breastfed counterparts, and serum levels of PFOA, PFNA, and PFHxS increased 8–11 percent per week of exclusive breastfeeding (Gyllenhammar et al., 2018). Where measured and estimated PFAS concentrations in breast milk in the United States have been compared with drinking water screening values of the Agency for Toxic Substances and Disease Registry (ATSDR), some exceedances have been observed, especially in communities impacted by PFAS contamination (LaKind et al., 2022) (see Figure 4-3).

Whether lactational exposure to PFAS can have adverse health effects in children has not been well studied to date. Formula feeding can also lead to PFAS exposure through either contaminated formula or formula reconstituted with contaminated drinking water. Given the increased exposures observed in breastfed versus formula-fed infants, it is not clear whether the benefits of breastfeeding outweigh the risks to the child among lactating persons with very high levels of PFAS exposure.

Guidance to breastfeed remains the best feeding advice for most infants given the many benefits of breastfeeding for both mothers and babies.[2] Even though PFAS exposures have been occurring for many years, research has consistently shown benefits of breastfeeding, providing confidence in the traditional guidance, although a more in-depth understanding of this exposure route is warranted to inform protection of such a vulnerable population.

MEDICAL INTERVENTIONS FOR POTENTIALLY REDUCING PFAS BODY BURDEN

There have been few studies overall and no clinical trials evaluating treatments to reduce PFAS body burden, even in cases of very high exposure. The few evaluations available have focused on the use of cholesterol-lowering medications and phlebotomy.

PFAS are secreted in the bile and have enterohepatic recirculation; therefore, researchers have been interested in medications that enhance bile sequestration as potential approaches for reducing PFAS body burden. Cholestyramine is a bile-sequestering agent that is used mainly to reduce low-density lipoprotein (LDL) cholesterol. In a cross-sectional study of C-8 Health Project participants, 36 of 56,175 adults were being treated with cholestyramine and were found to have lower levels of PFAS compared with those not taking this medication (Ducatman et al., 2021). Another medication, probenecid, was not

---

[2] See https://www.cdc.gov/breastfeeding/about-breastfeeding/why-it-matters.html (accessed May 23, 2022).

significantly associated with serum PFAS levels in this study. Cholestyramine has also been evaluated in a few small case studies of 1 to 20 individuals (Genius et al., 2010, 2013). Results of these studies suggest that cholestyramine may be an effective treatment to accelerate PFAS fecal excretion, but replication in studies with more participants is needed. No studies have assessed whether PFAS levels rebound when treatment with cholestyramine is discontinued.

Phlebotomy has been discussed as a way to reduce the body burden of PFAS. Genuis and colleagues (2014) asked six patients aged 16–53 years from a highly PFAS-exposed family to submit to routine blood draws (500 mL) for up to 5 years, resulting in a cumulative 2–12 L of blood drawn. The levels of PFOA, PFOS, and PFHxS decreased in these individuals over that time at a faster rate than expected according to first-order excretion kinetics. In a recent randomized controlled trial of 285 firefighters, serum PFOS and PFHxS concentrations were significantly reduced in subjects who regularly donated blood or plasma over 12 months compared with the control group; the decline was more pronounced in the plasma donation group (Gasiorowski et al., 2022). While these studies indicate that phlebotomy can be effective at reducing PFAS levels in blood, there are no established serum concentrations of PFAS at which the benefits of this intervention reasonably outweigh the harms, and the safety and utility of this approach are uncertain.

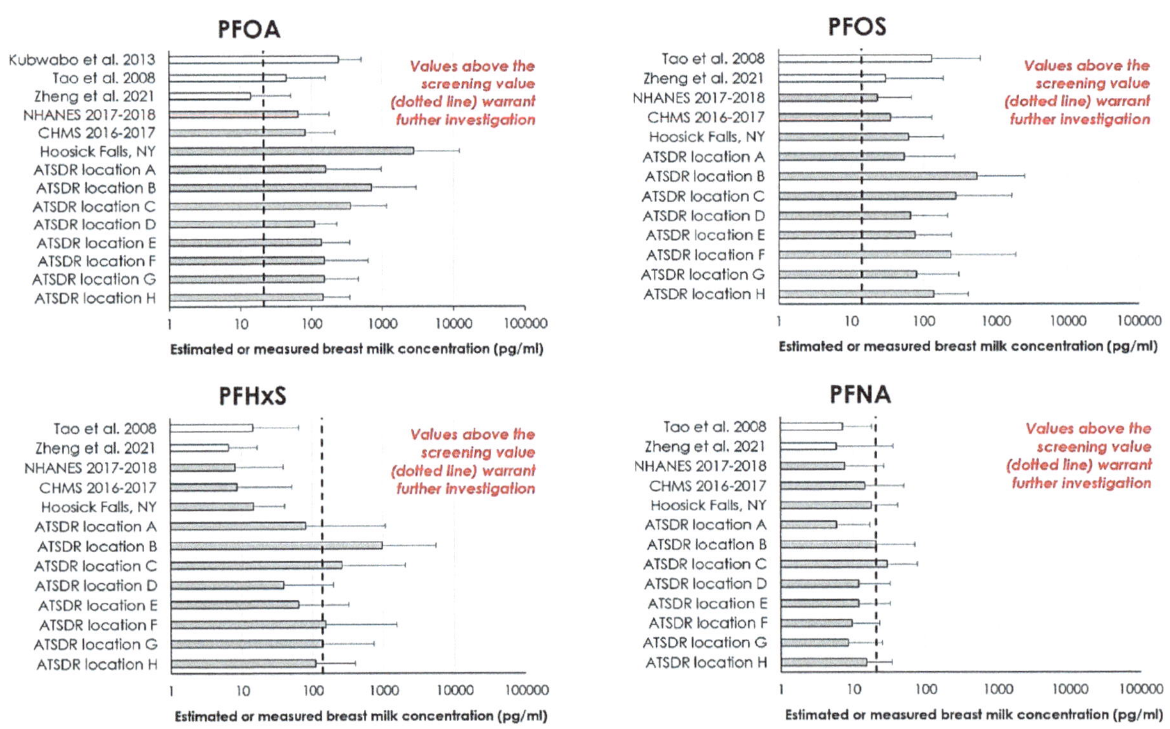

FIGURE 4-3 Measured (white bars) and estimated (gray bars) breast milk concentrations of per- and polyfluoroalkyl substances (PFAS) in the United States and Canada, in comparison with children's drinking water screening values (dotted line).

NOTES: Bars represent the mean measured breast milk levels for the data from Kubwabo and colleagues (2013) and Tao and colleagues (2008); the median measured breast milk levels from Zheng et al. (2021); and the geometric mean estimated breast milk levels for the National Health and Nutrition Examination Survey (NHANES), the Canadian Health Measures Survey (CHMS), and PFAS-contaminated communities. Error bars represent maximum measured concentrations in Kubwabo and colleagues (2013) and Tao and colleagues (2008), and 95th percentile estimated concentrations for NHANES, CHMS, and PFAS-contaminated sites. Of note, PFOS, PFHxS, and PFNA were not detected in breast milk samples from Kubwabo and colleagues (2013). The xx-axis is log-scale. Agency for Toxic Substances and Disease Registry children's drinking water screening values: PFOA (21 ppt), PFOS (14 ppt), PFHxS (140 ppt), and PFNA (21 ppt) (ATSDR, 2018).
SOURCE: LaKind et al., 2022.

EXISTING ADVICE ON PFAS EXPOSURE REDUCTION

Several federally funded academic projects and nonprofit organizations provide information about PFAS, including how to identify potential community exposure and reduce personal exposure. Because concerned individuals look to online resources for data and information, a few of the resources are summarized here, with the caveat that they have not been tested empirically, and the data they present may be incomplete, or the sources on which the data are based may be missing. Furthermore, in discussing these online resources, the committee is not endorsing them, nor do their content and conclusions necessarily represent the committee's views. In addition to the sources discussed below, many state health departments have state-level resources on PFAS that may be trusted sources of information.

PFAS-REACH

PFAS-REACH (Research, Education, and Action for Community Health) is a project funded by the National Institute of Environmental Health Sciences to develop guidance materials and data interpretation tools for use by communities impacted by PFAS-contaminated drinking water. The project is led by Silent Spring, Northeastern University, and Michigan State University, with collaboration from community partner organizations that include Testing for Pease, Massachusetts Breast Cancer Coalition, and Community Action Works. The project's online resource center, PFAS Exchange, provides factsheets and interactive maps; a factsheet on how to reduce one's exposure is most relevant to the discussion in this chapter.[3] Figure 4-4 shows the PFAS Exchange recommendations.

In your personal life:

✓ Avoid stain-resistant carpets and upholstery, as well as stain-resistant treatments and waterproofing sprays.

✓ Avoid products with the ingredient PTFE or other "fluoro" ingredients listed on the label.

✓ Choose cookware made of cast iron, stainless steel, glass, or enamel instead of Teflon.

✓ Filter your drinking water with an activated carbon or reverse osmosis filtration system.

✓ Eat more fresh foods to avoid take-out containers and other food packaging.

✓ Avoid microwave popcorn and greasy foods wrapped in paper.

✓ Look for nylon or silk dental floss that is uncoated or coated in natural wax.

In your community:

✓ Tell retailers and manufacturers you want products made without PFAS.

✓ Urge your local water utility to test for PFAS.

✓ Ask your state legislators to set up a statewide water and blood testing program.

✓ Encourage your state to follow the lead of other states in creating more health protective drinking water limits.

✓ Ask your elected officials to support restrictions on PFAS in consumer products and remediation of contaminated sites.

✓ Find out about local groups working to protect water quality by visiting:

www.pfas-exchange.org

FIGURE 4-4 Recommendations for reducing PFAS exposure available through the PFAS Exchange. SOURCE: PFAS-REACH project.

---

[3] See https://pfas-exchange.org/how-to-reduce-your-exposure-to-pfas (accessed June 17, 2022).

PFAS Project Lab

Northeastern University, one of the academic partners for the PFAS-REACH project, operates the PFAS Project Lab within its Social Science Environmental Health Research Institute (SSEHRI). Among the Lab's publicly available resources is the PFAS Sites and Community Resources Map.[4] The map interface provides the locations of known and suspected PFAS contamination sites (see Figure 4-5), as well as community resources and state action. The data were collected from government websites, news articles, and publicly available sources. The map began as a collaborative effort with the Environmental Working Group (EWG), an advocacy group (see below and Chapter 1, Figure 1-2). The SSEHRI map "aims to help affected residents and community groups to access information about data in their states and learn how to connect with other activists working on PFAS issues."[5]

Environmental Working Group

The EWG is included as a data source in the PFAS Project Lab. The EWG site also provides an interactive map that "serves to show the extent of PFAS water contamination as documented by states, the department of defense and EWG's testing," providing the locations of industrial and military sites with known PFAS contamination.[6] Additionally, the EWG provides a guide for avoiding exposure to PFAS chemicals. The EWG's recommendations are similar to those in the PFAS-REACH factsheet (see Figure 4-4) regarding consumer choices. The EWG has developed several consumer guides providing information on the chemicals (not just PFAS) present in a variety of commercial products, including sunscreen, cosmetics, personal care and beauty products, bug repellants, and household cleaners, among others.

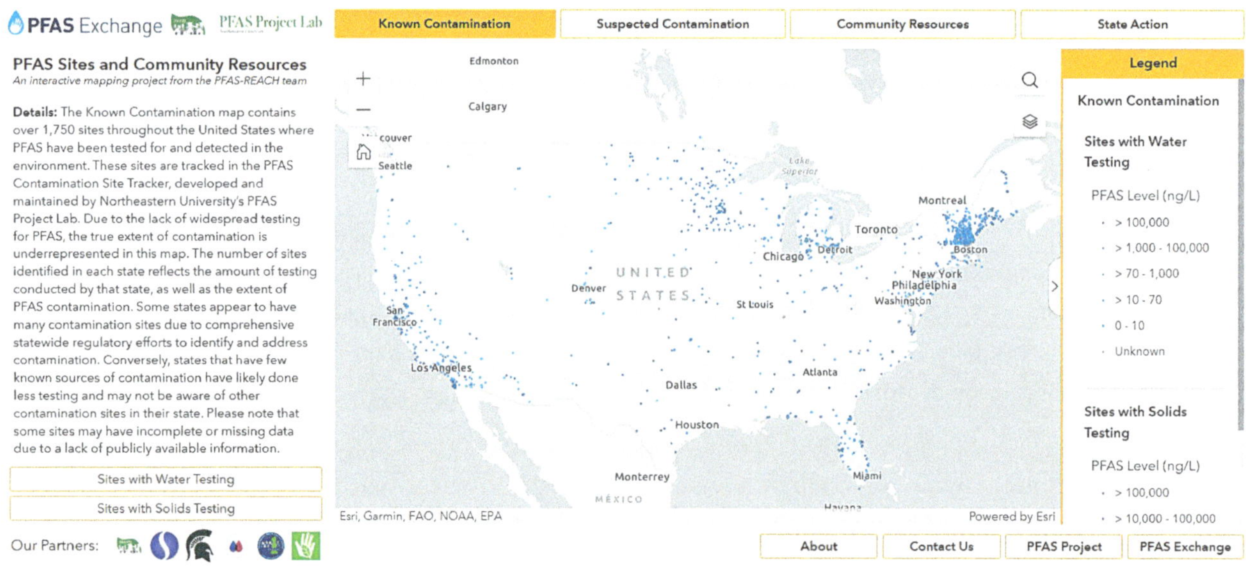

FIGURE 4-5 PFAS Project Lab map showing PFAS contamination sites in the continental United States. SOURCE: Social Science Environmental Health Research Institute (SSEHRI). https://experience.arcgis.com/experience/12412ab41b3141598e0bb48523a7c940 (accessed May 25, 2022).

---

[4] See https://experience.arcgis.com/experience/12412ab41b3141598e0bb48523a7c940 (accessed May 25, 2022).
[5] See https://www.ewg.org/news-insights/news/mapping-pfas-contamination-crisis (accessed June 15, 2022).
[6] See https://www.ewg.org/interactive-maps/pfas_contamination/#about (accessed June 17, 2022).

FINDINGS AND RECOMMENDATIONS

The committee found that few evidence-based recommendations can be made for reducing exposure to PFAS on an individual level.

Occupational exposures to PFAS may be much higher than community exposures. In accordance with the hierarchy of controls, methods for reducing workplace exposure can include replacing the chemical with a less hazardous one; engineering controls, such as ventilation to reduce inhalation of the chemical; administrative controls, such as rotating operations to reduce the amount of time an individual worker is around a chemical; or personal-level controls, such as personal protective equipment, including gloves and masks.

Ingestion is an important route of exposure to PFAS in the general population; thus it is important to reduce consumption of PFAS in drinking water and foods. Contamination of drinking water with PFAS is a widespread problem in the United States, and the extent of the contamination has not been completely characterized. Both municipal and private sources of drinking water (e.g., private wells) can be contaminated with PFAS as a result of fluorochemical manufacturing, use of firefighting foams, or discharge of landfill leachate to drinking water sources. If PFAS are in drinking water, switching to another source of water with lower PFAS concentrations will reduce exposure.

Consumption of game may also cause to exposure to PFAS. To date, 11 states have developed or are in the process of developing advisory guidelines for fish, wildlife, and other food products to protect human health from exposure to PFAS. These advisories offer guidance on limiting the quantity of consumption of these foods. These advisories are state-specific and range from do not eat (e.g., fish or deer in Michigan with PFOS concentrations over 300 parts per billion [ppb]) to no need to limit consumption (e.g., New Jersey fish with more than 0.56 nanograms per gram [ng/g] of PFOS). The Environmental Council of the States has compiled information from participating states on state PFAS standards, advisories, and guidance values (ECOS, 2020).

For clinicians, based on its review of the evidence on PFAS exposure reduction, the committee makes the following recommendations:

Recommendation 4-1[7]: Clinicians advising patients on PFAS exposure reduction should begin with a conversation aimed at first determining how they might be exposed to PFAS (sometimes called an environmental exposure assessment) and what exposures they are interested in reducing. This exposure assessment should include questions about current occupational exposures to PFAS (such as work with fluorochemicals or firefighting) and exposures to PFAS through the environment. Known environmental exposures to PFAS include living in a community with PFAS-contaminated drinking water, living near industries that use fluorochemicals, serving in the military, and consuming fish and game from areas with known or potential contamination.

Recommendation 4-2: If patients may be exposed occupationally, such as by working with fluorochemicals or as a firefighter, clinicians should consult with occupational health and safety professionals knowledgeable about the workplace practices to determine the most feasible ways to reduce that exposure.

Recommendation 4-3: Clinicians should advise patients with elevated PFAS in their drinking water that they can filter their water to reduce their exposure. Drinking water filters are rated by NSF International, an independent organization that develops public health standards for products. The NSF database can be searched online for PFOA to find

---

[7] The committee's recommendations are numbered according to the chapter of the main text in which they appear.

filters that reduce the PFAS in drinking water included in the committee's charge. Individuals who cannot filter their water can use another source of water for drinking.

Recommendation 4-4: In areas with known PFAS contamination, clinicians should advise patients that PFAS can be present in fish, wildlife, meat, and dairy products and direct them to any local consumption advisories.

There are fewer evidence-based exposure-reduction recommendations for patients without known sources of exposure:

Recommendation 4-5: Clinicians should direct patients interested in learning more about PFAS to authoritative sources of information on how PFAS exposure occurs and what mitigating actions they can take. Authoritative sources include the Pediatric Environmental Health Specialty Units (PEHSUs), the Agency for Toxic Substances and Disease Registry (ATSDR), and the U.S. Environmental Protection Agency (EPA).

Recommendation 4-6: When clinicians are counseling parents of infants on PFAS exposure, they should discuss infant feeding and steps that can be taken to lower sources of PFAS exposure. The benefits of breastfeeding are well known; the American Academy of Pediatrics, the American Academy of Family Physicians, and the American College of Obstetricians and Gynecologists support and recommend breastfeeding for infants, with rare exceptions. Clinicians should explain that PFAS can pass through breast milk from a mother to her baby. PFAS may also be present in other foods, such as the water used to reconstitute formula and infant food, and potentially in packaged formula and baby food. It is not yet clear what types and levels of exposure to PFAS are of concern for child health and development.

Additionally, there is a critical need for more data to understand PFAS exposure among breastfed infants:

Recommendation 4-7: Federal environmental health agencies should conduct research to evaluate PFAS transfer to and concentrations in breast milk and formula to generate data that can help parents and clinicians make shared, informed decisions about breastfeeding.

At this time, it not possible to eliminate all sources of PFAS exposure. There are some sources people can try to limit if they desire and have the resources to do so. If patients are resource-limited, it is most important that if PFAS contamination of their water is known or suspected, they use water filtration or another source of water for drinking that is lower in PFAS. In keeping with the principle of adaptability, it is also important to direct patients to reliable sources of information on PFAS, such as ATSDR, the U.S. Environmental Protection Agency, and state and local departments of public health so they can obtain accurate and up-to-date information.

REFERENCES

Alves, R. N., A. L. Maulvault, V. L. Barbosa, S. Cunha, C. J. A. F. Kwadijk, D. Álvarez-Muñoz, S. Rodríguez-Mozaz, O. Aznar-Alemany, E. Eljarrat, and D. Barceló. 2017. Preliminary assessment on the bioaccessibility of contaminants of emerging concern in raw and cooked seafood. *Food and Chemical Toxicology* 104:69–78.

Andrews, D. Q., and O. V. Naidenko. 2020. Population-wide exposure to per- and polyfluoroalkyl substances from drinking water in the United States. *Environmental Science & Technology Letters* 7(12):931–936. https://doi.org/10.1021/acs.estlett.0c00713.

Ao, J., T. Yuan, H. Xia, Y. Ma, Z. Shen, R. Shi, Y. Tian, J. Zhang, W. Ding, and L. Gao. 2019. Characteristic and human exposure risk assessment of per- and polyfluoroalkyl substances: A study based on indoor dust and drinking water in China. *Environmental Pollution* 254:112873.

ATSDR (Agency for Toxic Substances and Disease Registry). 2018. *ATSDR's Minimal Risk Levels (MRLs) and Environmental Media Evaluation Guides (EMEGs) for PFAS.* https://www.atsdr.cdc.gov/pfas/resources/mrl-pfas.html (accessed 21 May 2021).

ATSDR. 2021. *Toxicological profile for perfluoroalkyls.* Atlanta, GA: U.S. Department of Health and Human Services. https://www.atsdr.cdc.gov/toxprofiles/tp200.pdf (accessed July 1, 2022).

Bao, J., W. J. Yu, Y. Liu, X. Wang, Y. H. Jin, and G. H. Dong. 2019. Perfluoroalkyl substances in groundwater and home-produced vegetables and eggs around a fluorochemical industrial park in China. *Ecotoxicology and Environmental Safety* 171:199.

Bao, J., C. L. Li, Y. Liu, X. Wang, W. J. Yu, Z. Q. Liu, L. X. Shao, and Y. H. Jin. 2020. Bioaccumulation of perfluoroalkyl substances in greenhouse vegetables with long-term groundwater irrigation near fluorochemical plants in Fuxin, China. *Environmental Research* 188:109751.

Barbosa, V., A. L. Maulvault, R. N. Alves, C. Kwadijk, M. Kotterman, A. Tediosi, M. Fernández-Tejedor, J. J. Sloth, K. Granby, and R. R. Rasmussen. 2018. Effects of steaming on contaminants of emerging concern levels in seafood. *Food and Chemical Toxicology* 118:490–504.

Begley, T. H., K. White, P. Honigfort, M. L. Twaroski, R. Neches, and R. A. Walker. 2005. Perfluorochemicals: Potential sources of and migration from food packaging. *Food Additives and Contaminants* 22(10):1023–1031. https://doi.org/10.1080/02652030500183474.

Bhavsar, S. P., X. Zhang, R. Guo, E. Braekevelt, S. Petro, N. Gandhi, E. J. Reiner, H. Lee, R. Bronson, and S. A. Tittlemier. 2014. Cooking fish is not effective in reducing exposure to perfluoroalkyl and polyfluoroalkyl substances. *Environment International* 66:107–114.

Brown, J. B., J. M. Conder, J. A. Arblaster, and C. P. Higgins. 2020. Assessing human health risks from per- and polyfluoroalkyl substance (PFAS)-impacted vegetable consumption: A tiered modeling approach. *Environmental Science & Technology* 54(23):15202–15214. https://doi.org/10.1021/acs.est.0c03411.

Cariou, R., B. Veyrand, A. Yamada, A. Berrebi, D. Zalko, S. Durand, C. Pollono, P. Marchand, J. C. Leblanc, J. P. Antignac, and B. Le Bizec. 2015. Perfluoroalkyl acid (PFAA) levels and profiles in breast milk, maternal and cord serum of French women and their newborns. *Environment International* 84:71–81. htttps://doi.org/10.1016/j.envint.2015.07.014.

Carnero, A. R., A. Lestido-Cardama, P. V. Loureiro, L. Barbosa-Pereira, A. R. B. de Quirós, and R. Sendón. 2021. Presence of perfluoroalkyl and polyfluoroalkyl substances (PFAS) in food contact materials (FCM) and its migration to food. *Foods* 10(7):1443. https://doi.org/10.3390/foods10071443.

Chow, S. J., N. Ojeda, J. G. Jacangelo, and K. J. Schwab. 2021. Detection of ultrashort-chain and other per- and polyfluoroalkyl substances (PFAS) in U.S. bottled water. *Water Research* 201:117292. https://doi.org/10.1016/j.watres.2021.117292.

Curtzwiler, G. W., P. Silva, A. Hall, A. Ivey, and K. Vorst. 2021. Significance of perfluoroalkyl substances (PFAS) in food packaging. *Integrated Environmental Assessment and Management* 17(1):7–12. https://doi.org/10.1002/ieam.4346.

De Silva, A. O., J. M. Armitage, T. A. Bruton, C. Dassuncao, W. Heiger-Bernays, X. C. Hu, A. Kärrman, B. Kelly, C. Ng, A. Robuck, M. Sun, T. F. Webster, and E. M. Sunderland. 2021. PFAS exposure pathways for humans and wildlife: A synthesis of current knowledge and key gaps in understanding. *Environmental Toxicology and Chemistry* 40(3):631–657. https://doi.org/10.1002/etc.4935.

Death, C., C. Bell, D. Champness, C. Milne, S. Reichman, and T. Hagen. 2021. Per- and polyfluoroalkyl substances (PFAS) in livestock and game species: A review. *Science Of the Total Environment* 774:144795. https://doi.org/10.1016/j.scitotenv.2020.144795.

Del Gobbo, L., S. Tittlemier, M. Diamond, K. Pepper, B. Tague, F. Yeudall, and L. Vanderlinden. 2008. Cooking decreases observed perfluorinated compound concentrations in fish. *Journal of Agricultural and Food Chemistry* 56(16):7551–7559. https://doi.org/10.1021/jf800827r.

DeLuca, N. M., M. Angrish, A. Wilkins, K. Thayer, and E. A. Cohen Hubal. 2021. Human exposure pathways to poly- and perfluoroalkyl substances (PFAS) from indoor media: A systematic review protocol. *Environment International* 146:106308. https://doi.org/10.1016/j.envint.2020.106308.

Domingo, J. L., and M. Nadal. 2017. Per- and polyfluoroalkyl substances (PFASs) in food and human dietary intake: A review of the recent scientific literature. *Journal of Agricultural and Food Chemistry* 65(3):533–543. https://doi.org/10.1021/acs.jafc.6b04683.

Domingo, J. L., and M. Nadal. 2019. Human exposure to per- and polyfluoroalkyl substances (PFAS) through drinking water: A review of the recent scientific literature. *Environmental Research* 177:108648.

Ducatman, A., M. Luster, and T. Fletcher. 2021. Perfluoroalkyl substance excretion: Effects of organic anion-inhibiting and resin-binding drugs in a community setting. *Environmental Toxicology and Pharmacology* 85:103650. https://doi.org/10.1016/j.etap.2021.103650.

Emmett, E. A., F. S. Shofer, H. Zhang, D. Freeman, C. Desai, and L. M. Shaw. 2006. Community exposure to perfluorooctanoate: Relationships between serum concentrations and exposure sources. *Journal of Occupational and Environmental Medicine* 48(8):759–770.

ECOS (Environmental Council of the States). 2020. Processes and considerations for setting state PFAS standards. https://www.ecos.org/wp-content/uploads/2020/02/Standards-White-Paper-FINAL-February-2020.pdf (accessed March 25, 2020).

Egeghy, P. P., and M. Lorber. 2011. An assessment of the exposure of Americans to perfluorooctane sulfonate: A comparison of estimated intake with values inferred from NHANES data. *Journal of Exposure Science and Environmental Epidemiology* 21(2):150–168. https://doi.org/10.1038/jes.2009.73.

Emmett, E. A., F. S. Shofer, H. Zhang, D. Freeman, C. Desai, and L. Michael Shaw. 2006. Community exposure to perfluorooctanoate: Relationships between serum concentrations and exposure sources. *Journal of Occupational and Environmental Medicine/American College of Occupational and Environmental Medicine* 48(8):759.

Fenton, S. E., A. Ducatman, A. Boobis, J. C. DeWitt, C. Lau, C. Ng, J. S. Smith, and S. M. Roberts. 2021. Per- and polyfluoroalkyl substance toxicity and human health review: Current state of knowledge and strategies for informing future research. *Environmental Toxicology and Chemistry* 40(3):606–630. https://doi.org/10.1002/etc.4890.

Fraser, A. J., T. F. Webster, D. J. Watkins, M. J. Strynar, K. Kato, A. M. Calafat, V. M. Vieira, and M. D. McClean. 2013. Polyfluorinated compounds in dust from homes, offices, and vehicles as predictors of concentrations in office workers' serum. *Environment International* 60:128–136. https://doi.org/10.1016/j.envint.2013.08.012.

Fromme, H., S. Dietrich, L. Fembacher, T. Lahrz, and W. Völkel. 2015. Neutral polyfluorinated compounds in indoor air in Germany—The LUPE 4 study. *Chemosphere* 139:572–578. https://doi.org/10.1016/j.chemosphere.2015.07.024.

Gao, K., T. Zhuang, X. Liu, J. Fu, J. Zhang, J. Fu, L. Wang, A. Zhang, Y. Liang, M. Song, and G. Jiang. 2019. Prenatal exposure to per- and polyfluoroalkyl substances (PFASs) and association between the placental transfer efficiencies and dissociation constant of serum proteins-PFAS complexes. *Environmental Science & Technology* 53(1):6529–6538. https://doi.org/10.1021/acs.est.9b00715.

Gasiorowski, R., M. K. Forbes, G. Silver, Y. Krastev, B. Hamdorf, B. Lewis, M. Tisbury, M. Cole-Sinclair, B. P. Lanphear, R. A. Klein, N. Holmes, and M. P. Taylor. 2022. Effect of plasma and blood donations on levels of perfluoroalkyl and polyfluoroalkyl substances in firefighters in Australia: A randomized clinical trial. *JAMA Network Open* 5(4):e226257. https://doi.org/10.1001/jamanetworkopen.2022.6257.

Gellrich, V., H. Brunn, and T. Stahl. 2013. Perfluoroalkyl and polyfluoroalkyl substances (PFASs) in mineral water and tap water. *Journal of Environmental Science and Health, Part A*

Toxic/Hazardous Substances and Environmental Engineering 48(2):129–135. https://doi.org/10.1080/10934529.2013.719431. PMID: 23043333.

Genuis, S. J., D. Birkholz, M. Ralitsch, and N. Thibault. 2010. Human detoxification of perfluorinated compounds. *Public Health* 124(7):367–375. https://doi.org/10.1016/j.puhe.2010.03.002.

Genuis, S. J., L. Curtis, and D. Birkholz. 2013. Gastrointestinal elimination of perfluorinated compounds using cholestyramine and chlorella pyrenoidosa. *ISRN Toxicology* 657849. https://doi.org/10.1155/2013/657849.

Genuis, S. J., Y. Liu, Q. I. Genuis, and J. W. Martin. 2014. Phlebotomy treatment for elimination of perfluoroalkyl acids in a highly exposed family: A retrospective case-series. *PloS One* 9(12):e114295. https://doi.org/10.1371/journal.pone.0114295.

Gilliland, F. D. 1992. Fluorocarbons and human health: Studies in an occupational cohort. Thesis. University of Minnesota. https://static.ewg.org/reports/2019/pfa-timeline/1992_1-46-Gilliland-Thesis.pdf (accessed July 1, 2022).

Glüge, J., M. Scheringer, I. T. Cousins, J. C. DeWitt, G. Goldenman, D. Herzke, R. Lohmann, C. A. Ng, X. Trier, and Z. Wang. 2020. An overview of the uses of per- and polyfluoroalkyl substances (PFAS). *Environmental Science: Processes & Impacts* 22(12):2345–2373.

Gyllenhammar, I., J. P. Benskin, O. Sandblom, U. Berger, L. Ahrens, S. Lignell, K. Wiberg, and A. Glynn. 2018. Perfluoroalkyl acids (PFAAs) in serum from 2-4-month-old infants: Influence of maternal serum concentration, gestational age, breast-feeding, and contaminated drinking water. *Environmental Science & Technology* 52(12):7101–7110. https://doi.org/10.1021/acs.est.8b00770.

Gyllenhammar, I., J. P. Benskin, O. Sandblom, U. Berger, L. Ahrens, S. Lignell, K. Wiberg, and A. Glynn. 2019. Perfluoroalkyl acids (PFAAs) in children's serum and contribution from PFAA-contaminated drinking water. *Environmental Science and Technology* 53(19):11447–11457. https://doi.org/10.1021/acs.est.9b01746.

Harada, K. H., S. Hashida, T. Kaneko, K. Takenaka, M. Minata, K. Inoue, N. Saito, and A. Koizumi. 2007. Biliary excretion and cerebrospinal fluid partition of perfluorooctanoate and perfluorooctane sulfonate in humans. *Enviornmental Toxicology and Pharmacology* 24(2):134–139. https://doi.org/10.1016/j.etap.2007.04.003.

Haug, L. S., S. Huber, G. Becher, and C. Thomsen. 2011. Characterisation of human exposure pathways to perfluorinated compounds—Comparing exposure estimates with biomarkers of exposure. *Environment International* 37(4):687–693. https://doi.org/10.1016/j.envint.2011.01.011.

Heo, J. J., J. W. Lee, S. K. Kim, and J. E. Oh. 2014. Foodstuff analyses show that seafood and water are major perfluoroalkyl acids (PFAAs) sources to humans in Korea. *Journal of Hazardous Materials* 279:402–409. https://doi.org/10.1016/j.jhazmat.2014.07.004.

Herkert, N. J., J. Merrill, C. Peters, D. Bollinger, S. Zhang, K. Hoffman, P. L. Ferguson, D. R. U. Knappe, and H. M. Stapleton. 2020. Assessing the effectiveness of point-of-use residential drinking water filters for perfluoroalkyl substances (PFASs). *Environmental Science & Technology Letters* 7(3):178–184.

Herzke, D., S. Huber, L. Bervoets, W. D'Hollander, J. Hajslova, J. Pulkrabova, G. Brambilla, S. P. De Filippis, S. Klenow, G. Heinemeyer, and P. De Voogt. 2013. Perfluorinated alkylated substances in vegetables collected in four European countries; Occurrence and human exposure estimations. *Environmental Science and Pollution Research* 20:7930.

Hu, Y., C. Wei, L. Wang, Z. Zhou, T. Wang, G. Liu, Y. Feng, and Y. Liang. 2020. Cooking methods affect the intake of per- and polyfluoroalkyl substances (PFASs) from grass carp. *Ecotoxicology and Environmental Safety* 203:111003. https://doi.org/10.1016/j.ecoenv.2020.111003.

ITRC (Interstate Technology and Research Council). 2017. *History and use of per- and polyfluoroalkyl substances (PFAS).* https://pfas-1.itrcweb.org/wp-content/uploads/2017/11/pfas_fact_sheet_history_and_use__11_13_17.pdf (accessed March 25, 2020).

Iwabuchi, K., and I. Sato. 2021. Effectiveness of household water purifiers in removing perfluoroalkyl substances from drinking water. *Environmental Science and Pollution Research* 28(9):11665–11671.

Kärrman, A., I. Ericson, B. van Bavel, P. O. Darnerud, M. Aune, A. Glynn, S. Lignell, and G. Lindström. 2007. Exposure of perfluorinated chemicals through lactation: Levels of matched human milk and serum and a temporal trend, 1996–2004, in Sweden. *Environmental Health Perspectives* 115(2):226–230. https://doi.org/10.1289/ehp.9491.

Kim, M-J., J. Park, L. Luo, J. Min, J. Hoan Kim, H-D. Yang, Y. Kho, G. J. Kang, M-S. Chung, S. Shin, and B. Moon. 2020. Effect of washing, soaking, and cooking methods on perfluorinated compounds in mackerel (Scomber japonicus). *Food Science & Nutrition* 8(8):4399–4408. https://doi.org/10.1002/fsn3.1737.

Kim, S. K., K. T. Lee, C. S. Kang, L. Tao, K. Kannan, K. R. Kim, C. K. Kim, J. S. Lee, P. S. Park, Y. W. Yoo, J. Y. Ha, Y. S. Shin, and J. H. Lee. 2011. Distribution of perfluorochemicals between sera and milk from the same mothers and implications for prenatal and postnatal exposures. *Environmental Pollution (Barking, Essex: 1987)* 159(1):169–174. https://doi.org/10.1016/j.envpol.2010.09.008.

Koponen, J., K. Winkens, R. Airaksinen, U. Berger, R. Vestergren, I. T. Cousins, A. M. Karvonen, J. Pekkanen, and H. Kiviranta. 2018. Longitudinal trends of per- and polyfluoroalkyl substances in children's serum. *Environment International* 121(Pt 1):591–599. https://doi.org/10.1016/j.envint.2018.09.006.

Kubwabo, C., I. Kosarac, and K. Lalonde. 2013. Determination of selected perfluorinated compounds and polyfluoroalkyl phosphate surfactants in human milk. *Chemosphere* 91(6):771–777. https://doi.org/10.1016/j.chemosphere.2013.02.011.

LaKind, J. S., M-A. Verner, R. D. Rogers, H. Goeden, D. Q. Naiman, S. A. Marchitti, G. M. Lehmann, E. P. Hines, and S. E. Fenton. 2022. Current breast milk PFAS levels in the United States and Canada: After all this time, why don't we know more? *Environmental Health Perspectives* 130(2):025002.

Li, P., X. Oyang, Y. Zhao, T. Tu, X. Tian, L. Li, Y. Zhao, J. Li, and Z. Xiao. 2019. Occurrence of perfluorinated compounds in agricultural environment, vegetables, and fruits in regions influenced by a fluorine-chemical industrial park in China. *Chemosphere* 225:659.

Li, Y., T. Fletcher, D. Mucs, K. Scott, C. H. Lindh, P. Tallving, and K. Jakobsson. 2018. Half-lives of PFOS, PFHxS and PFOA after end of exposure to contaminated drinking water. *Occupational and Environmental Medicine* 75(1):46–51. https://doi.org/10.1136/oemed-2017-104651.

Liu, J., J. Li, Y. Liu, H. M. Chan, Y. Zhao, Z. Cai, and Y. Wu. 2011. Comparison on gestation and lactation exposure of perfluorinated compounds for newborns. *Environment International* 37(7):1206–1212. https://doi.org/10.1016/j.envint.2011.05.001.

Lorber, M., and P. P. Egeghy. 2011. Simple intake and pharmacokinetic modeling to characterize exposure of Americans to perfluoroctanoic acid, PFOA. *Environmental Science and Technology* 45(19):8006–8014. https://doi.org/10.1021/es103718h.

Luo, L., M-J. Kim, J. Park, H-D. Yang, Y. Kho, M-S. Chung, and B. Moon. 2019. Reduction of perfluorinated compound content in fish cake and swimming crab by different cooking methods. *Applied Biological Chemistry* 62(1):44. https://doi.org/10.1186/s13765-019-0449-x.

Manzano-Salgado, C. B., M. Casas, M. J. Lopez-Espinosa, F. Ballester, M. Basterrechea, J. O. Grimalt, A. M. Jiménez, T. Kraus, T. Schettgen, J. Sunyer, and M. Vrijheid. 2015. Transfer of perfluoroalkyl substances from parent to fetus in a Spanish birth cohort. *Environmental Research* 142:471–478. https://doi.org/10.1016/j.envres.2015.07.020.

MDH (Minnesota Department of Health). 2008, January. *MDH evaluation of point-of-use water treatment devices for perfluorochemical removal interim report.* Fact Sheet. https://wrl.mnpals.net/islandora/object/WRLrepository%3A1862/datastream/PDF/view (accessed March 21, 2021).

Morales-McDevitt, M. E., J. Becanova, A. Blum, T. A. Bruton, S. Vojta, M. Woodward, and R. Lohmann. 2021. The air that we breathe: Neutral and volatile PFAS in indoor air. *Environmental Science & Technology Letters* 8(10):897–902. https://doi.org/10.1021/acs.estlett.1c00481.

Olsen, G. W., J. M. Burris, D. J. Ehresman, J. W. Froelich, A. M. Seacat, J. L. Butenhoff, and L. R. Zobel. 2007. Half-life of serum elimination of perfluorooctanesulfonate, perfluorohexanesulfonate, and perfluorooctanoate in retired fluorochemical production workers. *Environmental Health Perspectives* 115(9):1298–1305. https://doi/org/10.1289/ehp.10009.

Patterson, C., J. Burkhardt, D. Schupp, E. R. Krishnan, S. Dyment, S. Merritt, L. Zintek, and D. Kleinmaier. 2019. Effectiveness of point-of-use/point-of-entry systems to remove per-and polyfluoroalkyl substances from drinking water. *AWWA Water Science* 1(2):e1131.

Rodgers, K. M., C. H. Swartz, J. Occhialini, P. Bassignani, M. McCurdy, and L. A. Schaider. 2022. How well do product labels indicate the presence of PFAS in consumer items used by children and adolescents? *Environmental Science & Technology* 56(10):6294–6304. https://doi.org/10.1021/acs.est.1c05175.

Schaider, L. A., S. A. Balan, A. Blum, D. Q. Andrews, M. J. Strynar, M. E. Dickinson, D. M. Lunderberg, J. R. Lang, and G. F. Peaslee. 2017. Fluorinated compounds in U.S. fast food packaging. *Environmental Science and Technology Letters* 4(3):105–111. https://doi.org/10.1021/acs.estlett.6b00435.

Schellenberger, S., C. Jonsson, P. Mellin, O. A. Levenstam, I. Liagkouridis, A. Ribbenstedt, A-C. Hanning, L. Schultes, M. M. Plassmann, and C. Persson. 2019. Release of side-chain fluorinated polymer-containing microplastic fibers from functional textiles during washing and first estimates of perfluoroalkyl acid emissions. *Environmental Science & Technology* 53(24):14329–14338.

Scher, D. P., J. E. Kelly, C. A. Huset, K. M. Barry, and V. L. Yingling. 2019. Does soil track-in contribute to house dust concentrations of perfluoroalkyl acids (PFAAs) in areas affected by soil or water contamination? *Journal of Exposure Science & Environmental Epidemiology* 29(2):218–226.

Serrano, L., L. Mª. Iribarne-Durán, B. Suárez, F. Artacho-Cordón, F. Vela-Soria, M. Peña-Caballero, J. A. Hurtado, N. Olea, M. F. Fernández, and C. Freire. 2021. Concentrations of perfluoroalkyl substances in donor breast milk in Southern Spain and their potential determinants. *International Journal of Hygiene and Environmental Health* 236:113796. https://doi.org/10.1016/j.ijheh.2021.113796.

Sunderland, E. M., X. C. Hu, C. Dassuncao, A. K. Tokranov, C. C. Wagner, and J. G. Allen. 2019. A review of the pathways of human exposure to poly- and perfluoroalkyl substances (PFASs) and present understanding of health effects. *Journal of Exposure Science & Environmental Epidemiology* 29:131.

Tao, L., J. Ma, T. Kunisue, E. L. Libelo, S. Tanabe, and K. Kannan. 2008. Perfluorinated compounds in human breast milk from several Asian countries, and in infant formula and dairy milk from the United States. *Environmental Science & Technology* 42 (22):8597–8602. https://doi.org/10.1021/es801875v.

Taylor, M. D., S. Nilsson, J. Bräunig, K. C. Bowles, V. Cole, N. A. Moltschaniwskyj, and J. F. Mueller. 2019. Do conventional cooking methods alter concentrations of per- and polyfluoroalkyl substances (PFASs) in seafood? *Food and Chemical Toxicology* 127:280–287.

Trudel, D., L. Horowitz, M. Wormuth, M. Scheringer, I. T. Cousins, and K. Hungerbühler. 2008. Estimating consumer exposure to PFOS and PFOA. *Risk Analysis* 28(2):251–269. https://doi.org/10.1111/j.1539-6924.2008.01017.x.

Vestergren, R., and I. T. Cousins. 2009. Tracking the pathways of human exposure to perfluorocarboxylates. *Environmental Science and Technology* 43(15):5565–5575. https://doi.org/10.1021/es900228k.

Washington, J. W., and T. M. Jenkins. 2015. Abiotic hydrolysis of fluorotelomer-based polymers as a source of perfluorocarboxylates at the global scale. *Environmental Science & Technology* 49(24):14129–14135. doi: 10.1021/acs.est.5b03686.

Young, A. S., E. H. Sparer-Fine, H. M. Pickard, E. M. Sunderland, G. F. Peaslee, and J. G. Allen. 2021.
    Per- and polyfluoroalkyl substances (PFAS) and total fluorine in fire station dust. *Journal of
    Exposure Science & Environmental Epidemiology* 31:930–942.
Zheng, G., E. Schreder, J. C Dempsey, N. Uding, V. Chu, G. Andres, S. Sathyanarayana, and A.
    Salamova. 2021. Per- and polyfluoroalkyl substances (PFAS) in breast milk: Concerning trends
    for current-use PFAS. *Environmental Science & Technology* 55(11):7510–7520.

5

PFAS Testing and Concentrations to Inform Clinical Care of Exposed Patients

The Centers for Disease Control and Prevention (CDC) and various other health agencies routinely conduct biomonitoring, including measuring of exposure biomarkers in human tissues, as an important part of environmental public health surveillance (Latshaw et al., 2017). With the exception of a few environmental exposures, however, such as children's blood lead measurements, biomonitoring is rarely included in routine clinical care. Biomonitoring integrates all sources and routes of exposure into measurements of internal exposure. In this regard it is advantageous because it is more reflective of aggregate exposure than are measures of external exposure, but as a consequence, it does not identify the specific source of an exposure. Nonetheless, when properly interpreted, biomonitoring data can be used to monitor exposure levels and trends, evaluate potential health risks connected to specific sites or populations, and inform public health decisions (Latshaw et al., 2017).

Biomonitoring can be based on the direct measurement of environmental chemicals or their reaction or breakdown products (metabolites) in human tissues and fluids, such as blood (serum, plasma, and whole blood), urine, hair, nails, and breast milk (CDC, 2021; NRC, 1987). Traditionally, biomarkers have been classified as biomarkers of exposure, effect, or susceptibility (see Figure 5-1).

This chapter addresses options and considerations to guide decision making for per- and polyfluoroalkyl substances (PFAS) testing in a patient's biological samples, strategies for interpreting biomonitoring data, and PFAS concentrations that could inform clinical care of exposed patients.

OPTIONS AND CONSIDERATIONS TO GUIDE DECISION MAKING FOR PFAS TESTING

PFAS Laboratory Methods

There are no standard methods for PFAS exposure biomonitoring; some, but not all, laboratories use methods similar to those used by the CDC. Unlike most clinical laboratories, laboratories that offer PFAS testing are not subject to measurement standardization through external proficiency testing programs that evaluate laboratory performance against preestablished criteria. Laboratories that offer PFAS testing also need not comply with clinical certification, such as Clinical Laboratory Improvement Amendments (CLIA) certification, for reporting of results to patients. To support the quality and integrity of results, PFAS testing should be conducted in laboratories that meet the following criteria:

- have an extensive quality assurance/quality control (QA/QC) program (Kannan et al., 2021);
- report National Institute of Standards and Technology Standard Reference Material (NIST-SRM)–traceable data (Kannan et al., 2021); and
- employ laboratory methods with relative standard deviations of less than 15 percent and with limits of detection (LODs) in the picogram/mL region, consistent with the LODs of the CDC and academic laboratories (FDA, 2018; SWGTOX, 2013).

119

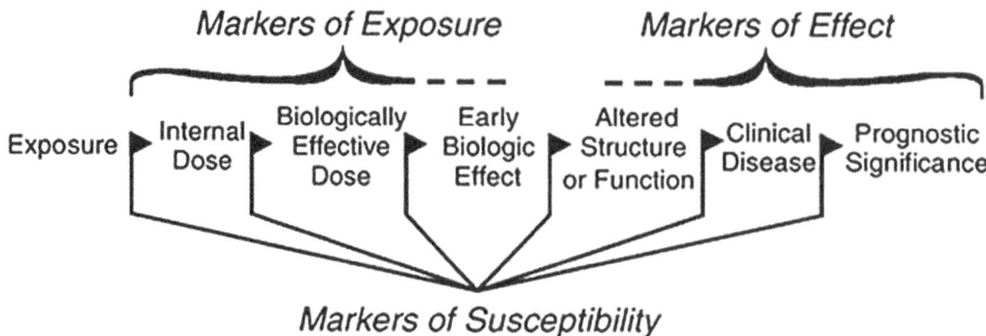

FIGURE 5-1 Simplified flow chart of classes of biomarkers.
SOURCE: NRC, 1987.

Which PFAS are most important to measure depends on the reasons for testing. A method comparable to that of the CDC that reports the linear and branched isomers of PFAS will allow comparison of individual results with those in the *National Report on Human Exposure to Environmental Chemicals* (CDC, 2015, 2021)—an informative method for individuals who wish to understand whether their exposure is high or low compared with background exposures in the U.S. population. For people with a past history of exposure to PFAS not included in the CDC panel, laboratory tests need to detect the compounds to which the person is suspected to have been exposed. However, adequate targeted exposure biomonitoring approaches may not currently exist for measuring many PFAS—especially those that are emerging or newly studied—that may be of concern in some communities. Most laboratories can test only for a limited number of PFAS; a typical PFAS panel will include fewer than 25 specific PFAS. SRMs exist for some but not all PFAS. NIST 1957 has reference values for 7 PFAS, and NIST 1958 includes 5 PFAS. The availability of SRMs for a broader range of PFAS would support higher-quality data for more PFAS.

Interpretation of PFAS biomonitoring data depends on the analytical methods being used, the matrix (blood, serum, plasma, urine, breast milk, etc.) in which the PFAS are measured, and the pharmacokinetics of the PFAS being measured.

Analytical Methods

Biological samples are generally analyzed using targeted analytical methods (i.e., for specific PFAS for which methods and standards are available), leaving a wide array of PFAS unaccounted for. Methods for measuring total extractable or adsorbable organofluorine can circumvent the limitations of targeted analytics, but these methods are also plagued by important challenges. Some non-PFAS are merged into the total organofluorine load. Moreover, clear health guidance values for total organofluorine have not been—and may very well never be—derived (see Box 5-1) because toxicological testing and risk assessments are typically conducted on a chemical-by-chemical basis so regulations can be derived for specific chemicals, although it is well known that single-chemical exposures are not reflective of how exposures occur in reality (NRC, 2009). Untargeted testing can also be used to assess exposure to PFAS, but is less standardized and precise than current targeted testing. Untargeted approaches, especially those that employ multiplexing techniques to account for high levels of endogenous chemicals, may provide a better overall picture of total PFAS exposure; as of now, however, those data would not be quantifiable and would provide only a qualitative (presence or absence) estimate of exposure (Guo et al., 2022).

BOX 5-1
Total Organofluorine Testing

In communities that could be exposed to many different PFAS, total organofluorine testing could be considered. This method measures all the organic compounds containing fluorine in a sample. In so doing, it addresses the challenge of the lack of a standard reference for so many PFAS. In a study of blood samples from the Swedish population, for example, 70 percent of organofluorine compounds measured in samples from women aged 16–44 using extractable organofluorine measurements were not accounted for by the 63 PFAS measured in targeted analyses (Aro et al., 2021). However, these measures of total organic fluorine include compounds that have a fluorine atom but are not typically considered PFAS (e.g., pesticides such as trifluralin, fluometuron, and benefin; or drugs such as Prozac). With untargeted mass spectroscopy techniques, then, it may be possible to detect many more PFAS, but confirming their identity and quantifying them can be challenging if authentic standards are not available for them, which is often the case. At this point, moreover, methods for measuring total organofluorine are not standardized and are used primarily in academic and research laboratories (Aro et al., 2022), and there are no reference- or risk-based concentrations with which to compare measurements, making interpretation difficult. Still, the committee believes total organofluorine measurements hold promise for addressing the vast number of PFAS compounds as a class and notes that at least one commercial laboratory has started using this testing.

Biological Matrix

The specific PFAS to which a patient is at risk of exposure may also dictate the biological samples (e.g., blood, urine) that should be collected. Most studies measure PFAS in serum, which is likely the best matrix for measuring PFAS with long biological half-lives (Calafat et al., 2019). Other matrices, such as whole blood or urine, may be better suited to the detection of PFAS with short biological half-lives or those whose distribution in the body differs from that of the most commonly studied PFAS, such as perfluorooctanoic acid (PFOA) and perfluorooctanesulfonic (PFOS) (Calafat et al., 2019; Poothong et al., 2017).

PFAS can be measured in multiple biological matrices, including but not limited to whole blood, serum, plasma, urine, breast milk, and hair (Alves et al., 2015). Overall, most studies have used either serum or plasma to evaluate PFAS exposure, and the measured concentrations in serum and plasma are comparable for PFOA, PFOS, and perfluorohexane sulfonate (PFHxS) (Ehresman et al., 2007). PFAS concentrations in other matrices can provide information on exposure during specific periods of life; for example, PFAS levels in segments of hair can help reconstruct historical exposures, and PFAS levels in breast milk make it possible to estimate infants' intake through breastfeeding (Zheng et al., 2021). On multiple occasions, this study's town hall participants indicated that they wanted to know not only their serum PFAS levels but also their breast milk levels, as these translate into early-life exposure for their children.

While urine, breast milk, and hair samples can be obtained relatively noninvasively, biomonitoring efforts in the United States have focused largely on serum, which is one of the most important biological compartments for the distribution of many PFAS in the body. Consequently, reference ranges and risk-based biomonitoring levels are limited or unavailable for other matrices. To date, the published literature provides measurements in breast milk for fewer than 200 people in the United States. If breast milk levels are measured, they can be interpreted by comparison with breast milk levels estimated using data from the National Health and Nutrition Examination Survey (NHANES), a series of surveys on the health status, health-related behaviors, and nutrition of the U.S. population (Lakind et al., 2022). Finally, much remains unknown about the distribution of the wide array of PFAS; for some of them, such traditional matrices as serum may not be optimal for estimating exposure to those PFAS whose distribution into these matrices is not significant.

Micro-samples are samples in which blood, usually from a pricked finger or heel stick, is collected into a small cuvette or spotted onto specially prepared filter paper and dried in open air under ambient conditions. Micro-samples have several advantages: they are minimally invasive, do not have to be collected by a phlebotomist, reduce biohazard risks, and in some cases do not require refrigeration or freezing (Freeman et al., 2018). On the other hand, the blood collected from a pricked finger is capillary blood, which, unlike venous blood, contains interstitial fluid, and research is needed to determine whether capillary blood concentrations are comparable to serum measurements. Capillary blood also requires validation, particularly if collected as dried blood spots because volume is difficult to estimate as a result of hematocrit and chromatographic effects, although K+ standardization is showing promise (Barr et al., 2021). Newborn blood spots have, however, been used successfully in research studies to evaluate exposures to PFAS and assess changes in exposures over time (Gross et al., 2020; Ma et al., 2013; Yeung et al., 2019).

Pharmacokinetics

Some PFAS, including PFOA and PFOS, have a relatively long biological half-life, and measurement of these PFAS in serum may represent chemical exposures that occurred in past years. For some other PFAS, half-lives are on the order of days. For example, GenX is estimated to have a biological half-life of 81 hours (Clark and TCS, 2021); as a result, its measurement in serum or urine likely represents exposure that occurred much more recently, such as in the last few weeks. In such cases, samples taken at different times may be needed to derive an adequate representation of long-term exposure. Alternative measures for estimating long-term exposure include exposure reconstruction of drinking water levels. Also, some physiological events may influence biological levels of PFAS. Pregnancy and breastfeeding can increase PFAS excretion, for example, so levels in mothers may not be reflective of levels prior to pregnancy (Kato et al., 2014).

Sample Contamination

The concentrations of PFAS measured in most human tissues can be considered trace concentrations, making sample contamination an issue. Contamination occurring during sample collection, transport, storage, or preparation can impact the ability to interpret a laboratory result. For example, PFAS are commonly used in medical collection equipment, such as Teflon-capped blood test tubes. Teflon should therefore be avoided in sample collection equipment intended for PFAS measurement (NASEM, 2018a), and field blanks may need to be performed to ensure that contamination is negligible.

Commercially Available PFAS Testing Laboratories

A few laboratories currently offer PFAS testing. NMS labs (Horsham, Pennsylvania) offers PFAS testing through a clinician request for out-of-pocket payment (Test Code 39307, CPT Code[s] 82542; reported by town hall speakers to cost more than $600) or reimbursement from insurance. NMS can measure six PFAS in serum: perfluorobutanesulfonic acid (as the linear isomer) (PFBS), perfluoroheptanoic acid (as the linear isomer) (PFHpA), perfluorohexanesulfonic acid (as the linear isomer) (PFHxS), perfluorononanoic acid (as the linear isomer) (PFNA), perfluorooctanesulfonic acid (as the linear isomer) (PFOS), and perfluorooctanoic acid (as the linear isomer) (PFOA).[1] It is important to note that although the NMS panel does not completely overlap with the specific PFAS the committee was charged to evaluate, it does include PFOA, PFOS, and PFNA, three of the PFAS that are commonly reported as detected in NHANES and therefore typically contribute the most to exposure. The NMS panel would be more comparable to NHANES if it included the branched isomers of PFOS and PFOA, and PFHxS.

---

[1] See https://www.nmslabs.com/tests/3427SP (accessed June 16, 2022).

EMPower DX (Framingham, Massachusetts), a subsidiary of Eurofins Scientific (Luxembourg City, Luxembourg), recently began offering direct-to-consumer testing for more than 40 PFAS via self-collected finger prick sample at a cost of $399.[2] The website for this lab reports that the test will "quantitatively determine PFAS concentrations … and positive or negative detections are reported with 99% confidence" (para. 4). In an email communication, company representatives reported that NIST reference standards are used, and that coefficients of variation are being developed for all PFAS in the panel.[3] As discussed above, method validation is needed for PFAS measurements from capillary blood to know how comparable they are to measurements from other PFAS tests.

Harms and Benefits of PFAS Testing

Based on the committee's principle of *proportionality* (see Chapter 2), it is necessary to consider the plausible harms and benefits associated with PFAS testing, although the weight of each will depend on the individual. Harms of PFAS testing include fear induced by blood draw, a small risk of injury or infection at the draw site, difficulties in interpreting results, and psychological stress that may occur when people who are tested learn that they or their family members have high levels of PFAS exposure. On the other hand, biomonitoring for PFAS blood levels may also alleviate fears associated with not knowing one's PFAS levels. Another potential benefit is increasing awareness of exposure so that exposures can be reduced. If sources of exposure are identified, actions taken to reduce these exposures, such as using a water filter, may also benefit family members in addition to the person who was tested. Community-level benefits may be associated with PFAS testing as well, such as empowering communities to respond to contamination and providing a baseline with which to evaluate the impact of community-level interventions to reduce exposure. Additionally, biomonitoring for PFAS in the context of epidemiologic studies could provide more information about PFAS-associated health effects.

To evaluate these potential harms and benefits associated with biomonitoring, the committee considered research studies evaluating the harms and benefits of reporting biomonitoring results to study participants (report-back). Report-back has been an active area of research by multidisciplinary teams, focused on whether and how to report study results to participants. These studies have varied in design, but methods used generally include interviews with study participants, investigators, and/or institutional review boards. They may also include focus groups; advisory councils; stakeholder workshops; observations at community meetings; one-on-one user testing of reports; and digital analytics, such as page views or time spent interacting with digital results pages. Investigators in these studies have identified harms associated with reporting of biomonitoring results, such as stress or concerns about the health effects associated with exposures, but the levels of stress reported have not been high enough to be considered extreme worry or panic (Emmett et al., 2009; Ohayon et al., 2017; Ramirez-Andreotta et al., 2016). Report-back studies have found further that when biomonitoring results are returned with information about how study participants could be exposed to the chemicals, what health effects are potentially associated with exposure, and what strategies could reduce exposure, participants express appreciation for receiving the results and feel empowered by the information (Adams et al., 2011; Altman et al., 2008; Brody et al., 2014; Giannini et al., 2018; Hernick et al., 2011; Perovich et al., 2018; Ramirez-Andreotta et al., 2016; Tomsho et al., 2019). Moreover, research studies have found that participants appreciate receiving test results even when the health implications of those results are unclear (Adams et al., 2011; Hernick et al., 2011; NASEM, 2018b). One study that compared the benefits of receiving individual versus aggregate biomonitoring results found that the former motivated participants to access information about environmental sources of chemical exposure and their health effects. The study also found that personal report-back increased engagement with exposure reports among Black participants.

---

[2] See https://empowerdxlab.com/products/product/pfas-exposure-test (accessed June 16, 2022).

[3] Personal communication, R. Mitzel, President Eurofins Air Toxics, LLC SVP Eurofins Specialty Services, December 6, 2021.

Report-back was associated with a small increase in psychological stress, but the authors suggest that this stress could motivate appropriate behavior change to reduce exposure (Brody et al., 2021).

In many cases, community members who provided testimony at the committee's town halls were strongly in favor of PFAS testing and described being frustrated with the numerous difficulties encountered in accessing the testing (see Appendix B). Kristen Mello of Westfield Residents Advocating for Themselves (WRAFT) said to the committee:

> You don't have a problem getting an insurance assessor when your car is hit, you don't have a problem getting an insurance assessor when you have a tornado, but this slow motion unfolding environmental and public health disaster … is intentionally keeping the information from us so that we cannot take action.

Furthermore, Emily Donovan of Clean Cape Fear, said:

> Sadly it feels like guinea pigs are treated better, because at least their exposures are thoroughly studied for the betterment of humanity.

And Cathy Wusterbarth of Need Our Water said:

> We've tested the fish; we've tested the deer; we've tested the groundwater, the waterways, and the foam. When are we going to test the people?… The only risk [of testing] is to the polluters who do not want us to link them to our exposure.

Community members stated that they want access to PFAS testing so that they can understand their personal level of exposure and, as suggested by Ayesha Khan of Nantucket PFAS Action Group, "help those who are exposed to be proactive in reducing exposure and managing risk." An example of the how PFAS testing can help people manage health risks was presented by Sandy Wynn-Stelt, who learned that the drinking water in her home was contaminated with PFAS at a level more than 1,000 times the current U.S. Environmental Protection Agency (EPA) health advisory and paid out of pocket for PFAS testing. She shared her results with her clinician, and they decided to conduct clinical follow-up for any of the conditions associated with PFAS, including impacts on thyroid hormone levels. The results of her thyroid tests led to other follow-up and ultimately to her diagnosis of thyroid cancer. Wynn-Stelt believes her PFAS testing led to an earlier diagnosis and better outcomes for her cancer.

Social and Ethical Implications of PFAS Testing

Many town hall speakers voiced frustration about not having access to testing for PFAS. Speakers mentioned the injustices related to having been exposed to PFAS without their consent or knowledge and their frustration with being unable to access testing. They shared that PFAS testing, when available, is expensive. Town hall speakers also suggested that discussion of PFAS testing should be conducted in a culturally sensitive manner, be available in a variety of languages, and be at a reading level that the average American can understand (e.g., 4th-grade level).

Another important social consideration is that people who are tested for PFAS may experience social and economic conflicts related to their testing. If a PFAS blood test identifies community contamination, business revenue and property values may be adversely affected (Harclerode et al., 2021). People who draw attention to the contamination may experience anger and social isolation from other community members. Indeed, one town hall speaker asked that their name be removed from the town hall agendas and the committee's website because they were worried about angering neighbors by talking about their contaminated well.

Art Schaap, a dairy farmer in New Mexico, discovered that his farm was unknowingly subjected to severe PFAS contamination due to use of military aqueous film-forming foam (AFFF). After learning of the contamination, Schaap voluntarily tested his herd for PFAS and discovered that his cows were highly exposed. The PFAS contamination on Schaap's farm devastated his livelihood because he had limited options available for getting rid of the contaminated animals in any profitable way. The dairy, beef, and rendering industries do not want PFAS-contaminated animals or products. The result for Schaap was the stranding of at least 4,000 cows, the death of 1,200, and the dumping of 1,500 loads of milk (see Appendix B for Schaap's complete testimony).

PFAS Testing: Findings and Recommendations

Applying the principle of *proportionality*, the committee believes that the benefits of PFAS testing for those who request it typically outweigh the harms (see Box 5-2). The harms reported in studies of report-back of biomonitoring results are worry about the harms of exposure, decreased property values, and potential social isolation or ostracism. In many cases, people who request testing are already worried about their exposure, which is why they are requesting the testing, and they may already be at risk of decreased property values associated with contamination. People in exposed communities have been "contaminated without consent" and "poisoned without permission."[4]

BOX 5-2
Potential Harms and Benefits of PFAS Testing

Potential Harms
- Fear of blood draw
- Small risk of injury or infection at draw site
- Difficulties in interpreting results
- Stress or concern about the health effects from exposure
- Decreased property values
- Social isolation
- Clinical consequences from medical follow-up as a result of exposure

Potential Benefits
- Increased awareness of exposure so that exposure can be reduced
- Empowerment of communities to respond to contamination
- Relief from the stress of not knowing one's exposure level
- Identification of the potential risk for health conditions associated with PFAS exposure to inform subsequent preventive care
- Help in monitoring whether efforts to reduce exposure are working through the conduct of baseline and follow-up tests.

An important element of the principle of *justice* is making PFAS testing easily accessible and readily available to all regardless of ability to pay, race, ethnicity, age, occupation, or location. Some populations are at increased risk of PFAS exposure and therefore may be at increased risk for a wide range of health conditions. These factors favor making exposure biomonitoring available to all who desire it. The principle of *autonomy* (i.e., respect for the ability of people to make their own health decisions) also favors allowing people who are likely to have a history of elevated exposure to PFAS and want PFAS testing to receive it and those who do not, to refuse it. These decisions require *shared decision making*[5] between patient and clinician and should include clarifying that exposure biomonitoring results

---

[4] Quotes from Andrea Amico, speaker at the committee's meeting on July 13, 2021.

[5] Barry and colleagues (2012) build on concepts in Charles (1997) and the Institute of Medicine's (2001) *Crossing the Quality Chasm: A New Health System for the 21st Century*, describing shared decision making as the

do not predict future health conditions and can only indicate the potential for an increased risk for certain conditions associated with exposures. Allowing people the opportunity to discuss with their clinicians whether they should undergo PFAS testing shows respect for patient values and is particularly important for people who have experienced environmental injustice as a result of PFAS contamination in their community.

The committee acknowledges that important factors need to be addressed. There are deficiencies in the current cost payment model, and the availability of PFAS testing may need to be addressed. Clinicians and health care facilities will also need to be made aware of reimbursement policies and laboratory codes (e.g., CPT; Reference Lab Order Code). New Hampshire recently required that health insurance cover PFAS blood testing.[6] PFAS testing can currently be ordered online without a provider, and the committee believes the testing and interpretation of its results are most beneficial if done with the guidance of a clinician.

The committee makes the following recommendations:

Recommendation 5-1: As communities with PFAS exposure are identified, government entities (e.g., Centers for Disease Control and Prevention [CDC]/Agency for Toxic Substances and Disease Registry [ATSDR], public health departments) should support clinicians with educational materials about PFAS testing so they can discuss testing with their patients. These educational materials should include the following information:

- **How people can be exposed to PFAS: Exposure routes include occupational exposures and work with fluorochemicals or as a firefighter; consumption of contaminated drinking water in communities that obtain their water from sources near commercial airports, military bases, fluorochemical manufacturing plants, wastewater treatment plants, landfills, or incinerators where PFAS-containing waste may have been disposed of or farms where sewage sludge may have been used; and consumption of contaminated fish or game if fishing or hunting occurs in contaminated areas. Individuals living near fluorochemical plants may also be exposed via inhalation of air emissions.**
- **Potential health effects of PFAS exposure and strategies for reducing exposure.**
- **Limitations of PFAS blood testing: PFAS blood testing does not identify the sources of exposure or predict future health outcomes; it only assesses body burden at the time of sample collection. For example, a person with low blood levels today may have had higher levels in the past.**
- **The benefits and harms of PFAS testing.**

Recommendation 5-2: Clinicians should offer PFAS testing to patients likely to have a history of elevated exposure. In all discussions of PFAS testing, clinicians should describe the potential benefits and harms of the testing and the potential clinical consequences (such as additional follow-up), related social implications, and limitations of the testing so that patient and clinician can make a shared, informed decision. Patients who are likely to have a history of elevated exposure to PFAS include those who have

---

pinnacle of patient-centered care: "The process by which the optimal decision may be reached ... is called shared decision-making and involves, at minimum, a clinician and the patient.... In shared decision-making, both parties share information: the clinician offers options and describes their risks and benefits, and the patient expresses his or her preferences and values. Each participant is thus armed with a better understanding of the relevant factors and shares responsibility in the decision about how to proceed" (p. 780).

⁶ See https://providernews.anthem.com/new-hampshire/article/coverage-for-pfas-and-pfc-blood-tests-for-new-hampshire-residents (accessed June 17, 2022).

- **had occupational exposure to PFAS (such as those who have worked with fluorochemicals or served as a firefighter);**
- **lived in communities where environmental and public health authorities (Centers for Disease Control and Prevention [CDC], Agency for Toxic Substances and Disease Registry [ATSDR], U.S. Environmental Protection Agency [EPA], state and local environmental or health authorities), or academic researchers have documented PFAS contamination; or**
- **lived in areas where PFAS contamination may have occurred, such as near facilities that use or have used fluorochemicals, commercial airports, military bases, wastewater treatment plants, farms where sewage sludge may have been used, or landfills or incinerators that have received PFAS-containing waste.**

When clinicians discuss results of PFAS with patients, the results will be most useful if accompanied by information on how exposure occurs, the potential associated health effects, and strategies that may reduce exposure—the same contextual information included in the report-back literature reviewed by the committee. Clinicians may also want to consider the frequency of PFAS testing based on the considerations in Box 5-3.

BOX 5-3
Considerations for Frequency of PFAS Testing

- Consider confirmatory retesting when the result is much higher or lower than anticipated given exposure history.
- Consider retesting if exposure changes because
  - public health actions (such as drinking water treatment programs or site cleanup are taken to reduce exposure);
  - the patient takes action(s) to reduce exposure (such as installing water filters, moving from a community with known high levels of PFAS in drinking water, or modifying occupational exposures); or
  - the patient moves into a community with known high levels of PFAS or otherwise has a suspected increase in exposure risk.
- For follow-up testing of PFAS with a long half-life, allow at least a year before retesting.
- Retesting is of no or limited value if initial serum levels are low and exposure does not change.

STRATEGIES FOR INTERPRETING BIOMONITORING DATA

There are two general strategies for interpreting chemical concentrations for exposure biomonitoring: reference-based (or descriptive) and risk-based (NRC, 2006). Both approaches can be useful to inform clinical care of exposed patients. Reference-based approaches can allow a clinician to understand whether the concentration of a chemical in the patient's biological sample is within or outside of the normal background range, while risk-based approaches can allow a clinician to determine whether a patient's biological concentration is below or above a value associated with tolerable, negligible, or minimal risk. Note that comparing concentrations of individual PFAS against reference- or risk-based levels may underestimate overall exposure to a complex mixture of PFAS and associated risk, such as in areas contaminated with PFAS used to replace other PFAS that have been phased out.

Figure 5-2 illustrates the distribution of biomarker concentrations in a generic reference population, expressed as cumulative frequency. As is commonly done in a clinical test, the 95th percentile of the distribution can be used to determine the upper-limit value of the test result. However, a different percentile may be chosen, depending on the circumstances, the characteristics of the reference population, the distribution of the results, and the purpose of the study. It is important to be aware that a particular cutpoint does not represent a level that separates the population into typical versus highly exposed (NRC, 2006). Box 5-4 describes the use of reference ranges to interpret other environmental exposures, such as exposures to lead and arsenic (see Box 5-4).

Reference-Based Approaches

Use of a reference-based approach to interpret biomonitoring data requires descriptive statistics from a reference population against which to compare the data. Typically, an individual's result would be compared with a statistical review of the reference data, typically in the form of a data distribution, such as 10th, 25th, 50th, 75th, and 95th percentiles within the reference population. The reference range offers a point of comparison, although some people or subgroups within that range may be subject to more or less exposure. This section describes a number of interpretive issues that arise with this approach. The validity and utility of biomonitoring values for use as reference ranges depend on the design of the studies from which those values were derived and the quality of the data, with special attention to the availability and comparability of data on the reference population in relation to the study population (NRC, 2006).

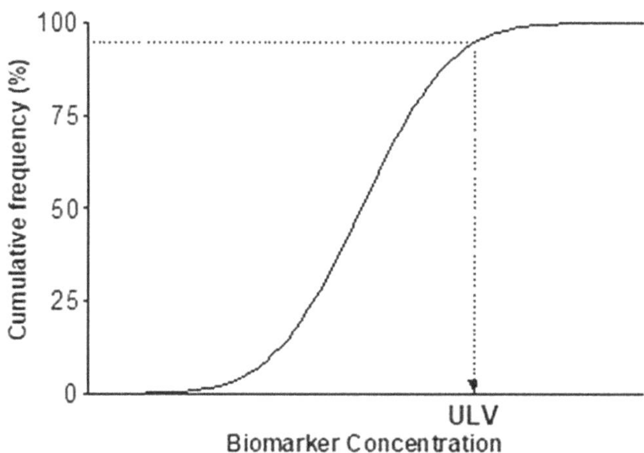

FIGURE 5-2 Distribution of biomarker concentrations in a generic reference population.
NOTE: ULV = upper limit value.
SOURCE: NRC, 2006.

BOX 5-4
Clinical Use of Reference Ranges

The examples of lead and arsenic may serve to illustrate the clinical use of biomonitoring results and reference ranges. In 2009, an elevated blood lead level (BLL) was defined by the Council of State and Territorial Epidemiologists as 10 micrograms per deciliter (µg/dL) or greater for children and adults. Subsequently, additional research documented health effects at BLLs below 10 µg/dL. In 2012, the Centers for Disease Control and Prevention (CDC) and the Advisory Committee on Childhood Lead Poisoning recommended that a BLL of 5 µg/dL or greater be defined as elevated. This reference range value was based on the 97th percentile of the BLL distribution for children aged 1–5 years, using 2007–2010 National Health and Nutrition Examination Survey (NHANES) data. This reference range value can be updated every 4 years using the two most recent NHANES surveys.[a] In May 2021, a BLL reference value of 3.5 µg/dL was set using the two most recent NHANES surveys to identify children with BLLs higher than those of most children. The adult elevated level was set at 5 µg/dL in 2012 on the basis of evidence of adverse health effects with no BLL threshold identified. The mean BLL at that

continued

BOX 5-4 *continued*

time was 1.09 µg/dL for U.S. adults older than age 20, with a 95th percentile value of 3.36 µg/dL. When a patient presents with a BLL above this reference value, the clinician, often in conjunction with public health officials, will initiate medical surveillance that includes repeated testing at intervals based on the level, history taking as to potential sources of the lead exposure, evaluation for signs and symptoms of lead toxicity, and treatment if necessary.[b]

Reference ranges are not used as commonly to interpret urinary arsenic levels. NHANES reports urinary total arsenic results (inorganic arsenic and methylated metabolites) to survey participants based on the following categories: ≤50 micrograms per liter (µg/l) is considered normal; >50 to <200 µg/L is considered high normal; and ≥200 µg/L is considered high. NHANES recommends that participants with a "high" result consult their medical provider. In a clinical setting, when the level of urinary arsenic is high normal (above 50 µg/dL or 100 µg/dL depending on the laboratory), the clinician will search for sources of exposure to arsenic and evaluate for signs and symptoms of arsenic toxicity. Diagnosis of arsenic toxicity is based on integration of exposure history, clinical findings, and laboratory results. In acute arsenic toxicity, total urine arsenic is typically greater than 1000 µg/L. In chronic arsenic toxicity, simple comparison with the "normal range" is not sufficient. It is important to evaluate for signs of arsenic toxicity and compare urine levels with a toxicity threshold from the literature (Baker et al., 2018).

[a] See https://www.cdc.gov/niosh/topics/ables/ReferenceBloodLevelsforAdults.html#_ftn3 (accessed June 17, 2022).
[b] See https://cdn.ymaws.com/www.cste.org/resource/resmgr/PS1/15-EH-01_revised_12.4.15.pdf (accessed June 17, 2022).

The reference range approach depends on data availability and data comparability for both the method used to measure PFAS and the population with which the results are being compared. Ideally, the reference population includes people similar in age, race or ethnicity, sex, and other demographic factors to the person whose PFAS testing result is being interpreted. Additionally, it is important that the PFAS be analyzed in the same tissues or fluids (blood, serum, breast milk, urine, etc.) and that the chemical analysis methods used are comparable and measure the same PFAS (NRC, 2006).

Note that interindividual variability in PFAS testing results may be a function of differences not only in exposure but also in pharmacokinetics with respect to excretory clearance. Such host factors as parity, breastfeeding status, menstrual status, age, genetic polymorphisms, concurrent acute or chronic disease, and medication use can affect pharmacokinetics.

Because NHANES is conducted in only a few communities each year, regional estimates or even urban versus rural comparisons cannot be made. Table 5-1 provides the most recent years of NHANES data (in serum ng/mL) for the total population sampled (ages 12 and older) for the four most predominant PFAS: PFOA, PFOS, PFHxS, and PFNA. These PFAS have been detected in almost all NHANES participants. Since the production and use of PFOS and PFOA were phased out, their levels have been declining (Brennan et al., 2021). Important limitations of NHANES data are that data for vulnerable populations, such as children younger than 12 years old and pregnant persons, are not always available. Children younger than 12 also are not included in the *National Report on Human Exposure to Environmental Chemicals*, although there are published estimates of PFAS exposure for children aged 3–11 years for 2013 and 2014 (Ye et al., 2018). Pregnant people are also not included in large numbers in NHANES, so multiple years of NHANES data will need to be combined to obtain a large enough sample for comparison (Watson et al., 2020); it may be more appropriate to compare pregnant people with the data for females in the same age range and time period. NHANES also does not specifically enroll participants in PFAS-exposed communities.

Distribution of PFAS Concentrations in Exposed Communities

While NHANES provides descriptive statistics for the general U.S. population, it is noteworthy that clinicians in exposed communities will likely encounter higher PFAS levels in their patients. Discussed below are examples from some of the contaminated communities throughout the United States. The first community identified as having known PFOA (C-8) exposure was residents living near the DuPont Teflon-manufacturing plant in Parkersburg, West Virginia. Contamination of six nearby public water districts and hundreds of private drinking water wells in West Virginia and Ohio was discovered, leading to public health concerns. As part of a settlement for a large class action lawsuit against DuPont, the C-8 Science Panel was established to determine potential health effects of PFOA exposure, and a 1-year cross-sectional survey (2005–2006), known as the C-8 Health Project, was conducted among approximately 70,000 residents with contaminated drinking water (Frisbee et al., 2009). The average measured serum PFOA level among residents in Little Hocking, Ohio, with the highest PFOA drinking water contamination was 227.6 ng/mL; for the entire C-8 Health Project survey, the average value for PFOA was 82.9 ng/mL. For comparison, the average PFOA serum level in the general U.S. population was 4.2 (ng/mL) in 2005 (CDC, 2015).

The 3M Company (Maplewood, Minnesota) produced PFAS at its Cottage Grove facility from the late 1940s until 2002. PFOA was a prominent PFAS made at this site. In late 2003, the Minnesota Pollution Control Agency discovered groundwater contamination near Cottage Grove and several other sites in the suburbs east of St. Paul, Minnesota (the 3M Chemolite site, 3M Woodbury site, and 3M Oakdale site, as well as the Washington County Landfill).[7] In 2006, water filtration systems for polluted public and private wells were installed to reduce PFAS exposure. The Minnesota Department of Health completed three projects to test blood levels of PFAS in people living in the east metro area of St. Paul (see Figure 5-3).

TABLE 5-1 Distributions of Serum PFAS Concentration (nanograms per milliliter [ng/mL]) in Four Cycles of the National Health and Nutrition Examination Survey (NHANES), 2011–2018

PFAS Chemical	NHANES Survey Years	Geometric Mean (95% confidence interval [CI])	50th Percentile (95% CI)	95th Percentile (95% CI)
PFOA	2011–2012	2.08 (1.95–2.22)	2.08 (1.96–2.26)	5.68 (5.02–6.49)
	2013–2014	1.94 (1.76–2.14)	2.07 (1.87–2.20)	5.57 (4.60–6.27)
	2015–2016	1.56 (1.47–1.66)	1.57 (1.47–1.77)	4.17 (3.87–4.67)
	2017–2018	1.42 (1.33–1.52)	1.47 (1.37–1.57)	3.77 (3.17–5.07)
PFOS	2011–2012	6.31 (5.84–6.82)	6.53 (5.99–7.13)	21.7 (19.3–23.9)
	2013–2014	4.99 (4.50–5.52)	5.20 (4.80–5.70)	18.5 (15.4–22.0)
	2015–2016	4.72 (4.40–5.07)	4.80 (4.40–5.30)	18.3 (15.5–22.7)
	2017–2018	4.25 (3.90–4.62)	4.30 (3.80–4.90)	14.6 (13.1–16.5)
PFNA	2011–2012	0.881 (0.801–0.968)	0.860 (0.750–0.960)	2.54 (2.28–2.89)
	2013–2014	0.675 (0.613–0.742)	0.700 (0.600–0.800)	2.00 (1.80–2.30)
	2015–2016	0.577 (0.535–0.623)	0.600 (0.500–0.600)	1.90 (1.50–2.20)
	2017–2018	0.411 (0.364–0.464)	0.400 (0.400–0.500)	1.40 (1.10–1.80)
PFHxS	2011–2012	1.28 (1.15–1.43)	1.27 (1.11–1.45)	5.44 (4.61–6.82)
	2013–2014	1.35 (1.20–1.52)	1.40 (1.20–1.60)	5.60 (4.70–7.10)
	2015–2016	1.18 (1.08–1.30)	1.20 (1.10–1.40)	4.90 (4.10–5.80)
	2017–2018	1.08 (0.996–1.18)	1.10 (1.00–1.20)	3.70 (3.30–5.60)

---

[7] See https://www.pca.state.mn.us/waste/pfas-investigation-and-clean-up (accessed June 17, 2022).

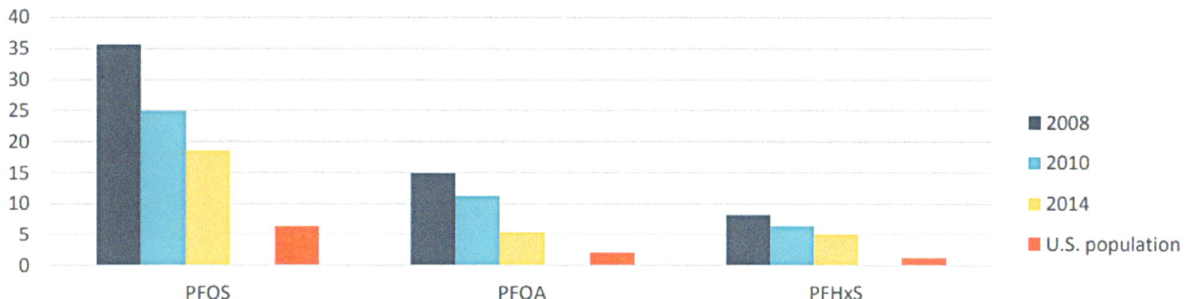

FIGURE 5-3 Geometric means of PFAS in blood from east metro St. Paul biomonitoring, in nanograms per milliliter (ng/mL).
SOURCE: U.S. population levels are from the 2011–2012 National Health and Nutrition Examination Survey (NHANES).

In Hoosick Falls, New York, high PFOA levels in water led to a public outcry, and a federal class action lawsuit was filed against Saint-Gobain Performance Plastics and Honeywell International. In 2016, the geometric mean PFOA serum level among participants in the Hoosick Falls Biomonitoring Study who used village water was 43.5 ng/mL (N = 1,640).[8] In the aftermath of Hoosick Falls, impacted community members voiced significant concern that the EPA's recommended health advisory level for drinking water at the time (400 ng/L[9]) provided insufficient protection. In January 2016, the EPA recommended that the community not drink water with PFOA in excess of 100 ng/L,[10] and in May 2016, after reviewing the existing body of PFOA and PFOS data,[11] it revised its health advisory level for drinking water to 70 ng/L for the sum of PFOA and PFOS. In New Hampshire, the state conducted biomonitoring in Merrimack Village District, where another Saint-Gobain plant had contaminated local drinking water. The average PFOA serum levels among participants was 3.9 ng/mL (N = 217).

In 2014, it was discovered that a public water supply well in Pease, New Hampshire, was contaminated with PFOS, PFOA, and PFHxS as a result of the use of firefighting foams at Pease Tradeport, a former Air Force Base. Beginning in April 2015, the New Hampshire Department of Health and Human Services conducted blood testing for people in communities where PFAS had been found in drinking water above lifetime health advisory levels. Between April and October 2015, 1,578 members of the Pease Tradeport community had their blood tested for PFAS. Results of biomonitoring from all individuals in the study showed that the geometric mean of PFAS exceeded that of the general U.S. population for PFOS (Pease: 8.59 ng/mL, NHANES: 6.31 ng/mL), PFOA (Pease: 3.09 ng/mL, NHANES: 2.08 ng/mL), and PFHxS (Pease: 4.12 ng/mL, NHANES: 1.28 ng/mL).[12]

Communities in Bucks and Montgomery counties in Pennsylvania have detected PFOS in their drinking water from use of firefighting foams during military activities. In response to this water contamination, the state health department conducted biomonitoring for PFOS exposure. The average PFOS serum level in this community was 10.2 ng/mL (N = 235).[13] A military base in Newburgh, New

---

[8] See https://www.health.ny.gov/environmental/investigations/hoosick/docs/infosheetgrouplong.pdf (accessed June 17, 2022).

[9] See https://www.epa.gov/sites/default/files/2015-12/documents/hoosickfallsmayorpfoa.pdf (accessed June 17, 2022).

[10] See https://www.epa.gov/sites/default/files/2016-01/documents/epa_statement_on_private_wells_in_the_town_of_hoosick.pdf (accessed June 16, 2022).

[11] See https://www.epa.gov/sites/default/files/2016-06/documents/drinkingwaterhealthadvisories_pfoa_pfos_updated_5.31.16.pdf (accessed June 16, 2022).

[12] See https://www.dhhs.nh.gov/dphs/documents/pease-pfc-blood-testing.pdf (accessed June 16, 2022).

[13] See https://www.health.pa.gov/topics/Documents/Environmental%20Health/PEATT%20Pilot%20Project%20Final%20Report%20April%2029%202019.pdf (accessed June 16, 2022).

York, also contaminated drinking water in that community, resulting in a median PFOS serum level of 16.3 ng/mL (N = 1,917).[14]

The primary manufacturer of PFOS, 3M, has also been implicated in drinking water contamination, in Decatur, Alabama. In a community study there, the median level was 39.3 ng/mL (N = 121) among participants with contaminated drinking water.[15]

Risk-Based Approaches

The most complex approaches for interpreting biomonitoring data are those that evaluate the risk associated with a biomonitoring result. Evaluation of risk may be desirable given the importance of the question of how risky a blood concentration is, as well as the fact that the descriptive approaches provide only relative information without assessing the risk to human health (NRC, 2006). Some exposure disease processes are thought to have a "threshold" or exposure level that needs to be exceeded before a response occurs, while others are thought to have a "nonthreshold" response such that there is no level at which no harm occurs. Examples of nonthreshold dose responses at a population level include lead, particulate matter, radon, and secondhand smoke. To set standards for nonthreshold effects, an acceptable risk needs to be determined (NRC, 2009). The standard reflects the expected daily intake associated with an acceptable risk for one or more adverse outcomes. Drinking water health advisories are risk-based standards that use dose estimates associated with daily consumption to predict long-term health effects. As a result, they should not be used to interpret biomonitoring results that represent an integration of multiple exposure sources over different periods depending on PFAS-specific half-lives (see Box 5-5).

BOX 5-5
PFAS Serum Levels Are Not Directly Comparable to PFAS Drinking Water Levels

In an effort to interpret PFAS testing results, members of the lay public sometimes compare PFAS levels in serum with PFAS drinking water advisory levels. This comparison may involve converting the units of the PFAS serum levels (usually given in nanograms per milliliter [ng/mL] of blood) to align with how PFAS drinking water advisory levels are generally reported (as nanograms per liter [ng/L] of water). The committee strongly discourages this comparison. Drinking water advisory levels reflect the levels of PFAS (in ng/L of water) that are assumed to represent a tolerable, negligible, or minimal risk for daily consumption in addition to other sources of exposure. In contrast, PFAS serum (ng/mL) levels are an integration of multiple exposure sources representing different periods of exposure, depending on PFAS-specific half-lives.

The committee also strongly recommends against translating a drinking water advisory level into a "safe" serum PFAS level. Using an online resource (https://www.ics.uci.edu/~sbartell/pfascalc.html [accessed June 16, 2022]; Bartell et al., 2017; Lu and Bartell, 2020), calculations can be made to answer such questions as "How long will it take for my serum PFOA levels to return to normal now that there's no more PFOA in my water?" or "My water has PFOA in it, but my blood hasn't yet been tested; do you expect my serum PFOA levels to be elevated?" The online calculator includes four PFAS chemicals (PFOA, PFOS, PFHxS, and PFNA) and an optional physiological adjustment to the serum half-life for menstruation. Although informative for some questions, the calculator does not allow users to include exposure sources other than drinking water. In contrast, agencies typically account for the contribution of drinking water to total exposure (e.g., 20 percent) when determining drinking water advisory levels. Serum PFAS levels reflect an integration of multiple exposure sources at a single time and are distinct from recommended levels in water for daily consumption.

Instead of comparing PFAS serum levels with PFAS drinking water levels, the committee recommends that individuals refer to health-based and reference-based serum levels.

---

[14] See https://www.health.ny.gov/environmental/investigations/newburgh/docs/infosheetgroupresults.pdf (accessed June 16, 2022).

[15] See https://www.atsdr.cdc.gov/hac/pha/Decatur/Perfluorochemical_Serum%20Sampling.pdf (accessed June 16, 2022).

There are several options for deriving serum or plasma concentrations that are consistent with health-based guidance values. In the most straightforward risk-based approach, epidemiologic studies with exposure-response relationships could be used to conduct biomonitoring-based risk assessment. As presented below, the German Human Biomonitoring Commission (HBM Commission) reviewed the epidemiologic literature to compile serum or plasma levels with and without observable effects. Contrasting patient serum PFAS levels with levels with or without effects in an epidemiologic study can facilitate understanding risk for the given response endpoint (Apel et al., 2020).

Another option is to derive serum or plasma concentrations that are consistent with health-based guidance values such as reference doses (RfDs) and tolerable daily intakes (TDIs), commonly referred to as biomonitoring equivalents (BEs) (Hays et al., 2008). Because these values are frequently based on animal experiments, some with no measure of serum concentrations, pharmacokinetic modeling is often used to translate a point of departure (e.g., the benchmark dose) into animal serum levels, which can be converted into BEs by adjusting for uncertainty factors. For this report, the committee focused on the human data to evaluate risk levels.

The committee reviewed the International Human Biomonitoring (i-HBM) Working Group dashboard to search for biomonitoring guidance values for PFAS and found that the only risk-based standards were the German HBM values for PFOS and PFOA. The committee also searched for RfDs based on human data in the tables in the *Environmental Council of the States White Paper on Setting State PFAS Standards* and in the authoritative reviews considered in the committee's review of PFAS health effects (see Chapter 3). The committee found that only the European Food Safety Authority (EFSA) scientific opinion contained a risk-based value estimated in humans.

German HBM Values

The German HBM Commission develops human biomonitoring (HBM) values for interpretation of the concentrations of environmental chemicals measured in biological samples. These values, which are derived for the general population (including vulnerable subgroups) with chronic exposure, represent concentrations below which no observed risk of adverse health effects is expected (HBM-I) or above which adverse health effects are possible (HBM-II) (Schulz et al., 2011).

In 2021, the HBM Commission published HBM-I values for PFOS and PFOA in plasma, which is assumed to have concentrations similar to those of serum (Ehresman et al., 2007; Hölzer et al., 2021). To derive HBM-I values, the HBM Commission conducted a literature review for studies of the health effects of PFOS and PFOA. It then selected epidemiological studies and identified points of departure for various health effects for use in quantifying an association between exposure and critical effects. The determination of these points of departure varied depending on the methods of each study. Where epidemiological analyses were based on quantiles of exposure, the point of departure represented the lower limit of the quantile for which significantly increased risk was observed. For studies using continuous measures of exposure, points of departure were based on either benchmark dose-response modeling or a qualitative assessment of the effects and dose-response relationship. Health outcomes in the reviewed studies included fertility (time to pregnancy), pregnancy-induced hypertension and diabetes, reduced birthweight, serum cholesterol concentrations, serum uric acid concentrations, reduced antibody response to vaccination, pubertal development, thyroid metabolism, and onset of menopause. Once points of departure for each compound, study, and outcome had been selected, HBM-I values were selected in the low range of points of departure for both chemicals. The points of departure identified represented a wide range of exposure values. It was determined that the HBM-1 values were 2 ng/mL for plasma PFOA and 5 ng/mL for plasma PFOS. The HBM Commission considers these HBM-I values to be precautionary. Exceedance of these levels should not be interpreted as increasing risk, but may warrant efforts to reduce exposure (Hölzer et al., 2021).

The HBM Commission also established HBM-II values in 2021, based on epidemiological studies on PFOA or PFOS and adverse health outcomes, including reduced birthweight, developmental effects, reduced fertility, reduced antibody response to vaccination, increased cholesterol concentrations (low-density lipoprotein [LDL] and total cholesterol), and type 2 diabetes. Because HBM-II values represent levels above which adverse health effects are possible, the points of departure for those values differed from those identified for HBM-I values. Where appropriate, benchmark dose-response modeling was used to determine a level for a given effect level. For studies with continuous exposure data in which benchmark dose-response modeling could not be performed, risk estimates from adjusted regression analyses were used. In the case of studies in which analyses were based on quantiles of exposure, the median value of the lowest quantile with a significant association with an adverse health outcome was used as the point of departure.

The points of departure chosen represent quantitatively defined changes (such as 5–10 percent, calculated with a confidence interval for a population) in certain target parameters (e.g., morbidity, laboratory values). In the evaluation of reduced birthweight, for example, the points of departure of 10 ng PFOA/mL and 15 ng PFOS/mL were determined from a meta-analysis that observed a reduction in birthweights by approximately 20 g per ng PFOA/mL and 20 g per ng PFOS/mL. Compiled points of departure were 3–10 ng/mL for PFOA and 1–30 ng/mL for PFOS. HBM-II values for plasma PFOA (10 ng/mL) and PFOS (20 ng/mL) were established as the mid- to high values in these ranges for the general population, excluding women of childbearing age. Lower values were derived for plasma PFOA (5 ng/mL) and PFOS (10 ng/mL) in women of childbearing age, mainly because studies indicated associations with developmental toxicity, reduced fertility, and increased incidence of gestational diabetes. However, the HBM Commission indicated that there is ample uncertainty around HBM-II values for PFOA and PFOS (Schümann et al., 2021).

European Food Safety Authority's Human Point of Departure

In 2020, EFSA published a scientific opinion on the derivation of a tolerable weekly intake for the sum of four predominant PFAS: PFOA, PFOS, PFHxS, and PFNA. Following a review of the animal and epidemiological literature on adverse health effects of PFAS, immunological endpoints were considered the most robust and sensitive for risk assessment. An epidemiological study in German children was used as the critical study for deriving the health-based guidance value. In this study, children's serum PFAS levels at 1 year of age were associated with lower antibody titers against diphtheria at a $BMDL_{10}$ (the lower one-sided confidence limit of the benchmark dose for a 10 percent response) value of 17.5 ng/mL for the sum of PFOS, PFOA, PFHxS, and PFNA (Abraham et al., 2020). Physiologically based pharmacokinetic (PBPK) modeling was used to translate this value in children's serum into a daily dose in their mothers, assuming that levels at 1 year of age in breastfed children reflect primarily the body burden acquired through placental and lactational transfer. To reach this $BMDL_{10}$ value in children's serum at the end of 12 months of breastfeeding, EFSA estimated that the maternal level at 35 years of age would need to be 6.9 ng/mL for the sum of the four PFAS, a level associated with an estimated maternal intake of 0.631 ng/kg body weight/day. Although EFSA did not specifically aim to determine acceptable serum PFAS levels, the 6.9 ng/mL serum concentration could be considered a serum level for women of reproductive age below which risk is negligible.

PFAS Concentrations That Could Inform Clinical Care: Findings and Recommendations

The HBM Commission has identified risk-based levels for two PFAS chemicals—PFOS and PFOA—while EFSA has established such values for the sum of PFOA, PFOS, PFHxS, and PFNA. No individual values are available for PFHxS and PFNA, and no values could be found for methyl-perfluorooctane sulfonamide (MeFOSAA), perfluorodecanoic acid (PFDA), and perfluoroundecanoic acid (PFUnDA). The lowest PFAS risk-based level is 2 ng/mL (HBM-I for PFOA), and the highest PFAS risk level is 20 ng/mL (HBM-II for PFOS in the general population), demonstrating that risks are unexpected below 2 ng/mL and that the risk of PFAS-associated effects at the population level is increased at 20 ng/mL. The risk-based levels for sensitive populations fall between 2 ng/mL and 20 ng/mL (see Figure 5-4).

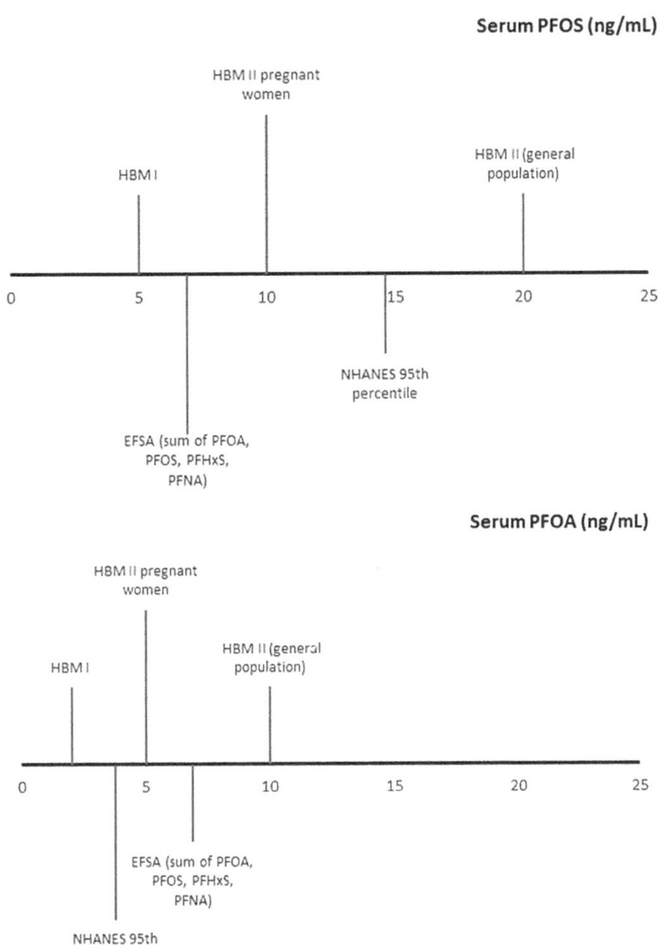

FIGURE 5-4 Reference- and risk-based serum PFOS and PFOA concentrations that could inform clinical assessments.
NOTES: Human biomonitoring (HBM) values are risk-based values derived from a compilation of epidemiological studies and represent levels at which no effect is expected (HBM-I) and above which effects are possible (HBM-II). The European Food Safety Authority (EFSA) risk-based value is a maternal serum PFAS concentration derived from an epidemiological study of children's prenatal and postnatal PFAS exposure and decreased antibody response to vaccines; maternal levels below this value are expected to have negligible impact on children's response to vaccines. The National Health and Nutrition Examination Survey (NHANES) reference-based serum concentration represents the 95th percentile of serum levels in a representative sample of the population aged 12 and older. PFHxS = perfluorohexanesulfonic acid; PFNA = perfluorononanoic acid; PFOA = perfluorooctanoic acid; PFOS = perfluorooctanesulfonic acid.

The committee used weighted data from NHANES to estimate PFAS exposures in the U.S. population. The HBM-I value, or the value below which effects are not expected, for PFOA (2 ng/mL) is the 73rd percentile, and the corresponding value for PFOS (5 ng/mL) is the 57th percentile. The HBM-II value, or the value above which effects may be expected for PFOA (10 ng/mL) is the 99th percentile and for PFOS (20 ng/mL) is the 98th percentile. For women of childbearing age (15–49), the HBM-II values for pregnant women for PFOS and PFOA are the 99th and 98th percentile, respectively. The committee observed that 25 percent of the U.S. population is exposed to PFAS above the EFSA point of departure (6.9 ng/mL sum of PFOA, PFOS, PFHxS, and PFNA).

Given the large number of PFAS and the fact that all humans are exposed to mixtures of PFAS, an approach that accounts for mixtures of PFAS would better inform clinical care than do single-chemical exposure values. Because the toxicities of different PFAS may not be equal, an approach using potency factors (e.g., dioxins) may be optimal for determining how a mixture of PFAS may exert its toxic effects. However, EFSA did not identify studies comparing the dose-response curves for different PFAS that would allow derivation of their potencies, and interspecies and sex differences also would complicate that effort. Thus EFSA assumed equal potencies for the four selected PFAS, which in humans share half-lives on the order of years. To facilitate easier comparison with estimated exposure, this calculation was performed on a weight rather than a molar basis. Bil and colleagues (2021) recently developed relative potency factors for several PFAS, which when applied result in the sum of PFOA equivalents in a mixture. Overall, the approach entails uncertainties, as the potency factors are derived from animal studies using mainly liver endpoints, which may correlate with effects in humans but to what degree is unknown. The additive approach used by EFSA has advantages. It is simple to apply and has been used in other efforts to regulate exposures to mixtures.[16] Also, there is evidence that many PFAS have similar toxic effects (Kwiatkowski et al., 2020). The committee believes the additive approach could be applied to the PFAS currently measured in the NHANES (MeFOSAA, PFHxS, PFOA [linear and branched isomers], PFDA, PFUnDA, PFOS [linear and branched isomers], and PFNA). Applying this approach of weight-based dose additivity, and using the HBM Commission's values, the committee makes the following recommendations (see Figure 5-5):

> **Recommendation 5-3: Clinicians should use serum or plasma concentrations of the sum of PFAS* to inform clinical care of exposed patients, using the following guidelines for interpretation:**
>
> - **Adverse health effects related to PFAS exposure are not expected at less than 2 nanograms per milliliter (ng/mL).**
> - **There is a potential for adverse effects, especially in sensitive populations, between 2 and 20 ng/mL.**
> - **There is an increased risk of adverse effects above 20 ng/mL.**
>
> *** Simple additive sum of MeFOSAA, PFHxS, PFOA (linear and branched isomers), PFDA, PFUnDA, PFOS (linear and branched isomers) and PFNA in serum or plasma. Caution is warranted when using capillary blood measurements as levels may differ from serum or plasma levels.**

The committee estimated that for the sum of PFAS in the NHANES, 2 ng/mL corresponds to the 2nd percentile, and 20 ng/mL corresponds to the 91st percentile, indicating that 89 percent of the U.S. population falls in the orange area in Figure 5-5, and 9 percent in the red area. Choosing cutoffs, as the committee has done, reflects ethical decisions and risks overstating or understating the risk of PFAS exposure given the uncertainty of the available information. Values suggested herein were derived from epidemiological studies evaluating associations at the population level, and the relevance of these values to interpret risk in individuals is uncertain. Also, these values were derived in part from studies in vulnerable populations (e.g., children, pregnant women), and so should account for many sensitive populations, but they may not protect all populations. Moreover, the assumption of weight-based dose additivity is likely an oversimplification. The molar sum may be more appropriate if the equal potency of all PFAS is assumed, and future research to produce toxic equivalency data or identify relevant potency factors could help refine the calculation (Bil et al., 2021). Furthermore, regardless of how the dose is calculated, there may not even be a level of PFAS exposure without some biological effect. Still, the increased risk from low levels of exposure is better addressed through population-health efforts than through individual action. The cutoff levels should be updated as more information becomes available.

---

[16] See https://www.epa.gov/sites/default/files/2014-11/documents/chem_mix_1986.pdf (accessed June 8, 2022).

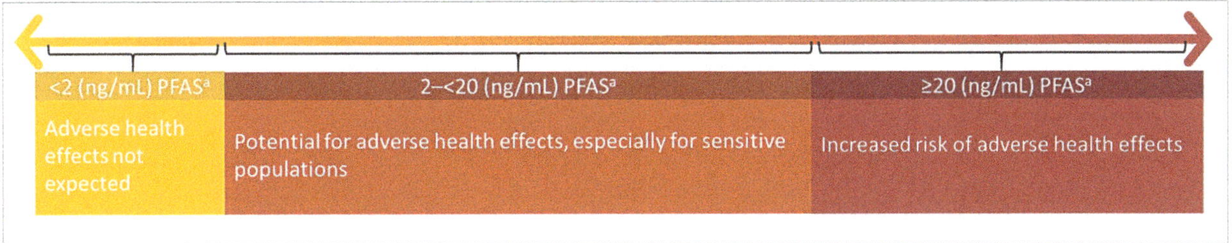

FIGURE 5-5 Graphical display of levels of PFAS to inform clinical care for the sum of MeFOSAA, PFHxS, PFOA (linear and branched isomers), PFDA, PFUnDA, PFOS (linear and branched isomers), and PFNA in serum or plasma.

The committee acknowledges that, in many communities, a large percentage of the population may have exposures to PFAS that would be considered high using the cutoffs presented here. This does not mean that health effects will occur, but likely relates to increased risk. More reference- and risk-based values are needed for other PFAS and other biological matrices, but given the expansiveness of the class, approaches based on relative potency factors may be more successful than developing risk-based levels for each PFAS in addressing this gap.

The committee also acknowledges the existence of data gaps with respect to reference populations. Appropriate PFAS reference populations may not be available for vulnerable populations such as pregnant people and young children, as pregnant people are included in low numbers in NHANES, and children younger than 12 are not routinely included in the publicly available data. Therefore, the committee makes the following recommendation:

Recommendation 5-4: The National Health and Nutrition Examination Survey should begin collecting and sharing more data on children younger than 12 years of age and pregnant people to generate reference populations for those groups.

CONCLUSION

Determining options and considerations to guide decision making for PFAS testing and PFAS concentrations that could inform clinical care of exposed patients will be beneficial in helping communities identify those who have elevated exposure to those chemicals. Testing for PFAS offers an opportunity to identify those who may be at increased risk of certain health outcomes. The recommendations in this chapter could expand PFAS testing among people already integrated into the health system (i.e., those with reliable insurance or other health care coverage). The committee acknowledges, however, that PFAS testing is expensive. Race, age, and other social and demographic characteristics already have disadvantaged many patients from accessing clinical preventive services. That disadvantage would be compounded by the expansion of PFAS testing services, as those services should be linked to counseling on steps for mitigating exposure and its impacts. Therefore, encouraging testing primarily among people with relatively stable access to care could have the unintended effect of aggravating disparities in exposure to PFAS, a severe disadvantage of encouraging testing without a funded, national PFAS testing program.

REFERENCES

Abraham, K., H. Mielke, H. Fromme, W. Völkel, J. Menzel, M. Peiser, F. Zepp, S. N. Willich, and C. Weikert. 2020. Internal exposure to perfluoroalkyl substances (PFASs) and biological markers in 101 healthy 1-year-old children: Associations between levels of perfluorooctanoic acid (PFOA) and vaccine response. *Archives of Toxicology* 94(6):2131–2147. https://doi.org/10.1007/s00204-020-02715-4.

Adams, C., P. Brown, R. Morello-Frosch, J. G. Brody, R. Rudel, A. Zota, S. Dunagan, J. Tovar, and S. Pattonand. 2011. Disentangling the exposure experience: The roles of community context and report-back of environmental exposure data. *Journal of Health and Social Behavior* 52(2):180–196. https://doi.org/10.1177/0022146510395593.

Altman, R. G., R. Morello-Frosch, J. G. Brody, R. Rudel, P. Brown, and M. Averick. 2008. Pollution comes home and gets personal: Women's experience of household chemical exposure. *Journal of Health and Social Behavior* 49(4):417–435. https://doi.org/10.1177/002214650804900404.

Alves, A., G. Jacobs, G. Vanermen, A. Covaci, and S. Voorspoels. 2015. New approach for assessing human perfluoroalkyl exposure via hair. *Talanta* 144:574–583. https://doi.org/10.1016/j.talanta.2015.07.009.

Apel, P., C. Rousselle, R. Lange, F. Sissoko, M. Kolossa-Gehring, and E. Ougier. 2020. Human biomonitoring initiative (HBM4EU)—Strategy to derive human biomonitoring guidance values (HBM-GVs) for health risk assessment. *International Journal of Hygiene and Environmental Health* 230:113622. https://doi.org/10.1016/j.ijheh.2020.113622.

Aro, R., U. Eriksson, A. Kärrman, and L. W. Y. Yeung. 2021. Organofluorine mass balance analysis of whole blood samples in relation to gender and age. *Environmental Science & Technology* 55(19):13142–13151. https://doi.org/10.1021/acs.est.1c04031.

Aro, R., U. Eriksson, A. Kärrman, K. Jakobsson, and L. W. Y. Yeung. 2022. Extractable organofluorine analysis: A way to screen for elevated per- and polyfluoroalkyl substance contamination in humans? *Environment International* 159:107035. https://doi.org/10.1016/j.envint.2021.107035.

Baker, B., V. Cassano, and C. Murray. 2018. Arsenic exposure, asessment, toxicity, diagnosis, and management: Guidance for occupational and environmental physicians. *Journal of Occupational and Environmental Medicine* 60(12):e634–e639. https://acoem.org/acoem/media/News-Library/Arsenic_Exposure_Assessment_Toxicity_Diagnosis.pdf (accessed June 17, 2022).

Barr, D. B., K. Kannan, Y. Cui, L. Merrill, L. Petrick, J. Meeker, T. Fennell, and E. Faustman. 2021. The use of dried blood spots for characterizing children's exposure to organic environmental chemicals. *Environmental Research* 195:110796. https://doi.org/10.1016/j.envres.2021.110796.

Barry, M. J., and S. Edgman-Levitan. 2012. Shared decision making—pinnacle of patient-centered care. *New England Journal of Medicine.* 366(9):780–781. https://doi.org/10.1056/NEJMp1109283.

Bil, W., M. Zeilmaker, S. Fragki, J. Lijzen, E. Verbruggen, and B. Bokkers. 2021. Risk assessment of per-and polyfluoroalkyl substance mixtures: A relative potency factor approach. *Environmental Toxicology and Chemistry* 40(3):859–870.

Brennan, N. M., A. T. Evans, M. K. Fritz, S. A. Peak, and H. von Holst. 2021. Trends in the regulation of per- and polyfluoroalkyl substances (PFAS): A scoping review. *International Journal of Environmental Research and Public Health* 18(20):10900. https://doi.org/10.3390/ijerph 182010900.

Brody, J. G., S. C. Dunagan, R. Morello-Frosch, P. Brown, S. Patton, and R. A. Rudel. 2014. Reporting individual results for biomonitoring and environmental exposures: Lessons learned from environmental communication case studies. *Environmental Health* 13:40. https://doi.org/10.1186/1476-069x-13-40.

Brody, J. G., P. M. Cirillo, K. E. Boronow, L. Havas, M. Plumb, H. P. Susmann, K. Z. Gajos, and B. A. Cohn. 2021. Outcomes from returning individual versus only study-wide biomonitoring results in an environmental exposure study using the Digital Exposure Report-Back Interface (DERBI). *Enviornmental Health Perspectives* 129(11):117005. https://doi.org/10.1289/ehp9072.

Calafat, A. M., K. Kato, K. Hubbard, T. Jia, J. C. Botelho, and L. Y. Wong. 2019. Legacy and alternative per- and polyfluoroalkyl substances in the U.S. general population: Paired serum-urine data from the 2013-2014 National Health and Nutrition Examination Survey. *Environment International* 131:105048. https://doi.org/10.1016/j.envint.2019.105048.

CDC (Centers for Disease Control and Prevention). 2021. *Fourth national report on human exposure to environmental chemicals, updated tables.* Atlanta, GA: U.S. Department of Health and Human Services.

CDC. 2022. *National report on human exposure to environmental chemicals.* Atlanta, GA: U.S. Department of Health and Human Services. https://www.cdc.gov/exposurereport/overview_ner.html (accessed June 28, 2022).

Charles, C., A. Gafni, and T. Whelan. 1997. Shared decision-making in the medical encounter: What does it mean? (or it takes at least two to tango). *Sociological Science in Medicine* 44:681–692. https://doi.org.10.1016/s0277-9536(96)00221-3.

Clark, D., and TCS (The Chemours Company). 2021, March 17. *Letter to EPA, Office of Pollution Prevention and Toxics regarding propanoic acid, 2,3,3,3-tetrafluoro-2-(1,1,2,2,3,3,3-heptafluoropropoxy)-CAS RN 13252-13-6* (also known as HFPO-DA). https://www.documentcloud.org/documents/21093350-chemours-tsca-fyi-letter-3-17-2021-v2 (accessed June 16, 2022).

Ehresman, D. J., J. W. Froehlich, G. W. Olsen, S-C. Chang, and J. L. Butenhoff. 2007. Comparison of human whole blood, plasma, and serum matrices for the determination of perfluorooctanesulfonate (PFOS), perfluorooctanoate (PFOA), and other fluorochemicals. *Environmental Research* 103(2):176–184. https://doi.org/10.1016/j.envres.2006.06.008.

Emmett, E. A., H. Zhang, F. S. Shofer, N. Rodway, C. Desai, D. Freeman, and M. Hufford. 2009. Development and successful application of a "Community-First" communication model for community-based environmental health research. *Journal of Occupational and Environmental Medicine* 51(2):146–156. https://doi.org/10.1097/JOM.0b013e3181965d9b.

FDA (U.S. Food and Drug Administration). 2018. *Bioanalytical method validation guidance for industry.* Washington, DC: U.S. Department of Health and Human Services.

Freeman, J. D., L. M. Rosman, J. D. Ratcliff, P. T. Strickland, D. R. Graham, and E. K. Silbergeld. 2018. State of the science in dried blood spots. *Clinical Chemistry* 64(4):656–679. https://doi.org/10.1373/clinchem.2017.275966.

Frisbee, S. J., A. P. Brooks, Jr., A. Maher, P. Flensborg, S. Arnold, T. Fletcher, K. Steenland, A. Shankar, S. S. Knox, C. Pollard, J. A. Halverson, V. M. Vieira, C. Jin, K. M. Leyden, and A. M. Ducatman. 2009. The C-8 health project: Design, methods, and participants. *Environmental Health Perspectives* 117(12):1873–1882. https://doi.org/10.1289/ehp.0800379.

Giannini, C. M., R. L. Herrick, J. M. Buckholz, A. R. Daniels, F. M. Biro, and S. M. Pinney. 2018. Comprehension and perceptions of study participants upon receiving perfluoroalkyl substance exposure biomarker results. *International Journal of Hygiene and Environmental Health* 221(7):1040–1046. https://doi.org/10.1016/j.ijheh.2018.07.005.

Gross, R. S., A. Ghassabian, S. Vandyousefi, M. J. Messito, C. Gao, K. Kannan, and L. Trasande. 2020. Persistent organic pollutants exposure in newborn dried blood spots and infant weight status: A case-control study of low-income Hispanic mother-infant pairs. *Environmental Pollution* 267:115427. https://doi.org/10.1016/j.envpol.2020.115427.

Guo, P., T. Furnary, V. Vasiliou, Q. Yan, K. Nyhan, D. P. Jones, C. H. Johnson, and Z. Liew. 2022. Non-targeted metabolomics and associations with per- and polyfluoroalkyl substances (PFAS) exposure in humans: A scoping review. *Environment International* 162(2022):107159.

Harclerode, M., S. Baryluk, H. Lanza, and J. Frangos. 2021. Preparing for effective, adaptive risk communication about per- and polyfluoroalkyl substances in drinking water. *AWWA Water Science* 3(5):e1236. https://doi.org/10.1002/aws2.1236.

Hays, S., L. Aylward, J. Lakind, M. Bartels, H. Barton, P. Boogaard, C. Brunk, S. Dizio, M. Dourson, D. Goldstein, J. Lipscomb, M Kilpatrick, D. Krewski, K Krishnan, M. Nordberg, M. Okino, Y-M. Tan, C. Viau, and J. Yager. 2008. Guidelines for the derivation of biomonitoring equivalents: Report from the Biomonitoring Equivalents Expert Workshop. *Regulatory Toxicology and Pharmacology* 51:S4–S15.

Hernick, A. D., M. Kathryn Brown, S. M. Pinney, F. M. Biro, K. M. Ball, and R. L. Bornschein. 2011. Sharing unexpected biomarker results with study participants. *Environmental Health Perspectives* 119(1):1–5. https://doi.org/10.1289/ehp.1001988.

Hölzer, J., H. Lilienthal, and M. Schümann. 2021. Human biomonitoring (HBM)-I values for perfluorooctanoic acid (PFOA) and perfluorooctane sulfonic acid (PFOS)—Description, derivation and discussion. *Regulatory Toxicology and Pharmacology* 121:104862. https://doi.org/10.1016/j.yrtph.2021.104862.

IOM (Institute of Medicine). 2001. *Crossing the quality chasm: A new health system for the 21st century.* Washington, DC: National Academy Press. https://doi.org/10.17226/10027.

Kannan, K., A. Stathis, M. J. Mazzella, S. S. Andra, D. Boyd Barr, S. S. Hecht, L. S. Merrill, A. L. Galusha, and P. J. Parsons. 2021. Quality assurance and harmonization for targeted biomonitoring measurements of environmental organic chemicals across the Children's Health Exposure Analysis Resource laboratory network. *International Journal of Hygiene and Environmental Health* 234:113741.

Kato, K., L. Y. Wong, A. Chen, C. Dunbar, G. M. Webster, B. P. Lanphear, and A. M. Calafat. 2014. Changes in serum concentrations of maternal poly- and perfluoroalkyl substances over the course of pregnancy and predictors of exposure in a multiethnic cohort of Cincinnati, Ohio pregnant women during 2003–2006. *Environmental Science and Technology* 48(16):9600–9608. https://doi.org/10.1021/es501811k.

Kwiatkowski, C. F., D. Q. Andrews, L. S. Birnbaum, T. A. Bruton, J. C. DeWitt, D. R. U. Knappe, M. V. Maffini, M. F. Miller, K. E. Pelch, A. Reade, A. Soehl, X. Trier, M. Venier, C. C. Wagner, Z. Wang, and A. Blum. 2020. Scientific basis for managing PFAS as a chemical class. *Environmental Science & Technology Letters* 7(8):532–543. https://doi.org/10.1021/acs.estlett.0c00255.

LaKind, J. S., M-A. Verner, R. D. Rogers, H. Goeden, D. Q. Naiman, S. A. Marchitti, G. M. Lehmann, E. P. Hines, and S. E. Fenton. 2022. Current breast milk PFAS levels in the United States and Canada: After all this time, why don't we know more? *Environmental Health Perspectives* 130(2):025002.

Latshaw, M. W., R. Degeberg, S. Sutaria Patel, B. Rhodes, E. King, S. Chaudhuri, and J. Nassif. 2017. Advancing environmental health surveillance in the US through a national human biomonitoring network. *International Journal of Hygiene and Environmental Health* 220(2):98–102.

Lu, S., and S. M. Bartell. 2020. *Serum PFAS calculator for adults, web-based software*, version 1.2. https://www.ics.uci.edu/~sbartell/pfascalc.html (accessed June 28, 2022).

Ma, W., K. Kannan, Q. Wu, E. M. Bell, C. M. Druschel, M. Caggana, and K. M. Aldous. 2013. Analysis of polyfluoroalkyl substances and bisphenol A in dried blood spots by liquid chromatography tandem mass spectrometry. *Analytical and Bioanalytical Chemistry* 405(12):4127–4138. https://doi.org/10.1007/s00216-013-6787-3.

NASEM (National Academies of Sciences, Engineering, and Medicine). 2018a. *Feasibility of addressing environmental exposure questions using department of defense biorepositories: Proceedings of a workshop—in brief.* Washington, DC: The National Academies Press. https://doi.org/10.17226/25287.

NASEM. 2018b. *Returning individual research results to participants: Guidance for a new research paradigm.* Washington, DC: The National Academies Press. https://doi.org/10.17226/25094.

NRC (National Research Council). 1987. Biological markers in environmental health research. *Environmental Health Perspectives* 74:3–9. https://doi.org/10.1289/ehp.74-1474499.

NRC. 2006. *Human biomonitoring for environmental chemicals.* Washington, DC: The National Academies Press. https://doi.org/10.17226/11700.

NRC. 2009. *Science and decisions: Advancing risk assessment.* Washington, DC: The National Academies Press. https://doi.org/10.17226/12209.

Ohayon, J. L., E. Cousins, P. Brown, R. Morello-Frosch, and J. G. Brody. 2017. Researcher and institutional review board perspectives on the benefits and challenges of reporting back biomonitoring and environmental exposure results. *Environmental Research* 153:140–149. doi: https://doi.org/10.1016/j.envres.2016.12.003.

Perovich, L. J., J. L. Ohayon, E. M. Cousins, R. Morello-Frosch, P. Brown, G. Adamkiewicz, and J. G. Brody. 2018. Reporting to parents on children's exposures to asthma triggers in low-income and public housing, an interview-based case study of ethics, environmental literacy, individual action, and public health benefits. *Environmental Health: A Global Access Science Source* 17(1):48. https://doi.org/10.1186/s12940-018-0395-9.

Poothong, S., C. Thomsen, J. A. Padilla-Sanchez, E. Papadopoulou, and L. S. Haug. 2017. Distribution of novel and well-known poly- and perfluoroalkyl substances (PFASs) in human serum, plasma, and whole blood. *Environmental Science & Technology* 51(22):13388–13396. https://doi.org/10.1021/acs.est.7b03299.

Ramirez-Andreotta, M. D., J. G. Brody, N. Lothrop, M. Loh, P. I. Beamer, and P. Brown. 2016. Improving environmental health literacy and justice through environmental exposure results communication. *International Journal of Environmental Research and Public Health* 13(7):690. https://doi.org/10.3390/ijerph13070690.

Schulz, C., M. Wilhelm, U. Heudorf, and M. Kolossa-Gehring. 2011. Update of the reference and HBM values derived by the German Human Biomonitoring Commission. *International Journal of Hygiene and Environmental Health* 215(1):26–35. https://doi.org/10.1016/j.ijheh.2011.06.007.

Schümann, M., H. Lilienthal, and J. Hölzer. 2021. Human biomonitoring (HBM)-II values for perfluorooctanoic acid (PFOA) and perfluorooctane sulfonic acid (PFOS)—Description, derivation and discussion. *Regulatory Toxicology and Pharmacology* 121:104868. https://doi.org/10.1016/j.yrtph.2021.104868.

SWGTOX (Scientific Working Group for Forensic Toxicology). 2013. Scientific Working Group for Forensic Toxicology (SWGTOX) standard practices for method validation in forensic toxicology. *Journal of Analytical Toxicology* 37(7):452–474.

Tomsho, K. S., C. Schollaert, T. Aguilar, R. Bongiovanni, M. Alvarez, M. K. Scammell, and G. Adamkiewicz. 2019. A mixed methods evaluation of sharing air pollution results with study participants via report-back communication. *International Journal of Environmental Research and Public Health* 16(21):4183. https://doi.org/10.3390/ijerph16214183.

Watson, C. V., M. Lewin, A. Ragin-Wilson, R. Jones, J. M. Jarrett, K. Wallon, C. Ward, N. Hilliard, and E. Irvin-Barnwell. 2020. Characterization of trace elements exposure in pregnant women in the United States, NHANES 1999–2016. *Environmental Research* 183:109208. https://doi.org/10.1016/j.envres.2020.109208.

Ye, X., K. Kato, L. Y. Wong, T. Jia, A. Kalathil, J. Latremouille, and A. M. Calafat. 2018. Per- and polyfluoroalkyl substances in sera from children 3 to 11 years of age participating in the National Health and Nutrition Examination Survey 2013–2014. *International Journal of Hygiene and Environmental Health* 221(1):9–16. https://doi.org/10.1016/j.ijheh.2017.09.011.

Yeung, E. H., E. M. Bell, R. Sundaram, A. Ghassabian, W. Ma, K. Kannan, and G. M. Louis. 2019. Examining endocrine disruptors measured in newborn dried blood spots and early childhood growth in a prospective cohort. *Obesity (Silver Spring, MD)* 27(1):145–151. https://doi.org/10.1002/oby.22332.

Zheng, G., E. Schreder, J. C. Dempsey, N. Uding, V. Chu, G. Andres, S. Sathyanarayana, and A. Salamova. 2021. Per- and polyfluoroalkyl substances (PFAS) in breast milk: Concerning trends for current-use PFAS. *Environmental Science & Technology* 55(11):7510–7520.

6

Guidance for Clinicians on Exposure Determination, PFAS Testing, and Clinical Follow-Up

Despite continued uncertainty about the exact nature of risks from PFAS exposure, clinicians will need to advise and make decisions with patients regarding their exposure. While there is evidence of an association of PFAS with several health outcomes, the likelihood that a particular individual will have any specific adverse health outcome following exposure to PFAS cannot currently be determined with great specificity. There are also gaps in knowledge about how individuals can reduce any potential risks related to PFAS exposure. Despite these gaps, however, many individuals and communities expect clinicians to address PFAS-associated risks as part of routine health care delivery. Although close monitoring and exposure mitigation might prevent or lessen the severity of health effects for those exposed to PFAS, aggressive clinical follow-up could lead to unnecessary treatment, with attendant risk of treatment-related adverse effects; increase patients' anxiety; and even provide false assurance. For these reasons, trust and clear communication between clinicians and patients are of the utmost importance as they face the task of making decisions that consider all options and incorporate informed preferences, although how best to include children, especially adolescents, in shared decision making is a complicated matter and an active focus of research (Boland et al., 2019). Ongoing and future research should eventually guide clinicians in predicting patient risk and provide an understanding of the benefits and harms of interventions designed to avoid adverse health outcomes.

CRITERIA FOR SCREENING

Screening is the process of testing to identify individuals at high risk for developing a clinical condition or those who have a condition for which signs or symptoms may not be evident. Population screenings are one type of clinical preventive services recommended by health and medical professional agencies and organizations such as the American Academy of Family Physicians (AAFP), the American College of Obstetricians and Gynecologists (ACOG), the American Academy of Pediatrics (AAP), the American Heart Association (AHA), the Health Resources and Services Administration (HRSA), and the U.S. Preventive Services Task Force (USPSTF). These organizations usually recommend population-level screenings as part of routine clinical care when it is clear that they offer a net benefit (compared with potential harms of the screening itself and any subsequent treatment). Although there are different frameworks for determining when to adopt population-level screening, the criteria articulated by Wilson and Jungner (1968) are a common platform for these frameworks:

- The condition should be an important public health concern.
- There should be a treatment for the condition.
- Facilities for diagnosis and treatment should be available.
- There should be a latent stage of the condition.
- There should be a test or examination for the condition.
- The test should be acceptable to the population.
- The natural history of the disease should be adequately understood.
- There should be an agreed-upon policy on whom to treat.

- The total cost of finding a case should be economically balanced in relationship to medical expenditure as a whole.
- Case finding should be a continuous process.

The degree to which these criteria are applied and how they are weighted must be tailored to the particular clinical issue and the perspective of those developing the preventive service. For example, the degree to which the evidence base must be clear can vary based on the urgency of the clinical context, and cost is often not explicitly considered because cost/benefit data from the societal perspective are rarely available. Despite this variation, a key common theme across all frameworks is the need to assess whether the expected benefits of population screening exceed the potential harms. This assessment can be challenging when significant scientific uncertainty exists, as is the case for population screening for PFAS exposure. The committee faced challenges in making population-level recommendations related to the following issues:

- PFAS testing could identify risks for many diverse health outcomes, as opposed to the usual case in which screening identifies one condition or a group of related conditions. Assessment of benefit and harm is difficult if the various potential health outcomes differ in this regard. Furthermore, developing recommendations for clinical follow-up after PFAS exposure is challenging because of the heterogeneity of potential health outcomes.
- The benefits of screening might accrue to individuals other than those who would be screened. For example, determining that an individual had a harmful environmental exposure might not help that person but could lead to broader public health measures that could protect the community.

When there are gaps in knowledge about the benefits and harms of screening or when benefits and harms are closely matched, clinicians should assess individual patient preferences. Informed decision making can be challenging given the above gaps and the limited time clinicians and patients have together. Because standard screening criteria have important limitations in settings of substantial scientific uncertainty, such as PFAS-related health effects, this chapter offers recommendations for basing screening decisions on the principles the committee proposes in Chapter 2.

PFAS-ASSOCIATED HEALTH OUTCOMES

The committee identified several health outcomes associated with PFAS exposure (see Chapter 5). Many are common diseases in the general population, and all have multiple known risk factors (Schrager, 2018). The committee believed it was important to categorize the strength of the evidence for each outcome, but concluded that all conditions with an association should be considered for patient follow-up, as acknowledging the potential risk may make doctors and patients more likely to prioritize screenings. If a patient has a known or suspected exposure to PFAS, the committee encourages clinicians to prioritize screenings for those conditions related to PFAS when relevant and possible. The committee encourages clinicians to use evidence-based best practices and strategies when speaking with patients to support shared decision making and clear health risk communication. Resources such as the Consumer Assessment of Healthcare Providers and Systems (CAHPS) Ambulatory Care Improvement Guide can provide evidence-based trainings, tools, and strategies for shared decision making and clear communication (AHRQ, 2020). The committee did not conduct a meta-analysis to determine the level of increased risk posed by PFAS exposure for each health outcome. Risks vary by exposure level, life stage, and whether patients have other risk factors for developing a health effect. These uncertainties make it infeasible to determine the optimal screening tests and their frequency.

The committee found sufficient evidence of increased risk for the following health outcomes with exposure to PFAS:

- reductions in birthweight;
- dyslipidemia in children and adults;
- kidney cancer in adults; and
- decreased antibody response in children and adults, but with insufficient evidence of an increase in risk or severity of infection or differences in vaccine effectiveness.

The committee found limited suggestive evidence of increased risk for the following health outcomes with exposure to PFAS:

- breast cancer in adults,
- pregnancy-induced hypertension (gestational hypertension and preeclampsia),
- liver enzyme elevations (in children and adults),
- testicular cancer (in adults),
- thyroid dysfunction (in adults), and
- ulcerative colitis (in adults).

For many of these adverse health effects, however, including cancers, it is unclear what clinicians and individuals can do to monitor for their development and intervene to lower the risk related to PFAS exposure. Guidance for clinicians' engagement in sharing decision making with their patients regarding follow-up care for PFAS-associated health outcomes is included later in this chapter.

Clinical Practice Guidelines for PFAS-Associated Health Outcomes

The committee reviewed clinical practice guidelines for the health effects associated with PFAS exposure, as well as the recommendations from the C-8 Medical Panel (see Table 6-1). The C-8 Medical Panel developed a medical monitoring protocol for a community with high PFAS exposure surrounding Parkersburg, West Virginia. The protocol specifies follow-up for human diseases for which the C-8 Science Panel found a probable link: dyslipidemia, kidney cancer, pregnancy-induced hypertension (gestational hypertension and preeclampsia), testicular cancer, thyroid disfunction, and ulcerative colitis.

Outcomes with Sufficient Evidence of Association

Dyslipidemia in Children and Adults

In the United States, blood testing for lipid and cholesterol levels is recommended throughout the life course based on age and risk factors. AAP recommends that all children be screened once between ages 9 and 11 years and again between 17 and 21 years, with selective screening for children over 2 years of age with a family history of lipid and cholesterol disorders or heart disease (Richerson et al., 2017). AHA recommends screening every 4–6 years for people aged 20 or older who are at low risk for cardiovascular disease (Grundy et al., 2019). The C-8 Medical Panel provided cholesterol screening starting as early as 2 years of age, which is similar to the age recommended for cholesterol screening for children with a family history of lipid disorders (C-8 Medical Panel, 2013).

Reductions in Birthweight

Birthweight is an important and well-established pregnancy outcome. Clinical prevention guidance for reductions in birthweight is to screen for risk factors that lead to reductions in birthweight. Common prenatal exposures associated with a risk of reduced birthweight in full-term newborns include use of tobacco, alcohol, or other drugs.

TABLE 6-1 An Overview of Screening Recommendations for the Health Effects Associated with Exposure to PFAS

Recommendations for the General Population[a]	C-8 Medical Panel Recommendation[b,c] for Class Members
Lower birthweight	
Screen for risk factors for low birthweight in pregnant persons during prenatal well visit[d]	No screening recommended for C-8 class members[e]
Dyslipidemia (in adults and children)	
Screen all children once between ages 9 and 11 years and again between ages 17 and 21; among those with a familial history, begin screening at age 2 years and follow up yearly[f]	Screen children for cholesterol levels at age 2 years and older unless already screened during the prior 5 years, already diagnosed, or receiving treatment
Screen adults aged 40–75 with no history of cardiovascular disease (CVD), one or more CVD risk factors, and a calculated 10-year CVD event risk of 10% or greater[g]	
American Heart Association recommends all adults aged 20 or older have their cholesterol and other traditional risk factors checked every 4–6 years as long as their risk remains low[h]	
Kidney cancer (in adults)	
No routine screening recommended	Screen individuals aged 20–39 years with a symptom questionnaire; follow up with abdominal exam and urine test if symptoms present
	Screen individuals aged 40 or older with a symptom questionnaire, abdominal exam, and urine testing
Decreased antibody response (in adults and children)	
No routine screening recommended	No screening recommended for C-8 class members[e]
Breast cancer in adults	
U.S. Preventive Services Task Force (USPSTF): Screen average-risk women aged 50–74 with mammography every 2 years[i]	No screening recommended for C-8 class members[e]
Women's Preventive Services Initiative (WPSI): Screen average-risk women with mammography; mammography should be initiated no earlier than age 40 and no later than age 50 and occur annually or biennially until age 74[j]	
Pregnancy-induced hypertension (gestational hypertension and preeclampsia)	
Screen pregnant persons for hypertension and preeclampsia throughout pregnancy[k,l]	Screen pregnant persons for hypertension and proteinuria throughout pregnancy
Liver enzyme alterations in children and adults	
No routine screening recommended	No screening recommended for C-8 class members[e]
Testicular cancer in adults	
American Academy of Family Physicians (AAFP) and USPSTF recommend against screening asymptomatic adolescents and adults for testicular cancer[m]	Screen for testicular cancer with a physical exam and questionnaire
Thyroid dysfunction in adults	
Screen newborns as part of recommended uniform screening panel[n]	Screen adults' serum thyroid-stimulating hormone; otherwise test based on signs and symptoms of thyroid dysfunction
No routine screening recommended[o]	
Ulcerative colitis in adults	
No routine screening recommended	Screen for symptoms with questionnaire starting at age 15 years

[a] In cases in which conflicting or differing recommendations for population-level screenings have been issued, the clinical practice guideline developers have been identified for clarity.

[b] C-8 Medical Panel, 2013.
[c] The C-8 Medical Panel recommended that members of the class action lawsuit, *Jack W. Leach et al. v. E.I. du Pont de Nemours & Company* (no. 01-C-608 W.Va., Wood County Circuit Court, filed April 10, 2002), be screened once in 2013; the panel has since updated its guidance, and now recommends that class members be screened three times, 3 years apart (C-8 Medical Panel, 2022). See Chapter 1 for more details about the lawsuit.
[d] Hagan et al., 2017.
[e] The C-8 Medical Panel was allowed to develop recommendations only for the six conditions identified by the C-8 Science Panel in 2013: pregnancy-induced hypertension, kidney cancer, testicular cancer, thyroid disease, ulcerative colitis, and hypercholesterolemia. The C-8 Medical Panel did not deliberate about screenings for lower birthweight, decreased antibody response, breast cancer, or liver enzyme functions (C-8 Medical Panel, 2013).
[f] Richerson et al., 2017.
[g] USPSTF, 2016b.
[h] Grundy et al., 2019.
[i] USPSTF, 2016a.
[j] WPSI, 2019.
[k] USPSTF, 2021.
[l] USPSTF, 2017.
[m] USPSTF, 2011.
[n] HRSA, 2018.
[o] USPSTF, 2015.

Kidney Cancer in Adults

There are no effective screening approaches for identifying kidney cancer early in its course, and therefore no authoritative clinical screening recommendations. Urinalysis is effective at finding blood in the urine, which can be a sign of advanced kidney cancer but is also indicative of other disorders, such as infections and kidney stones. The C-8 Medical Panel recommends that clinicians ask about family history, symptoms of kidney cancer (gross hematuria, chronic abdominal pain, recent involuntary weight loss, unexplained fever for 1 week), conduct a physical exam for abdominal mass, and check a urine dipstick for blood (C-8 Medical Panel 2013, 2022). Although palpation for an abdominal mass is not harmful, it is unlikely to lead to early detection and may provide false reassurance.

Decreased Antibody Response in Children and Adults

The predictive value of antibody titers after vaccination or infection is usually unclear, and the evidence for an association between PFAS exposure and infection risk or severity is insufficient. Nor are there any evidence-based recommendations or other clinical guidance for addressing decreased antibody response. This is an important area for additional research.

Outcomes with Limited Suggestive Evidence of Association

Breast Cancer in Adults

The USPSTF recommends biennial mammographic screening for all average-risk women aged 50–74 (USPSTF, 2016a). For women aged 40–49, the USPSTF states that the decision to start screening mammography should be an individual one, and that women who place a higher value on the potential benefit than on the potential harms may choose to begin biennial screening. For women aged 75 and older, the USPSTF makes no recommendation because of insufficient evidence (USPSTF, 2016a). The HRSA-supported Women's Preventive Services Initiative recommends that average-risk women initiate "mammography screening no earlier than age 40 and no later than age 50," that "screening mammography should occur at least biennially and as frequently as annually," and that it "should continue through at least age 74 and age alone should not be the basis to discontinue screening" (WPSI, 2019).

Pregnancy-Induced Hypertension

Pregnancy-induced hypertension, including gestational hypertension and preeclampsia, affects about 1 in every 12–17 pregnant persons aged 20–44 in the United States, and poses serious health risks for pregnant persons and fetuses during pregnancy (Bateman et al., 2012). As part of standard prenatal care, pregnant persons seeing a clinician will have their blood pressure monitored routinely throughout their pregnancy to reduce various risks to both parent and fetus associated with high blood pressure (Kilpatrick, 2017). This standard measurement is feasible, and its benefits outweigh its harms. The C-8 Medical Panel reinforces the standard blood pressure monitoring recommendations (C-8 Medical Panel 2013, 2022).

Elevated Liver Enzymes in Children and Adults

Elevated liver enzymes do not represent a health outcome by themselves, although they generally indicate a level of liver inflammation. Elevations are often found incidentally with multiphasic blood test panels administered as part of routine medical care or as part of an assessment of patient symptoms or concerns. Elevated liver enzymes have many causes, the most common of which include use of over-the-counter pain medications (particularly acetaminophen) and certain prescription medications (including statin drugs), alcohol consumption, heart failure, viral hepatitis, fatty liver disease, and obesity. There are no authoritative recommendations for screening for liver enzyme alterations in otherwise healthy patients who are not being monitored for liver dysfunction, and follow-up recommendations for elevations depend on which enzymes are affected, the degree of elevation, and characteristics of the individual patient.

Testicular Cancer in Adults

Testicular cancer, though rare, is the most common cancer in American males aged 15–25 (NCI, n.d.). Most cases of testicular cancer are discovered incidentally by patients or their partners. The USPSTF found that there is inadequate evidence that screening by clinician examination or patient self-examination has a higher yield or greater accuracy for detecting testicular cancer at earlier stages. It is also not known whether earlier detection would lead to better health outcomes. The USPSTF concludes that there is no benefit of screening for testicular cancer in the general population and recommends against it (USPSTF, 2011). The C-8 Medical Panel recommends a risk questionnaire and clinical testicular exam, considering ultrasound if additional risk factors are identified (C-8 Medical Panel, 2013, 2022).

Thyroid Disease and Dysfunction in Adults

There are well-defined approaches for identifying and treating thyroid hormone dysfunction, especially among older adults and women (CDC, 2014). The standard clinical practice for identifying thyroid dysfunction is to order blood testing for levels of thyroid-stimulating hormone (TSH) when there are signs or symptoms that could be attributable to hypo- or hyperthyroidism (NIDDK, 2017). Standard approaches exist for diagnosis and treatment of thyroid disorders. In its review, however, the USPSTF concluded that the current evidence is insufficient to assess the balance of benefits and harms of screening for thyroid dysfunction in nonpregnant, asymptomatic adults. The C-8 Medical Panel recommends that adults in PFAS-contaminated communities receive TSH-level screening in addition to testing based on signs and symptoms (C-8 Medical Panel 2013, 2022).

Ulcerative Colitis in Adults

Ulcerative colitis is a chronic bowel disease with a prevalence of 1.3 percent among U.S. adults that results in inflammation and sores in the lining of the large intestine (colon) and rectum (Dahlhamer et al., 2016). It can be debilitating and lead to life-threatening complications. In addition to such generalized symptoms as fatigue, fever, and weight loss, affected individuals experience abdominal pain; blood or pus in stool; rectal bleeding; and frequent, recurring diarrhea. Diagnosis requires endoscopy and laboratory testing (Hanauer, 2004). There are no authoritative recommendations for screening for ulcerative colitis. The C-8 Medical Panel recommends that clinicians administer a questionnaire to elicit symptoms of the condition beginning at age 15 (C-8 Medical Panel, 2013, 2022); however, no information is available regarding the accuracy of the questionnaire.

RECOMMENDATIONS FOR PATIENT FOLLOW-UP

Where appropriate, the committee developed recommendations for patient follow-up for PFAS-associated health outcomes that should be offered to patients based on shared, informed decision making between patient and clinician. The clinical practice guidelines for standard medical care and the C-8 Medical Panel recommendations served as the basis of these recommendations. Some conditions associated with PFAS exposure have no established clinical prevention guidance, while other clinical prevention recommendations are beneficial when applied in cases of increased risk. There is value in specifying follow-up that is part of the standard of care as defined by authoritative clinical professional groups because clinicians often are unable to deliver all recommended preventive services during primary care visits (Privett and Guerrier, 2021) and must decide with patients which services to focus on considering both evidence and patient values and preferences (USPSTF et al., 2022).

The committee used its three established cutoff levels for PFAS in serum or plasma (detailed in Chapter 5) to determine follow-up based on PFAS exposure level, although the risks are not the same within each of these three categories. The risks from PFAS likely increase with increased exposure thus PFAS levels of 3 nanograms per milliliter (ng/mL) and 19 ng/mL do not represent the same risk. PFAS blood testing measures burden at the time of sample collection. For example, a person with low blood levels today may have had higher levels in the past. Clinicians should use judgment and shared decision making in making follow-up decisions based on PFAS exposure and other risk factors. Figure 6-1 suggests that clinicians engage in shared, informed decision making with their patients regarding follow-up care for PFAS-associated health outcomes.

> **Recommendation 6-1: Clinicians should treat patients with serum PFAS concentration below 2 nanograms per milliliter (ng/mL) with the usual standard of care.**

> **Recommendation 6-2: For patients with serum PFAS concentration of 2 nanograms per milliliter (2 ng/mL) or higher and less than 20 ng/mL, clinicians should encourage PFAS exposure reduction if a source of exposure is identified, especially for pregnant persons. Within the usual standard of care clinicians should:**

> - **Prioritize screening for dyslipidemia with a lipid panel (once between 9 and 11 years of age, and once every 4 to 6 years over age 20) as recommended by the American Academy of Pediatrics (AAP) and the American Heart Association (AHA).**
> - **Screen for hypertensive disorders of pregnancy at all prenatal visits per the American College of Obstetricians and Gynecologists (ACOG).**
> - **Screen for breast cancer based on clinical practice guidelines based on age and other risk factors such as those recommended by the U.S. Preventive Services Task Force (USPSTF).**

≥20 (ng/mL) PFAS*

Encourage PFAS exposure reduction if a source of exposure is identified, especially for pregnant persons.

In addition to the usual standard of care, clinicians should:

- Prioritize screening for dyslipidemia with a lipid panel (for patients over age 2) following American Academy of Pediatrics (AAP) recommendations for high-risk children and American Heart Association (AHA) guidance for high-risk adults.
- At all well visits:
  - Conduct thyroid function testing (for patients over age 18) with serum thyroid stimulating hormone (TSH),
  - Assess for signs and symptoms of kidney cancer (for patients over age 45), including with urinalysis, and
  - For patients over age 15, assess for signs and symptoms of testicular cancer and ulcerative colitis.

2–<20 (ng/mL) PFAS*

Encourage PFAS exposure reduction if a source has been identified, especially for pregnant persons.

Within the usual standard of care clinicians should:

- Prioritize screening for dyslipidemia with a lipid panel (once between 9 and 11 years of age, and once every 4 to 6 years over age 20) as recommended by the AAP and AHA.
- Screen for hypertensive disorders of pregnancy at all prenatal visits per the American College of Obstetricians and Gynecologists (ACOG).
- Screen for breast cancer based on clinical practice guidelines based on age and other risk factors such as those recommended by U.S. Preventive Services Task Force (USPSTF).

<2 (ng/mL) PFAS*

Provide usual standard of care

* Simple additive sum of MeFOSAA, PFHxS, PFOA (linear and branched isomers), PFDA, PFUnDA, PFOS (linear and branched isomers), and PFNA in serum or plasma

FIGURE 6-1 Clinical guidance for follow-up with patients after PFAS testing.
NOTE: MeFOSAA = methylperfluorooctane sulfonamidoacetic acid; PFDA = perfluorodecanoic acid; PFHxS = perfluorohexane sulfonic acid, PFNA = perfluorononanoic acid; PFOA = perfluorooctanoic acid; PFOS = perfluorooctanesulfonic acid; PFUnDA = perfluoroundecanoic acid.

Recommendation 6-3: For patients with serum PFAS concentration of 20 nanograms per milliliter (ng/mL) or higher, clinicians should encourage PFAS exposure reduction if a source of exposure is identified, especially for pregnant persons. In addition to the usual standard of care, clinicians should:

- **Prioritize screening for dyslipidemia with a lipid panel (for patients over age 2) following American Academy of Pediatrics (AAP) guidelines for high-risk children and American Heart Association (AHA) guidance for high-risk adults.**
- **At all well visits:**
  - **conduct thyroid function testing (for patients over age 18) with serum thyroid stimulating hormone (TSH),**
  - **assess for signs and symptoms of kidney cancer (for patients over 45), including with urinalysis, and**
  - **for patients over 15, assess for signs and symptoms of testicular cancer and ulcerative colitis.**

APPLYING THE COMMITTEE'S EXPOSURE, TESTING, AND CLINICAL FOLLOW-UP RECOMMENDATIONS

The Committee created a flow chart summarizing PFAS education, exposure assessment, and clinical follow-up (see Figure 6-2). In communities where PFAS exposure has been identified, the Agency of Toxic Substances and Disease Registry (ATSDR) and other government entities should support local clinicians with educational materials about PFAS exposure and testing. Clinicians should then determine whether a particular patient is likely to have a history of elevated exposure to PFAS. If so, the clinician should offer PFAS testing and make a shared, informed decision on that testing. If testing is chosen, the labs should be ordered (Test Code 39307, Current Procedural Terminology [CPT] Code 82542). Test results should be interpreted by summing the concentrations of MeFOSAA, PFHxS, PFOA (linear and branched isomers), PFDA, PFUnDA, PFOS (linear and branched isomers), and PFNA. The laboratory may not report results for all PFAS considered by the committee or may include different PFAS in their panel. In that case, the sum of PFAS should include only the PFAS in the analyte list considered by the committee. For example, if the lab tests for PFOA, PFOS, PFHxS, PFNA, and PFBS, the summation should include PFOA, PFOS, PFHxS, and PFNA. Differing analyte lists may cause some variation in test results. Still, as long as PFOA, PFOS, PFHxS, and PFNA are included in the analyte list, the results may not vary too greatly, as these four analytes are the ones most commonly detected in the United States. If any analyte is below the limit of detection, the clinician should calculate the analyte limit of detection divided by the square root of 2 and use this value in the summation. The summation should then be compared against Figure 6-2 to determine an appropriate clinical follow-up plan based on shared, informed decision making between patient and clinician.

CONCLUSION

The committee believes its clinical follow-up recommendations may be helpful to clinicians who have been asked to address PFAS-associated risks as part of routine health care delivery. There are potential harms both from aggressive clinical follow-up and from ignoring the risks of PFAS exposure. Clinicians and patients should decide which screening practices and services to pursue through a process of shared, informed decision making, along with consideration of the patients' level of PFAS exposure and other risk factors they may have. For young children, these discussions will likely take place with the parents; for adolescents, shared, informed decision making is complicated and an active area of research (Boland et al., 2019). The committee's patient follow-up recommendations should be updated as clinical practice guidelines change and as more is learned about the health effects of PFAS.

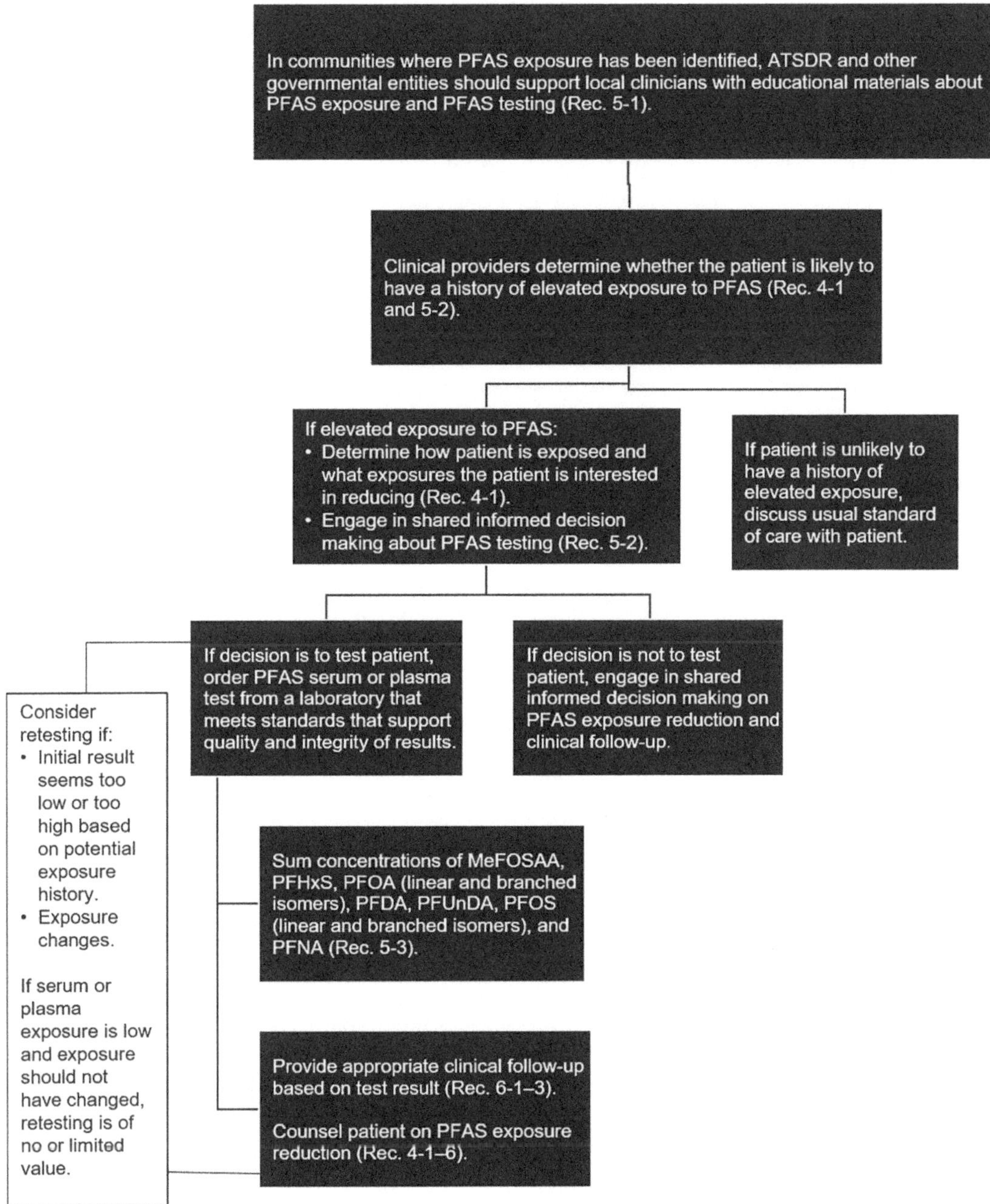

FIGURE 6-2 Flow chart on how the committee's recommendations work together in a clinical setting. NOTE: ATSDR = Agency for Toxic Substances and Disease Registry; MeFOSAA = methylperfluorooctane sulfonamidoacetic acid; PFDA = perfluorodecanoic acid; PFHxS = perfluorohexane sulfonic acid, PFNA = perfluorononanoic acid; PFOA = perfluorooctanoic acid; PFOS = perfluorooctanesulfonic acid; PFUnDA = perfluoroundecanoic acid.

REFERENCES

AHRQ (Agency for Healthcare Research and Quality). 2020. *The CAHPS ambulatory care improvement guide: Practical strategies for improving patient experience.* https://www.ahrq.gov/cahps/quality-improvement/improvement-guide/improvement-guide.html (accessed July 1, 2022).

Bateman, B. T., K. M. Shaw, E. V. Kuklina, W. M. Callaghan, E. W. Seely, and S. Hernández-Díaz. 2012. Hypertension in women of reproductive age in the United States: NHANES 1999–2008. *PLOS ONE* 7(4):e36171. https://doi.org/ 10.1371/journal.pone.0036171.

Boland, L., I. D. Graham, F. Légaré, K. Lewis, J. Jull, A. Shephard, M. L. Lawson, A. Davis, A. Yameogo, and D. Stacey. 2019. Barriers and facilitators of pediatric shared decision-making: A systematic review. *Implementation Science* 14(1):7. https://doi.org/10.1186/s13012-018-0851-5.

C-8 Medical Panel. 2013. *Information on the C-8 (PFOA) Medical Monitoring Program screening tests prepared by the medical panel for the C-8 class members.* http://www.c-8medicalmonitoring program.com/docs/med_panel_education_doc.pdf (accessed January 28, 2022).

C-8 Medical Panel. 2022. *C-8 Medical Monitoring Program.* http://www.c-8medicalmonitoringprogram.com (accessed January 28, 2022).

CDC (Centers for Disease Control and Prevention). 2014. *HTDS Guide—About thyroid disease: Section summary.* https://www.cdc.gov/nceh/radiation/hanford/htdsweb/guide/thyroid.htm (accessed January 28, 2022).

Dahlhamer, J. M., E. P. Zammitti, B. W. Ward, A. G. Wheaton, and J. B. Croft. 2016. Prevalence of inflammatory bowel disease among adults aged ≥18 years—United States, 2015. *Morbidity and Mortality Weekly Report* 65(42):1166–1169.

Grundy, S, M., N. J. Stone, A. L. Bailey, C. Beam, K. K. Birtcher, R. S. Blumenthal, L. T. Braun, S. de Ferranti, J. Faiella-Tommasino, D. E. Forman, R. Goldberg, P. A. Heidenreich, M. A. Hlatky, D. W. Jones, D. Lloyd-Jones, N. Lopez-Pajares, C. E. Ndumele, C. E. Orringer, C. A. Peralta, J. J. Saseen, S. C. Smith, L. Sperling, S. S. Virani, and J. Yeboah. 2019. 2018 AHA/ACC/AACVPR/ AAPA/ABC/ACPM/ADA/AGS/APhA/ASPC/NLA/PCNA Guideline on the Management of Blood Cholesterol: A Report of the American College of Cardiology/American Heart Association Task Force on Clinical Practice Guidelines. *Circulation* 139(25):e1082–e1143. https://doi.org/10.1161/CIR.0000000000000625.

Hagan, J. F., J. S. Shaw, and P. M. Duncan (Eds.). 2017. *Bright Futures guidelines for health supervision of infants, children, and adolescents: Adolescence visits: 11 through 21 years.* https://brightfutures.aap.org/Bright%20Futures%20Documents/BF4_InfancyVisits.pdf (accessed February 9, 2022).

Hanauer, S. B. 2004. Update on the etiology, pathogenesis and diagnosis of ulcerative colitis. *Nature Clinical Practice Gastroenterology & Hepatology* 1(1):26–31.

HRSA (Health Resources and Services Administration). 2018. *Recommended uniform screening panel.* https://www.hrsa.gov/advisory-committees/heritable-disorders/rusp/index.html (accessed January 28, 2022).

Kilpatrick, S. J., L-A. Papile, and G. A. Macones (Eds.). 2017. *Guidelines for perinatal care* (8th ed.). Elk Grove Village, IL: American Academy of Pediatrics.

NCI (National Cancer Institute). n.d. *Cancer stat facts: Testicular cancer.* https://seer.cancer.gov/statfacts/html/testis.html (accessed January 28, 2022).

NIDDK (National Institute of Diabetes and Digestive and Kidney Diseases). 2017. *Thyroid tests.* https://www.niddk.nih.gov/health-information/diagnostic-tests/thyroid (accessed June 30, 2022).

Privett, N., and S. Guerrier. 2021. Estimation of the time needed to deliver the 2020 USPSTF preventive care recommendations in primary care. *American Journal of Public Health* 111:145–149. https://doi.org/10.2105/AJPH.2020.305967.

Richerson, J. E., G. R. Simon, J. J. Abularrage, A. D. Arauz Boudreau, C. N. Baker, G. A. Barden, O. Brown, J. M. Hackell, and A. P. Hardin. 2017. Bright Futures guidelines. *Pediatrics* 139(4)85–87.

Schrager, S. 2018. Five ways to communicate risks so that patients understand. *Family Practice Management* 25(6):18–31.

USPSTF (U.S. Preventive Services Task Force). 2011. *Testicular cancer: Screening.* https://www.uspreventiveservicestaskforce.org/uspstf/recommendation/testicular-cancer-screening (accessed January 28, 2022).

USPSTF. 2015. *Thyroid dysfunction: Screening.* https://www.uspreventiveservicestaskforce.org/uspstf/recommendation/thyroid-dysfunction-screening (accessed January 28, 2022).

USPSTF. 2016a. *Breast cancer: Screening.* https://www.uspreventiveservicestaskforce.org/uspstf/recommendation/breast-cancer-screening (accessed February 9, 2022).

USPSTF. 2016b. *Statin use for the primary prevention of cardiovascular disease in adults: Preventive medication.* https://www.uspreventiveservicestaskforce.org/uspstf/recommendation/statin-use-in-adults-preventive-medication (accessed February 9, 2022).

USPSTF. 2017. *Preeclampsia: Screening.* https://www.uspreventiveservicestaskforce.org/uspstf/recommendation/preeclampsia-screening (accessed January 28, 2022).

USPSTF. 2021. *Hypertension in adults: Screening.* https://www.uspreventiveservicestaskforce.org/uspstf/recommendation/hypertension-in-adults-screening (accessed January 28, 2022).

USPSTF. 2022. Collaboration and shared decision-making between patients and clinicians in preventive health care decisions and U.S. Preventive Services Task Force recommendations. *JAMA* 327(12):1171–1176. http://doi.org/10.1001/jama.2022.3267.

Wilson, J. M., and Y. G. Jungner. 1968. [Principles and practice of mass screening for disease]. *Boletín de la Oficina Sanitaria Panamericana* 65(4):281–393.

WPSI (Women's Preventive Services Initiative). 2019. *Women's preventive services guidelines.* https://www.hrsa.gov/womens-guidelines-2019 (accessed February 9, 2022).

7

Revising ATSDR's PFAS Clinical Guidance

Since the potential for harmful effects of per- and polyfluoroalkyl substances (PFAS) exposure were made known to the public in 2000, clinicians have increasingly needed guidance on advising their communities regarding sources, routes, and effects of exposure, even as significant uncertainty about the health effects of exposure remains (see Chapter 3). The C-8 Medical Panel was the first body to offer clinical guidance regarding PFAS exposure, but the guidance was limited to members of the class action lawsuit (C-8 Medical Panel, 2013).[1] In response to this increasing need from clinicians for advice about responding to this environmental hazard, the Agency for Toxic Substances and Disease Registry (ATSDR) published clinical guidance regarding PFAS exposure in December 2019 (ATSDR, 2019).

RECOMMENDATIONS FOR CHANGES TO ATSDR'S CLINICAL GUIDANCE

In accordance with its Statement of Task (see Chapter 1), the committee in this chapter recommends several changes to ATSDR's guidance to ensure consistency with the conclusions, findings, and recommendations in this report. The Statement of Task specifies three main considerations regarding changes to the guidance. The first two—decision making for PFAS testing and PFAS concentrations informing clinical care of exposed patients—are discussed in Chapter 5; the third—clinical follow-up care specific to PFAS exposure—is addressed in Chapter 6.

The following recommendations illustrate the potential use of the information presented in previous chapters of this report:

> **Recommendation 7-1: The Agency for Toxic Substances and Disease Registry (ATSDR) should update its PFAS clinical guidance to make it more succinct and accord with the review of PFAS-associated health effects, exposure reduction considerations, PFAS testing recommendations and interpretation, and recommendations for clinical follow-up presented in this report. When describing the health effects of PFAS, ATSDR should avoid using terms typically used to categorize toxicants, such as "endocrine disrupter" or "neurotoxin," because they are vague and not necessarily clinically meaningful. When discussing the strength of the association between PFAS and a health outcome, ATSDR should use standard categories of association (such as sufficient evidence of an association, limited or suggestive evidence of an association, inadequate or insufficient evidence of an association, and limited or suggestive evidence of no association).**

> **Recommendation 7-2: The Agency for Toxic Substances and Disease Registry (ATSDR) should incorporate a reader-centered approach when developing its guidance, with the knowledge that many different audiences will turn to its clinical guidance document to prepare for discussions with their clinicians. ATSDR should also solicit feedback on the guidance from a variety of stakeholders, such as**

---

[1] See Chapter 1 for a brief overview of the lawsuit, *Jack W. Leach et al. v. E.I. du Pont de Nemours & Company* (no. 01-C-608 W.Va., Wood County Circuit Court, filed April 10, 2002), and Chapter 6 for details regarding the clinical guidance for class members.

community groups, practicing clinicians, and medical associations. In addition, ATSDR should encourage clinicians to use evidence-based organizational health literacy strategies to support shared, informed decision making; patient-centered care; cultural humility; and accessible language when communicating with patients about potential health risks.

Recommendation 7-3: The Agency for Toxic Substances and Disease Registry (ATSDR) should develop a process for updating its PFAS guidance that adheres to criteria for making guidelines trustworthy, such as being based on a thorough, transparent, unbiased review of the evidence and being developed by a knowledgeable panel of experts free from strong biases and conflicts of interest. A review of the evidence on the health effects of PFAS should be completed by an authoritative neutral party every 2 years, and the clinical guidance should be updated every 5 years or sooner if warranted by the evidence on the health effects of PFAS. Clinicians and members of communities with elevated PFAS exposure should be engaged to inform the problem and review updated guidance.

In addition to these considerations, the committee believed it would be useful to address some more technical aspects of the ATSDR guidance document, including its writing and design and its dissemination and implementation. To this end, the committee referred to existing clinical guidance on PFAS exposure, as well as materials on health literacy and health communication, including the Agency for Healthcare Research and Quality's (AHRQ's) Health Literacy Universal Precautions Toolkit (Brega, 2015) and the Centers for Medicare & Medicaid Services' (CMS's) Toolkit for Making Written Material Clear and Effective (McGee, 2010). Given the Centers for Disease Control and Prevention's (CDC's) long-standing commitment to effective health communication (Donovan, 1995; Gagen and Kreps, 2019; Roper, 1993; Tinker and Silberberg, 1997), the committee decided not to focus on the finer details of these considerations but on their high-level aspects.

The committee reviewed several other materials published by other organizations to aid clinicians in addressing exposure to PFAS and determining options for clinical follow-up. Table 7-1 describes the clinical guidance documents published by the Pediatric Environmental Health Specialty Units (PEHSUs), and PFAS Research, Education, and Action for Community Health (PFAS-REACH) and identifies the strengths of each.

TABLE 7-1 Description of PFAS Clinical Guidance Documents

Clinical Guidance Document	Description of Content	Strengths of Document
How to Conduct a Clinical Visit with Patients Concerned About PFAS (PEHSU, 2021)	Includes a numbered list of five strategies for clinicians to use in discussing PFAS exposure with concerned patients and provides some detail on how to apply them. Also provides some references to other materials with further detail.	Focuses on empathy and shared decision making and the potential need for exposure reduction. Offers advice only for overall approaches to navigating discussions with patients, and directs clinicians to supplementary Pediatric Environmental Health Specialty Units (PEHSUs) and Agency for Toxic Substances and Disease Registry (ATSDR) resources.
PFAS Exposure: Information for Patients and Guidance for Clinicians to Inform Patient and Clinician Decision Making: For Clinicians (PFAS-REACH, 2021)	Summarizes clinical services recommended by other trustworthy groups for adults and children with above-average PFAS exposure. Categorizes information by type of service—laboratory tests, exams, or counseling topics	Accessible summary of material produced by others. Does not claim to provide rigorous analysis of others' findings. Does provide more information than the PEHSU document in an accessible format.

WRITING AND DESIGN OF ATSDR'S CLINICAL GUIDANCE

The CMS Toolkit for Making Written Material Clear and Effective provides advice about paying attention to content, organization, writing style, engagement, and motivation (McGee, 2010):

- "Content" relates to what readers want and need to know and whether the information provided is accurate and up to date.
- The "organization" of the written material involves how well the material paces readers by grouping information in meaningful sections.
- "Writing style" refers to whether the text is conversational and uses the active voice, is specific and concrete, and uses familiar and culturally appropriate terms.
- "Engagement and motivation" refers to whether the material has a positive and friendly tone, offers trustworthy information sources, provides relatable statistics, or offers information about how readers can learn more.

The CMS Toolkit also includes "design" principles related to such matters as

- overall design and page layout;
- font, size of print, and contrast;
- headings, lists, and blocks of text;
- use of color;
- photographs and illustrations; and
- tables, charts, and diagrams (McGee, 2010).

As noted above, the committee limited its attention to design intricacies and overall layout, other than noting that the ATSDR guidance document is likely too long for the target audience of clinicians. Best practices for guidance (AAFP, 2021) include providing recommendations that are specific, offer clear direction, and are succinct—principles consistent with comments made to the committee by community members and clinicians (see Appendix B). Incorporating such best practices and using a reader-centered approach will allow ATSDR's clinical guidance to reach and support as many clinicians (and other community members) as possible.

DISSEMINATING AND IMPLEMENTING ATSDR'S CLINICAL GUIDANCE

Process for Updating the Guidance

In addition to updates to the ATSDR guidance stemming from this report, the committee proposes the establishment of a process for regularly updating the guidance (see Figure 7-1). The first step in that process is engaging with impacted communities to inform understanding of the issue at hand. It will also be important for ATSDR to update its reviews regarding PFAS-associated health effects, as well as to catalyze future research by identifying gaps in the evidence. This process should include both a review of guidelines issued by other authoritative bodies reflecting decisions about the health effects of PFAS and a review of the epidemiologic literature to identify any new studies that may warrant updating or revising ATSDR's own authoritative guidance. These reviews should be conducted by a neutral party every 2 years, or sooner if a watershed study, such as a large cohort or nested case-control study, on PFAS exposure and health effects is published before the next review is scheduled. AHRQ, the U.S. Preventive Services Task Force (USPSTF), and other organizations use similar processes that may be informative. One such process includes conducting "living" systematic reviews to continually update an existing review as new evidence becomes available (Elliott et al., 2017). It will be important in this process to incorporate reviews conducted by the U.S. Environmental Protection Agency, the National

Toxicology Program, and other authoritative bodies. The committee proposes that ATSDR revise its guidance within 5 years of its 2021 analysis, consistent with the timelines for updating of the National Guideline Clearinghouse (NGC) and the USPSTF recommendations. The process for updating the literature review should encompass studies on PFAS exposure reduction, including those evaluating behavior change, interventions, or clinical measures, as well as studies on risk-based levels of PFAS to inform clinical care. It will also be important to review recommendations on standard care, as clinical follow-up recommendations would change if, for example, a beneficial screening test existed for a PFAS-associated health outcome. Finally, as noted above, the updating process should incorporate approaches for assembling feedback from clinicians and community members prior to each review. It should also include processes for updated documents to be reviewed by clinicians, in consultation with community members impacted by PFAS (see Figure 7-1 for an overview of this proposed updating process).

Transparency and Trustworthiness

Transparency enhances the trust clinicians and others place in clinical guidance. To advise ATSDR in this regard, the committee turned to the Institute of Medicine's (IOM's) standards for developing trustworthy clinical practice guidelines (CPGs) (IOM, 2011). The 2011 IOM report *Clinical Practice Guidelines We Can Trust* defines CPGs as

> statements that include recommendations intended to optimize patient care. They are informed by a systematic review of evidence and an assessment of the benefits and harms of alternative care options. (p. 15)

FIGURE 7-1 Suggested framework for updating the Agency for Toxic Substances and Disease Registry's clinical guidance based on new evidence.

The definition also includes six attributes required to make CPGs trustworthy, including that they

- are based on a systematic review of the existing evidence;
- are developed by a knowledgeable, multidisciplinary panel of experts and representatives from key affected groups;
- consider important patient subgroups and patient preferences, as appropriate;
- are based on an explicit and transparent process that minimizes distortions, biases, and conflicts of interest;
- provide a clear explanation of the logical relationships between alternative care options and health outcomes, and provide ratings of both the quality of evidence and the strength of recommendations; and
- are reconsidered and revised as appropriate when important new evidence warrants modifications of recommendations (p. 26).

The above definition of and standards for trustworthiness are intended to distinguish CPGs as transparent and methodologically rigorous, evidence-based guidelines as compared with other important forms of clinical guidance, such as expert advice, consensus statements, and practice bulletins. CPGs provide recommendations based on the balance of benefits and harms of different interventions for preventing or treating disease. Although ATSDR may be unable to issue CPGs regarding general PFAS exposure, as the complexity of the data may make a formal systematic review to inform such CPGs difficult, it should still strive to ensure that its clinical guidance reflects these criteria when possible.

CONCLUSION

The need for clear clinical guidance regarding PFAS exposure has only increased, despite the remaining uncertainties regarding the specific health risks of exposure. Yet it is difficult to distill the necessary information for effective clinical practice with respect to PFAS exposure, its possible health effects, and options for shared decision making once exposure has been confirmed. In addition to the distillation of the information itself, moreover, the writing, design, and general implementation of the clinical guidance are as important and as complex as the distillation of the information itself. This chapter offers recommendations that, if implemented, may assist ATSDR in its efforts to offer the guidance needed by clinicians. The committee believes ATSDR can continue to offer up-to-date, useful, and trustworthy clinical guidance by incorporating the updated data on the health effects of PFAS exposure presented in earlier chapters of this report, as well as adhering to the important risk communication and health literacy principles discussed above.

REFERENCES

AAFP (American Academy of Family Physicians). 2021. *Clinical practice guidelines.* https://www.aafp.org/family-physician/patient-care/clinical-recommendations/definitions/cpg.html (accessed December 28, 2021).

ATSDR (Agency for Toxic Substances and Disease Registry). 2019. *PFAS: An overview of the science and guidance for clinicians on per- and polyfluoroalkyl substances (PFAS).* https://www.atsdr.cdc.gov/pfas/docs/ATSDR_PFAS_ClinicalGuidance_12202019.pdf (accessed March 25, 2020).

Brega, A. G., J. Barnard, N. M. Mabachi, B. D. Weiss, D. A. DeWalt, C. Brach, M. Cifuentes, K. Albright, and D. R. West. 2015. *AHRQ health literacy universal precautions toolkit, second edition.* Rockville, MD. AHRQ Publication No. 15-0023-EF. https://www.ahrq.gov/health-literacy/improve/precautions/toolkit.html (acessed July 1, 2022).

C-8 Medical Panel. 2013. *Information on the C-8 (PFOA) Medical Monitoring Program screening tests prepared by the medical panel for the C-8 class members.* http://www.c-8medicalmonitoring program.com/docs/med_panel_education_doc.pdf (accessed January 28, 2022).

Donovan, R. J. 1995. Steps in planning and developing health communication campaigns: A comment on CDC's framework for health communication. *Public Health Reports* 110(2):215–217. https://stacks.cdc.gov/view/cdc/64183 (accessed July 1, 2022).

Elliott, J. H., A. Synnot, T. Turner, M. Simmond, E. Akl, S. McDonald, G. Salanti, J, Meerpohl, H. MacLehose, J. Hilton, D. Tovey, I, Shemilt, and J. Thomas, on behalf of the Living Systematic Review Network. 2017. Living systematic review: 1. Introduction—the why, what, when, and how. *Journal of Clinical Epidemiology* 91:23–30. https://doi.org/10.1016/j.jclinepi.2017.08.010.

Gagen, D. M., and G. L. Kreps. 2019. An examination of the clarity of government health websites using the Centers for Disease Control and Prevention's Clear Communication Index. *Frontiers in Communication* 4. https://doi.org/10.3389/fcomm.2019.00060.

IOM (Institute of Medicine). 2011. *Clinical practice guidelines we can trust.* Washington, DC: The National Academies Press. https://doi.org/10.17226/13058.

McGee, J. 2010. Toolkit for making written material clear and effective. Baltimore, MD: Centers for Medicare & Medicaid Services. https://www.cms.gov/Outreach-and-Education/Outreach/WrittenMaterialsToolkit (accessed July 1, 2022).

PEHSU (Pediatric Environmental Health Specialty Unit). 2021. *How to conduct a clinical visit with patients concerned about PFAS exposure.* https://deohs.washington.edu/pehsu/sites/deohs.washington.edu.pehsu/files/PFAS%20Clinical%20Visit%20tips%20for%20clinicians_AK_Jan%202021.pdf (accessed June 30, 2022).

PFAS-REACH (PFAS Research, Education, and Action for Community Health). 2021. *PFAS exposure: Information for patients and guidance for clinicians to inform patient and clinician decision making—For people in PFAS-impacted communities.* https://wordpress.silentspring.org/wp-content/uploads/2021/06/PFAS-REACH-Medical-screening-guidance_PFAS-impacted-communities.pdf (accessed June 30, 2022).

Roper, W. L. 1993. Health communication takes on new dimensions at CDC. *Public Health Reports* 108(2):179. https://www.ncbi.nlm.nih.gov/pmc/articles/PMC1403358.

Tinker, T. L., and P. G. Silberberg. 1997. *An evaluation primer on health risk communication programs and outcomes.* https://www.atsdr.cdc.gov/risk/evalprimer/index.html (accessed June 28, 2022).

8

Implementing the Committee's Recommendations to Improve Public Health

Improving public health requires the use of multifaceted approaches to emerging health issues. In environmental health—the subset of public health focused on environmental factors—mitigation of potential harms associated with chemical exposures is often complicated because there is no exposure surveillance system for most chemicals. To address per- and polyfluoroalkyl substances (PFAS) effectively as a public health issue, it will be important to identify the people and communities with high exposures to PFAS, improve environmental health education for both clinicians and the public, and prevent exposures. Thus, the committee's recommendations would be most effective if implemented as part of a larger national effort toward increased PFAS-focused biomonitoring, exposure surveillance, and education around environmental health.

To promote a coordinated response to PFAS exposures, the committee makes the following recommendations:

> **Recommendation 8-1: Laboratories conducting PFAS testing of serum or plasma should report the results to state public health authorities, following the respective states' statutes and reporting regulations. This reporting would improve PFAS exposure surveillance; it could be linked with the Centers for Disease Control and Prevention's environmental public health tracking network and help build capacity for improvements in the state-based national biomonitoring network.**

BIOMONITORING AND SURVEILLANCE

PFAS exposure is nearly ubiquitous, but individual levels of exposure are not consistently measured or documented (CDC, 2021; NASEM, 2020, 2021a). Given the persistence of most PFAS in the human body, biomonitoring will likely produce meaningful exposure data, especially for those PFAS with longer half-lives. Collecting these data is critical to identifying communities and other areas with elevated PFAS exposure. Failing to identify such communities magnifies existing disparities for those that already have limited access to health care, live in areas of high exposure, and do not know that they may be at elevated risk for PFAS-associated health outcomes. Using biomonitoring data to establish exposure surveillance programs while also ensuring that medical professionals and the public have access to clear, accurate information about the risks of PFAS exposure will improve public health overall while also empowering people and communities to safeguard their own health.

According to the 2006 report by the National Research Council *Human Biomonitoring for Environmental Chemicals*:

> Identifying, controlling, and preventing population exposures to potentially harmful environmental chemicals have been cornerstones of U.S. environmental health efforts. Biomonitoring has become a tool that is central to these efforts. (NRC, 2006, p. 1)

While aggregate PFAS biomonitoring data can be used to promote population health, individual PFAS biomonitoring can provide people with information about their personal exposures, which is important in informing clinical care (see Chapter 5). Therefore, access to PFAS blood tests for patients likely to have a history of elevated exposure as well as the results of those tests can help individuals better understand and, if possible, address their personal risk from PFAS exposure. Additionally, while biomonitoring studies are important to identify exposed communities and study potential health outcomes, PFAS testing intended to inform an individual's medical care should be coordinated by clinicians to ensure appropriate care and follow-up.

Improving access to PFAS blood testing for individuals could also allow for the collection of aggregate biomonitoring data if the results were compiled into a database. As discussed in Chapter 5, public health professionals can use those data to determine reference ranges of PFAS exposure in the general population. Understanding those ranges of exposure would support prioritization and allocation of funding and resources for preventive medicine or public health actions. Other high-profile cases demonstrate this powerful use of biomonitoring data to improve public health practice, as occurred in 2015 when Dr. Mona Hanna-Attisha began reviewing her patients' medical records and realized that the percentage of patients with elevated blood lead levels had increased after the city of Flint, Michigan, changed its water source.[1]

In addition to informing clinical care, biomonitoring data can be used in establishing exposure surveillance programs. Biomonitoring-based exposure surveillance enables public health practitioners and policy makers to identify exposure trends, mitigate ongoing exposures, and study associations between exposures and disease outcomes (Goldman et al., 1995; Thacker et al., 1996). In contrast with agencies' ongoing monitoring of environmental contaminants in air, soil, and water, there are few comprehensive databases or collection strategies for measuring and tracking levels in exposed people (Latshaw et al., 2017). The data that are currently collected, such as measurements from the National Health and Nutrition Examination Survey (NHANES),[2] often provide a snapshot of exposure that is useful in tracking nationwide trends in exposure over time, but cannot inform a localized public health response or help identify potential exposure "hot spots." Additionally, unlike the National Health Interview Survey,[3] the NHANES does not list urban versus rural as a publicly available variable, which limits researchers' ability to identify potential location-based disparities in PFAS exposure. Without collecting and compiling comprehensive PFAS biomonitoring data, federal agencies are less able to prioritize highly exposed communities in their mitigation strategies. Additionally, a lack of biomonitoring data limits studies on putative health effects, resulting in continued uncertainty that hinders public health response. As Kristen Mello stated during the committee's April 7, 2021, town hall, "Don't avoid taking our data and then complain about data gaps."

An effective strategy for PFAS exposure surveillance would involve multiple levels of public health response (Colles et al., 2021; Eatman and Strosnider, 2017). As with U.S. programs on pediatric monitoring for lead exposure, laboratories would report the biomonitoring results, zip code, and demographic information to state health departments, which in turn would report the data to the federal government. Data collected by states could be more geographically granular than the data typically accessible through NHANES, thereby enabling localized public health responses for highly contaminated areas (Nassif et al., 2021). Additionally, hierarchical data collection and reporting processes might include repeated stopgaps and promote oversight to ensure that clinicians, researchers, and public health agencies were using the data as effectively as possible.

ENVIRONMENTAL HEALTH EDUCATION

If the use of biomonitoring data and environmental surveillance programs is to be most effective at preventing or at least minimizing clinical disease associated with exposure, clinicians and the public

---

[1] See ww.npr.org/sections/health-shots/2018/06/25/623126968/pediatrician-who-exposed-flint-water-crisis-shares-her-story-of-resistance (accessed June 17, 2022).

[2] See https://www.cdc.gov/nchs/nhanes/index.htm (accessed June 17, 2022).

[3] See https://www.cdc.gov/nchs/nhis (accessed June 17, 2022).

must have access to the information and know what to do with it. From a clinical perspective, providers need a better working knowledge of environmental health and chemical exposures. At the committee's town halls, speakers described their frustration with trying to obtain medical care from practitioners who were unfamiliar with PFAS and did not understand the environmental health contexts of the communities in which they practiced. At the town hall on April 7, 2021, Hope Grosse said:

> When I would go to the doctors and tell them about some of the exposures of over 50 chemicals that I was exposed to, the doctors would laugh and say no. Clearly, they didn't have any information about environmental components [of disease]. They made me feel small; they made me feel stupid and embarrassed even just asking the question.

The National Academies has previously acknowledged the issue of limited clinical education in environmental health. The 1991 Institute of Medicine (IOM) report *Addressing the Physician Shortage in Occupational and Environmental Medicine: Report of a Study* notes that all levels of medical education, from undergraduate to graduate and continuing education programs, provide limited, if any, training in occupational and environmental medicine. The report recommends that occupational and environmental medicine concepts be introduced early and continuously during medical education (IOM, 1991). Likewise, *Environmental Medicine: Integrating a Missing Element into Medical Education* notes a need for more environmentally literate physicians (IOM, 1995a). Decades later, however, Green-McKenzie and colleagues (2021) found that only 70 percent of medical students had heard of occupational and environmental medicine, and most of them had received only one lecture on the topic. After medical school, residency programs align with specialty requirements, which may not include further training in environmental and occupational medicine. In addition to the more formal medical training doctors receive in medical school and residency programs, professional associations such as the American Medical Association,[4] the American College of Preventive Medicine,[5] and the American College of Occupational and Environmental Medicine[6] offer continuing medical education modules on environmental health. However, these modules are optional, meaning that physicians may not learn about environmental and occupational medicine after medical school, either.

Physicians are not alone in their limitations with respect to environmental health education. According to the 1995 IOM report *Nursing, Health, and the Environment: Strengthening the Relationship to Improve the Public's Health*, "environmental health currently receives scant attention in nursing education and research" (IOM, 1995b, p. 14). The report calls for more nurses with education in environmental health, given that nurses represent the largest proportion of the health care workforce. Similarly, a recent commentary on the nursing profession notes that most nurses do not see themselves as environmental health practitioners or scientists (McCauley and Hayes, 2021). To address this shortcoming, environmental health is now a recommended domain for nursing education (NASEM, 2021b); in addition, nurses may need additional continuing education on environmental health topics.

Continuing education of clinicians is important to address knowledge gaps in environmental health. Increasing the opportunities for clinician education in environmental and occupational medicine; promoting collaborative, interdisciplinary environmental health networks; and framing environmental health as a tenet of preventive medicine and primary care are essential to ensure proper medical care for those facing environmental exposures and to assist families in reducing exposures when possible. One resource that can be leveraged to help respond to environmental exposures is the Pediatric Environmental Health Specialty Units (PEHSUs). ATSDR funds the PEHSUs with support from EPA. The PEHSUs are a "national network of experts in the prevention, diagnosis, management, and treatment of health issues that arise from environmental exposures from preconception through adolescence."[7] While the primary

---

[4] See https://www.ama-assn.org (accessed June 17, 2022).
[5] See https://www.acpm.org (accessed June 17, 2022).
[6] See https://acoem.org (accessed June 17, 2022).
[7] See https://pehsu.net (accessed July 27, 2022).

focus is reproductive and children's health, clinicians will work with the whole family in determining exposure pathways and reducing these exposures. The PEHSU network includes nurses, physicians, medical toxicologists, and public health professionals and promotes collaboration among these diverse groups of practitioners. The network provides continuing education opportunities and information on PFAS for both health care professionals and community members. PEHSU-affiliated practitioners have proposed concrete environmental health competency areas for medical education programs (Goldman et al., 2021). Similarly, the Alliance of Nurses for Healthy Environments (ANHE)[8] addresses environmental exposures and health, including PFAS and related clinical care, in various online resources and engages with universities to integrate environmental health into their nursing education curricula. Other organizations, such as the Children's Environmental Health Network,[9] also provide educational materials in environmental health for health care providers.

Many public health practitioners have received training in environmental health because until recently, it was a required component of master of public health (M.P.H.) programs; however, the curriculum for public health is changing. In 2016, the Council on Education for Public Health (CEPH) released new accreditation criteria, aimed at offering schools of public health greater flexibility in course offerings and students a more applied, practice-based education.[10] With these more flexible coursework requirements, the CEPH removed the explicit requirement for education in environmental health sciences, replacing it and the other core disciplines with a series of competencies. This change may inadvertently lead to reduced environmental health training for public health professionals.[11] Indeed, by December 2019, just 91 percent of accredited M.P.H. programs continued to offer a concentration in environmental health sciences.[12] Addressing such challenges as PFAS requires a public health workforce with competency in environmental health sciences, toxicology, and epidemiology, which is why the Association for Prevention Teaching and Research calls on schools and programs in public health to maintain existing environmental health content and develop such content if previously eliminated or never offered (Levy et al., 2022).[13]

The public's environmental health literacy complements clinician education in environmental health by promoting individuals' personal health decision making while galvanizing communities to demand population-level public health responses. Historically, health literacy as a concept centered on individual decision making based on an understanding of health information. In Healthy People 2030, however, the U.S. Department of Health and Human Services (HHS) expanded the definition to encompass organizations' responsibility to ensure that individuals have access, understanding, and services to inform their health decisions, a concept HHS refers to as organizational health literacy.[14]

The term *health literacy* as used in this report focuses on Nutbeam's (2000) second and third levels of health literacy—interactive and critical literacy—which emphasize empowerment, autonomy, and action. More specifically, National Institute of Environmental Health Sciences (NIEHS) authors Finn and O'Fallon (2017) describe environmental health literacy as

> a philosophical perspective, a public health policy to improve literacy and health literacy
> in the general public, and a set of strategies to empower individuals and communities to
> exert control over the environmental exposures that may lead to adverse health outcomes.
> (p. 495)

---

[8] See https://envirn.org (accessed June 17, 2022).

[9] See https://cehn.org (accessed June 17, 2022).

[10] See https://media.ceph.org/documents/Environmental_Health.pdf (accessed June 17, 2022).

[11] See http://www.publichealthnewswire.org/?p=environmental-health-in-education (accessed June 17, 2022).

[12] See https://media.ceph.org/documents/Environmental_Health.pdf (accessed June 17, 2022).

[13] See https://www.aptrweb.org/page/ClimateChange (accessed June 17, 2022).

[14] See https://health.gov/healthypeople/priority-areas/health-literacy-healthy-people-2030/history-health-literacy-definitions (accessed June 17, 2022).

Building on the concept of health literacy, environmental health literacy has three dimensions: (1) awareness and knowledge of an environmental exposure, (2) skills and self-efficacy that allow individuals to make decisions to protect their own health, and (3) community change based on actions that reduce environmental exposures and protect public health (Gray, 2018). The committee recognizes that many exposures to PFAS are beyond individual control, and placing the onus for action on individuals often imposes an undue burden. Nevertheless, environmental health literacy can empower people to understand their own risks and ask their providers informed questions. As demonstrated by the community liaisons and town hall speakers who contributed to this study, many individuals have already worked especially hard to access information and educate themselves on PFAS.

In keeping with the principle of reducing the burden on exposed community members, PFAS-related environmental health literacy should center on making PFAS research publicly available and understandable for general audiences. Open access publication platforms are more accessible than those that are behind paywalls. Summaries written in plain language for the lay public, as blogs, news articles, and podcasts would be even better for increasing awareness about PFAS. Such resources as ATSDR's ToxFAQs[15]; Northeastern University's PFAS-TOX Database[16]; the PFAS Research, Education, and Action for Community Health PFAS Exchange[17]; and the Environmental Working Group's PFAS Contamination in the U.S. map[18] distill key information for the public and may serve as models for other environmental health literacy work. Other online resources, such as Purdue University's Online Writing Lab, describe considerations for evaluating the credibility of online sources that can serve as a guide for both those developing and those using the resources.[19] People must have access to information that can help them make health decisions and advocate for the health of their communities.

BARRIERS TO IMPLEMENTATION

Barriers to the committee's proposed implementation approach include difficulties with PFAS testing, lack of data standardization, poor coordination among experts from different disciplines, and a lack of funding for and availability of education in environmental and occupational medicine and environmental health literacy activities. Few laboratories currently have the capability to test for PFAS. The testing methodology is complex, and not all PFAS can be detected in serum or plasma. Data standardization is also a challenge (Latshaw et al., 2017) given the variability in testing methods, participant demographics, and exposure sources. As noted in Chapter 5 of this report, PFAS testing is expensive, and limiting the availability of testing based on access to consistent clinical care and insurance coverage could exacerbate health disparities in exposed communities. Furthermore, comprehensive PFAS exposure surveillance requires collaboration among a range of experts across different fields (Nassif et al., 2021); training in environmental and occupational medicine and environmental health is lacking; and support for PFAS health literacy is limited.

Given widespread PFAS exposure and the putative health effects associated with these chemicals, public health authorities would do well to prioritize addressing these barriers. Laboratory capacity to test for PFAS could be increased through coordination with the Association of Public Health Laboratories. Biomonitoring data and exposure surveillance programs could be standardized as long as they were developed with structures in place to ensure that the data were collected with the intent of their being comparable. The Network Steering Committee of the National Biomonitoring Network provides a template for successful implementation (Nassif et al., 2021).

---

[15] See https://wwwn.cdc.gov/TSP/ToxFAQs/ToxFAQsLanding.aspx (accessed June 17, 2022).

[16] See https://pfasproject.com/pfas-toxic-database (accessed June 17, 2022).

[17] See https://pfas-exchange.org (accessed June 17, 2022).

[18] See https://www.ewg.org/interactive-maps/pfas_contamination/map (accessed June 17, 2022).

[19] See https://owl.purdue.edu/owl/research_and_citation/conducting_research/evaluating_sources_
of_information/evaluating_digital_sources.html (accessed June 17, 2022).

Finally, continuing education opportunities such as "train-in-place" programs allow clinicians to learn environmental and occupational medicine concepts during the course of their ongoing practice (Green-McKenzie et al., 2021). Optimizing these programs and other innovative education strategies, as well as planning communication efforts as part of biomonitoring studies (NRC, 2006), can increase environmental health knowledge without requiring unattainable funding allocations.

MOVING FORWARD

Based on the recommendations and conclusions of this report, ATSDR, NIEHS, environmental health organizations, and bodies responsible for clinical education can take specific actions to address issues of PFAS contamination and clinical care, which will ultimately lead to improved public health (see Figure 8-1). By updating its PFAS clinical guidance according to the recommendations in Chapter 7, ATSDR will lay the groundwork for clinicians to participate in shared decision making and provide informed care for individuals exposed to PFAS. Access to informed care will allow communities with elevated exposure to PFAS to receive relevant medical advice, testing, screening, and treatment. In addition to improved guidance on clinical care, coordination between ATSDR and the National Biomonitoring Network will allow for the expansion of testing and surveillance. Additional communities identified as having elevated PFAS exposures can then receive informed care.

Additionally, NIEHS can use the information in this report to translate the research gaps identified in Chapter 3 into research needs, prioritizing research on potential associations between PFAS exposure and adverse health outcomes where the evidence is limited. As the research gaps narrow, ATSDR can update its clinical guidance to ensure that clinicians' knowledge and monitoring of PFAS-associated health outcomes remain current.

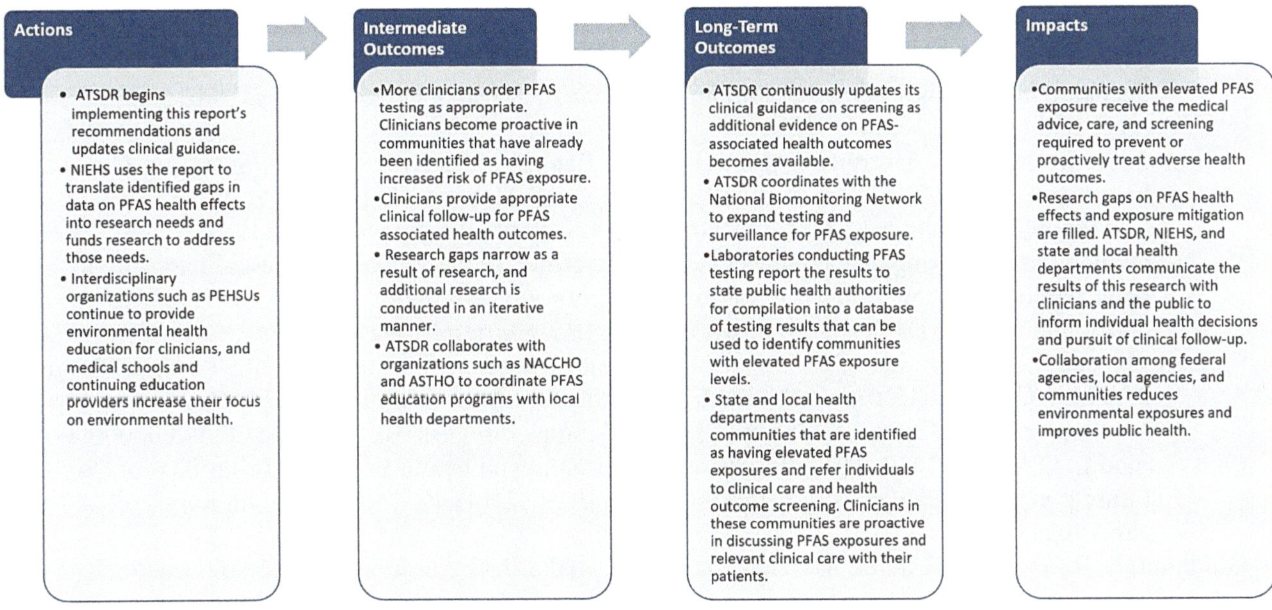

FIGURE 8-1 Recommended approach to mitigating PFAS exposure and adverse health outcomes.
NOTE: ASTHO = Association of State and Territorial Health Officials; ATSDR = Agency for Toxic Substances and Disease Registry; NACCHO = National Association of County and City Health Officials; NIEHS = National Institute of Environmental Health Sciences; PEHSU = Pediatric Environmental Health Specialty Unit.

Environmental health education for clinicians and the public is key to reducing PFAS exposure and ensuring shared decision making and appropriate follow-up. Introductory environmental health training provides the foundation for PFAS-specific clinical education. Given this, the Liaison Committee on Medical Education,[20] the Accreditation Council for Graduate Medical Education,[21] the American Board of Medical Specialties,[22] the Accreditation Commission for Education in Nursing,[23] and other bodies responsible for curricula and accreditation of clinician education programs may consider incorporating environmental health into clinician training. Input from such interdisciplinary organizations as PEHSUs and ANHE can support these efforts. State and local health departments may serve as a link between the public and clinicians and proactively provide current information about environmental exposures to the communities they serve. Health departments could receive support for this work through liaison organizations funded by the Centers for Disease Control and Prevention, including the National Association of County and City Health Officials[24] and the Association of State and Territorial Health Officials.[25] Intentional collaboration at every level—from clinician education; to community partnerships; to local, state, and federal government initiatives—will begin closing the gaps in environmental health infrastructure and training, facilitating the overdue coordinated response and clinical care that people with elevated PFAS exposures deserve.

These efforts in environmental exposure surveillance, environmental health literacy, and improved environmental health tracking will aid in continuing to identify communities impacted by PFAS, which in turn will support exposure mitigation and effective preventive medicine responses. As the committee looks forward, it sees a pressing need for a robust environmental health infrastructure to continue to respond to PFAS, as well as to address other complex emerging and persistent environmental challenges.

REFERENCES

CDC (Centers for Disease Control and Prevention). 2021. *Fourth national report on human exposure to environmental chemicals, updated tables.* Atlanta, GA: U.S. Department of Health and Human Services. https://www.cdc.gov/biomonitoring/pdf/fourthreport_updatedtables_feb2015.pdf (accessed June 28, 2022).

Colles, A., D. Coertjens, B. Morrens, E. Den Hond, M. Paulussen, L. Bruckers, E. Govarts, A. Covaci, G. Koppen, K. Croes, V. Nelen, N. Van Larebeke, S. De Henauw, T. Fierens, G. Van Gestel, H. Chovanova, M. Mampaey, K. Van Campenhout, I. Loots, W. Baeyens, and G. Schoeters. 2021. Human biomonitoring data enables evidence-informed policy to reduce internal exposure to persistent organic compounds: A case study. *International Journal of Environmental Research and Public Health* 18(11):5559. https://doi.org/10.3390/ijerph18115559.

Eatman, S., and H. M. Strosnider. 2017. CDC's National Environmental Public Health Tracking Program in action: Case studies from state and local health departments. *Journal of Public Health Management and Practice* 23(Suppl 5):S9–S17. https://doi.org/10.1097/phh.0000000000000640.

Finn, S., and L. O'Fallon. 2017. The emergence of environmental health literacy—From its roots to its future potential. *Environmental Health Perspectives* 125(4):495–501. https://doi.org/10.1289/ehp.1409337.

Goldman, L. R., H. Anton-Culver, M. Kharrazi, and E. Blake. 1995. Banking of human tissue for biomonitoring and exposure assessment: Utility for environmental epidemiology and surveillance. *Environmental Health Perspectives* 103:31–34. https://doi.org/10.2307/3432557.

---

[20] See https://lcme.org (accessed June 17, 2022).
[21] See https://www.acgme.org (accessed June 17, 2022).
[22] See https://www.abms.org (accessed June 17, 2022).
[23] See https://www.acenursing.org (accessed June 17, 2022).
[24] See https://www.naccho.org (accessed June 17, 2022).
[25] See https://www.astho.org (accessed June 17, 2022).

Goldman, R. H., L. Zajac, R. J. Geller, and M. D. Miller. 2021. Developing and implementing core competencies in children's environmental health for students, trainees, and healthcare providers: A narrative review. *BMC Medical Education* 21:503. https://doi.org/10.3390/ijerph15030466.

Gray, K. M. 2018. From content knowledge to community change: A review of representations of environmental health literacy. *International Journal of Environmental Research and Public Health* 15:466. https://doi.org/10.3390/ijerph15030466.

Green-McKenzie, J., U. Savanoor, H. Duran, C. Jones, D. Vearrier, P. Malak, E. Emmett, and F. Shofer. 2021. Outcomes of a survey-based approach to determine factors contributing to the shortage of occupational medicine physicians in the United States. *Journal of Public Health Management & Practice* 27(3 Suppl):S200–S205. htps://doi.org/10.1097/PHH.0000000000001315.

IOM (Institute of Medicine). 1991. *Addressing the physician shortage in occupational and environmental medicine: Report of a study.* Washington, DC: National Academy Press. https://doi.org/10.17226/9494.

IOM. 1995a. *Environmental medicine: Integrating a missing element into medical education.* Washington, DC: National Academy Press. https://doi.org/10.17226/4795.

IOM. 1995b. *Nursing, health, and the environment.* Washington, DC: National Academy Press. https://doi.org/10.17226/4986.

Latshaw, M. W., R. Degeberg, S. S. Patel, B. Rhodes, E. King, S. Chaudhuri, and J. Nassif. 2017. Advancing environmental health surveillance in the US through a national human biomonitoring network. *International Journal of Hygiene and Environmental Health* 220(2):98–102.

Levy, C. R., L. M. Phillips, C. J. Murray, L. A. Tallon, and R. M. Caron. 2022. Addressing gaps in public health education to advance environmental justice: Time for action. *American Journal of Public Health* 112(1):69–74. https://doi.org/10.2105/ajph.2021.306560.

McCauley, L., and R. Hayes. 2021. From Florence to fossil fuels: Nursing has always been about environmental health. *Nursing Outlook* 69(5):720–731. https://doi.org/10.1016/j.outlook.2021.06.007.

NASEM (National Academies of Sciences, Engineering, and Medicine). 2020. *Understanding, controlling, and preventing exposure to PFAS: Proceedings of a workshop—in brief.* Washington, DC: The National Academies Press. https://doi.org/10.17226/25856.

NASEM. 2021a. *Federal government human health PFAS research workshop: Proceedings of a workshop—in brief.* Washington, DC: The National Academies Press. https://doi.org/10.17226/26054.

NASEM. 2021b. *The future of nursing 2020–2030: Charting a path to achieve health equity.* Washington, DC: The National Academies Press. https://doi.org/10.17226/25982.

Nassif, J., A. M. Calafat, and K. M. Aldous. 2021. The U.S. national biomonitoring network—Enhancing capability and capacity to assess human chemical exposures. *International Journal of Hygiene and Environmental Health* 237:113828. https://doi.org/10.1016/j.ijheh.2021.113828.

NRC (National Research Council). 2006. *Human biomonitoring for environmental chemicals.* Washington, DC: The National Academies Press. https://doi.org/10.17226/11700.

Nutbeam, D. 2000. Health literacy as a public health goal: A challenge for contemporary health education and communication strategies into the 21st century. *Health Promotion International* 15(3):259–257. https://doi.org/10.1097/PHH.0000000000001315.

Thacker, S. B., D. F. Stroup, R. G. Parrish, and H. A. Anderson. 1996. Surveillance in environmental public health: Issues, systems, and sources. *American Journal of Public Health* 86(5):633–638. https://doi.org/10.2105/ajph.86.5.633.

Appendix A

Committee Member, Staff, and Community Liaison Biographies

COMMITTEE MEMBERS

Bruce N. Calonge, M.D., M.P.H. (NAM) (*Chair*), is an associate professor of family medicine at the Colorado School of Medicine, University of Colorado Denver, and an associate professor of epidemiology at the Colorado School of Public Health. He most recently served as the president and chief executive officer of The Colorado Trust, a private grant-making foundation dedicated to achieving health equity for all Coloradans. Nationally, Dr. Calonge chairs the Centers for Disease Control and Prevention's (CDC's) Task Force on Community Preventive Services; chairs the National Academies of Sciences, Engineering, and Medicine's Board on Population Health and Public Health Practice; and is a member of the National Academies' Roundtable on the Promotion of Health Equity. He is the past chair of the U.S. Preventive Services Task Force, the past chair of the CDC's Evaluation of Genomic Applications in Practice and Prevention Working Group, and a past member of the Secretary of Health and Human Services' Advisory Committee on Heritable Disorders in Newborns and Children. Prior to coming to The Colorado Trust, Dr. Calonge was the chief medical officer of the Colorado Department of Public Health and Environment. He received his B.A. in chemistry from Colorado College; his M.D. from the University of Colorado; and his M.P.H. from the University of Washington, where he also completed his preventive medicine residency. Additionally, Dr. Calonge completed his family medicine residency at the Oregon Health & Science University.

Laura Anderko, Ph.D., RN, is an environmental health nurse consultant and the co-director of the Federal Region 3 Pediatric Environmental Health Specialty Unit (PEHSU) of the Mid-Atlantic Center for Children's Health and the Environment at the M. Louise Fitzpatrick College of Nursing, Villanova University. Prior to joining Villanova University, she held the Robert and Kathleen Scanlon Chair in Values Based Health Care and was a professor at the School of Nursing & Health Studies at Georgetown University. Dr. Anderko is an educator, clinician, and scholar in public health, nursing, and environmental health. Over the past 8 years in her role with the Region 3 PEHSU, she has consulted with and educated local governments, communities, families, and health professionals about potential health impacts and risk-reduction strategies for PFAS. She has published extensively on PFAS, focusing on the clinicians' role in addressing health impacts and strategies for reducing risks. Dr. Anderko has served on several federal advisory committees, including the Children's Health Protection Advisory Committee for the U.S. Environmental Protection Agency, and is a founding member and steering committee member of the Alliance of Nurses for Healthy Environments. She currently serves as a special advisor for the National Environmental Health Partnership Council, coordinated by the American Public Health Association. Dr. Anderko was recognized by the Obama White House as a Champion of Change for her efforts in climate change and public health. She received her M.S. in nursing from Northern Illinois University and her Ph.D. in public health from the University of Illinois Chicago.

Erin M. Bell, Ph.D., is a professor in the Department of Epidemiology and Biostatistics and the Department of Environmental Health Sciences in the School of Public Health of the University at Albany. She joined the faculty at the University at Albany after completing her postdoctoral training at the National Cancer Institute's Occupational and Environmental Epidemiology Branch. Dr. Bell has taught epidemiology and environmental health to undergraduate and graduate students for almost 20 years, serving most recently as the faculty director of the undergraduate program in public health at the University at Albany. She is also an affiliate with the Center for Social and Demographic Analysis. Her research studies focus on environmental exposures as they relate to reproductive, immune, and cancer outcomes. Dr. Bell is currently the co–principal investigator of two cohort studies: the Upstate KIDS study, which follows more than 6,000 children to identify potential risk factors for developmental health effects, and the Health Study of New York State Communities Exposed to PFAS Contaminated Drinking Water, funded by the Agency for Toxic Substances and Disease Registry as one of seven sites participating in the Multi-site PFAS Health Study. She has served on several committees with the National Academies of Sciences, Engineering, and Medicine, including the Committee to Review the Health Effects in Vietnam Veterans of Exposure to Herbicides: Seventh, Eighth, and Tenth Biennial Updates. Dr. Bell received her M.S. in epidemiology and biostatistics from the University of Massachusetts Amherst and her Ph.D. in epidemiology from the University of North Carolina at Chapel Hill.

Dana Boyd Barr, Ph.D., is a research professor at the Emory University Rollins School of Public Health, where she co-directs the Laboratory for Exposure Assessment and Development in Environmental Research and leads research within the National Institutes of Health–funded Human Exposome Research Center: Understanding Lifetime Exposures (HERCULES). Her research focuses on using analytical chemistry techniques to assess exposure to toxicants, primarily related to maternal and child health. Prior to joining the faculty at Emory, Dr. Boyd Barr worked for 23 years at the Centers for Disease Control and Prevention (CDC), where she devoted much of her time to the development of methods for assessing human exposure to a variety of environmental toxicants. She is the past president of the International Society of Exposure Science (ISES), former editor-in-chief of the *Journal of Exposure Science and Environmental Epidemiology*, and the current deputy editor for *Environmental Health Perspectives*. Dr. Boyd Barr has received many awards for her research, including ISES's Daisy Award for Outstanding Investigator, two awards from the Secretary of Health and Human Services for exposure-health investigations involving diethylene glycol and methyl parathion poisoning, the 2004 Federal Scientific Employee of the Year award, the CDC's Mackel Award for outstanding collaboration among epidemiology and laboratory, and the U.S. Environmental Protection Agency's Silver Medal for outstanding service to environmental health. In 2014, 2015, and 2016, she was designated by Thomson Reuters as a Highly Cited Scientist in environment/ecology, representing the top 1 percent of scientists in her field. She recently became a member of the National Academies of Sciences, Engineering, and Medicine's Division on Earth and Life Studies' Board on Environmental Studies and Toxicology. Dr. Boyd Barr received her Ph.D. in analytical chemistry from Georgia State University.

Kevin C. Elliott, Ph.D., is a professor at Michigan State University (MSU) with joint appointments in the Lyman Briggs College, the Department of Fisheries and Wildlife, and the Department of Philosophy. His research lies at the intersection of the philosophy of science, research ethics, and environmental ethics, exploring the roles of ethical and social values in scientific research on environmental pollution, as well as ethical issues related to science communication, science policy, and team science. Dr. Elliott has authored a wide range of articles and book chapters and has published two books with Oxford University Press: *Is a Little Pollution Good for You?: Incorporating Societal Values in Environmental Research* (2011) and *A Tapestry of Values: An Introduction to Values in Science* (2017). He has also co-edited two books: *Exploring Inductive Risk: Case Studies of Values in Science* (with Ted Richards, Oxford University Press, 2017) and *Current Controversies in Values and Science* (with Dan Steel, Routledge Press, 2017). Dr. Elliott served as the program committee chair for the 2018 meeting of the Philosophy of Science Association and is the associate editor for the journal *Philosophy of Science*. He served as a member of the advisory council for the National Institute of Environmental Health Sciences from 2014 to 2018 and is currently a member of the MSU Center for PFAS Research. Dr. Elliott has served on organizing committees and workshops with the National Academies of Sciences, Engineering, and

Medicine, including the Committee on the Use of Emerging Science for Environmental Health Decisions. He received his Ph.D. and M.A. in history and philosophy of science from the University of Notre Dame.

Melissa Gonzales, Ph.D., M.S., is a professor and the chair in the Department of Environmental Health Sciences at the Tulane University School of Public Health and Tropical Medicine. She formerly was a professor in the Division of Epidemiology, Biostatistics, and Preventive Medicine at the University of New Mexico (UNM) School of Medicine. At UNM she was the co-director of UNM's Center for Native Environmental Health Equity Research; director of evaluation for the vice chancellor's Office of Diversity; and associate vice chancellor for research and evaluation in the Office of Diversity, Equity and Inclusion at the UNM Health Sciences Center. As an environmental health scientist with expertise in exposure assessment and epidemiology, Dr. Gonzales's community-engaged, translational research focuses on understanding the contribution of environmental exposure to health disparities, rural and minority health, and the translation of research for improved health equity through informed policy. Her research and leadership roles at UNM include serving in the Metal Exposure Toxicity Assessment on Tribal Lands in the Southwest Superfund Center, and the Transdisciplinary Research, Equity, and Engagement Center for Advancing Behavioral Health. Her work includes the Albuquerque Hispanic Moms Study, the Zuni Exposure Study, the Colorectal Disease Prevention Study, and the UNM– University of Texas at El Paso ARCH study of asthma and air pollution among children living on the U.S.-Mexico border. Dr. Gonzales has previously served on two committees with the National Academies of Sciences, Engineering, and Medicine: the Committee to Review the Health Effects in Vietnam Veterans of Exposure to Herbicides: Tenth Biennial Update, and the Committee to Evaluate the Potential Exposure to Agent Orange/TCDD Residue and Level of Risk of Adverse Health Effects for Aircrew of Post-Vietnam C-123 Aircraft. She received her M.S. in toxicology/industrial hygiene at the University of Arizona College of Pharmacy and her Ph.D. in environmental health at the University of California, Berkeley, School of Public Health.

Erin N. Haynes, Ph.D., is the Kurt W. Deuschle Professor in Preventive Medicine and Environmental Health and the chair of the Departments of Epidemiology and Preventive Medicine and Environmental Health within the University of Kentucky (UK) College of Public Health. She is the deputy director of the National Institutes of Health's (NIH's) National Institute of Environmental Health Sciences' (NIEHS's) UK Center for Appalachian Research in Environmental Sciences; associate director of the UK Center for Clinical and Translational Science; co-chair of the NIEHS Disaster Research Response Committee; and senior associate editor of the *Journal of Appalachian Health*. Her current research, funded by NIEHS, focuses on investigating the effects of neurotoxicant exposure, particularly metals, on neurodevelopment in rural adolescents, and development and validation of a real-time, lab-on-a-chip sensor for blood metals detection. Dr. Haynes works with community members to address environmental health issues and develops citizen science tools to enable environmental health research. She also serves on the NIH/NIEHS Environmental Health Sciences study section and as an academic counselor for the International Society of Exposure Science. Dr. Haynes received her M.S. in environmental genetics and molecular toxicology from the University of Cincinnati, and her Dr.P.H. in environmental health sciences from the University of Michigan.

Jane Hoppin, S.M., Sc.D., is a professor of biological sciences and the deputy director of the Center for Human Health and the Environment at North Carolina State University. Her research focuses on how epidemiologic evidence can inform mechanisms of disease from environmental exposures. Dr. Hoppin is currently leading a National Institutes of Health–funded study of PFAS exposures in the Cape Fear River basin, as well as a study evaluating the impact of pesticide exposure to residents in the banana-growing region of Costa Rica. She has authored more than 200 publications related to exposure assessment for chemicals in the environment and health outcomes. Dr. Hoppin previously spent almost 15 years at the National Institute of Environmental Health Sciences, working on the Agricultural Health Study. She received her S.M. and Sc.D. in environmental health and epidemiology from the Harvard T.H. Chan School of Public Health.

Tamarra James-Todd, Ph.D., M.P.H., is the Mark and Catherine Winkler Assistant Professor of Environmental Reproductive and Perinatal Epidemiology at the Harvard T.H. Chan School of Public Health, an associate epidemiologist at Brigham and Women's Hospital, and an instructor in medicine at Harvard Medical School. Her research expertise is in reproductive and developmental outcomes following exposure to environmental risk factors. Specifically, she evaluates the impact of pregnancy as a sensitive period of environmental chemical exposure for women's cardiometabolic health across the reproductive life course. Dr. James-Todd is the principal investigator (PI) of the Environmental Reproductive and Glucose Outcomes (ERGO) Study, which evaluates the impact of environmental chemicals and glucose metabolism during the perinatal period. She is also the PI of a study evaluating the effects of PFAS exposure on maternal cardiometabolic health across the reproductive life course within the Project Viva pregnancy/birth cohort. Dr. James-Todd has a particular focus on racial and ethnic disparities in environmental exposures and women's reproductive health outcomes. She has served on the U.S. Environmental Protection Agency's Scientific Advisory Board for the Chemical Assessment Advisory Committee and on the National Academies of Sciences, Engineering, and Medicine's Committee for Gulf War and Health: Volume 11: Generational Health Effects of Serving in the Gulf War. Dr. James-Todd received her M.P.H. in international health from the Boston University School of Public Health and her Ph.D. in epidemiology from Columbia University.

Alex R. Kemper, M.D., is the division chief of primary care pediatrics at Nationwide Children's Hospital and a professor of pediatrics at The Ohio State University College of Medicine. The focus of his research is the evaluation of preventive services delivered in the primary care setting. Dr. Kemper currently serves as the chair of the Evidence Review Workgroup for the Secretary of Health and Human Services' Advisory Committee on Heritable Disorders in Newborns and Children and as the deputy editor of *Pediatrics.* He is a former member of the U.S. Preventive Services Task Force and has served on the National Academies of Sciences, Engineering, and Medicine's Planning Committee on Examining Special Nutritional Requirements for Disease States. Dr. Kemper received his M.D. from Duke University, where he also completed his pediatric residency training, followed by combined fellowship training in health services research and medical informatics, and residency training in preventive medicine at the University of North Carolina.

Brian Linde, M.D., M.P.H., is an assistant professor in the Yale Occupational and Environmental Medicine Program at the Yale School of Medicine, where he oversees medical education and training and has designed curricula for physicians in training. He is also the chief of occupational health services at the Veterans Affairs Connecticut Healthcare System, where he oversees more than 3,000 employees, offering services including medical surveillance, work injury evaluation and management, infection prevention, and employee health and well-being. As a board-certified occupational and environmental physician, Dr. Linde also provides patient consultations to evaluate health effects of occupational and environmental exposures. He received his M.P.H. in occupational and environmental medicine from the Yale School of Public Health and his M.D. from Albert Einstein College of Medicine, Yeshiva University.

Marc-André Verner, Ph.D., M.S., is an associate professor in the Department of Occupational and Environmental Health, School of Public Health, Université de Montréal (Canada). He is also a member of the Centre for Public Health Research. Dr. Verner's expertise is in human health risk assessment, toxicology, biological modeling, and environmental epidemiology. His current research projects focus primarily on developmental exposure to environmental chemicals in the womb and postnatally through breastfeeding in the context of epidemiologic studies and risk assessment, including estimating gestational and lactational exposure to PFAS. Namely, he participated in the Minnesota Department of Health reevaluation of water guidance values for perfluorooctanoic acid and perfluorooctane sulfonic acid. In 2016, Dr. Verner received the Joan M. Daisey Outstanding Young Scientist Award for his studies using pharmacokinetic modeling to assess exposures during hypothesized windows of susceptibility to contaminants. He received his M.S. and Ph.D. from the Université du Québec à Montréal.

Veronica M. Vieira, D.Sc., M.S., is a professor in and the chair of the Department of Environmental and Occupational Health at the University of California, Irvine (UC Irvine), Program in Public Health. Her current work includes evaluating birth defects and infant morbidity in relation to air pollution using generalized additive models, and she has worked with the Boston University Superfund Research Program. She works extensively with reconstructing historic environmental exposures using geographic information systems and has experience with groundwater modeling and perfluorooctanoic acid. Dr. Vieira collaborated on the C-8 Health Project, contributing to several health and exposure studies, and is currently an investigator on the UC Irvine PFAS Health Study, part of a multisite study sponsored by the Centers for Disease Control and Prevention and the Agency for Toxic Substances and Disease Registry. She has served on the National Academies of Sciences, Engineering, and Medicine's Committee to Review Possible Toxic Effects from Past Environmental Containment at Fort Detrick. Dr. Vieira received her M.S. in environmental engineering from Stanford University and her D.Sc. in environmental health from the Boston University School of Public Health.

Xiaobin Wang (NAM), M.D., M.P.H., is the Zanvyl Krieger Professor in Children's Health, the director of the Center on the Early Life Origins of Disease at the Johns Hopkins Bloomberg School of Public Health, a professor of pediatrics at the Johns Hopkins University School of Medicine, and an American Board of Pediatrics–certified pediatrician at Johns Hopkins Hospital. Her work unites biomarkers, clinical medicine, epidemiology, and disease prevention. Dr. Wang served as the principal investigator in a number of molecular epidemiological studies funded by the National Institutes of Health and led multi-institution, multidisciplinary teams to investigate environmental, nutritional, genetic, and epigenetic factors during critical developmental windows (preconception, in utero, infancy, and childhood). Her team has conducted a series of studies in three unique study cohorts (Boston Birth Cohort, Chicago Family Cohort, and Chinese Twin Cohort). Dr. Wang has served on two committees with the National Academies of Sciences, Engineering, and Medicine: the Committee on Food Allergies: Global Burden, Causes, Treatment, Prevention, and Public Policy, and the Committee on Understanding Premature Birth and Assuring Healthy Outcomes. She is also a member of the National Academy of Medicine. Dr. Wang received her M.D. from Beijing Medical University, her M.P.H. from the School of Public Health and Tropical Medicine at Tulane University, and her Sc.D. from the Johns Hopkins School of Hygiene and Public Health. She completed a 3-year research fellowship in environmental epidemiology at the Harvard T.H. Chan School of Public Health and a residency in pediatrics at the Boston University Medical Center.

Chris J. Wiant, Ph.D., M.P.H., is the founding president and the former chief executive officer of the Caring for Colorado Foundation, serving from 2000 until his retirement in 2020. Previously the executive director of Colorado's Tri-County Health Department, Dr. Wiant is experienced in risk assessment and communication, exposure science, and environmental policy, as well as in collaborating with communities dealing with environmental contamination. He also served as the chief of the Environmental Chemistry Section of the Illinois Department of Public Health. Dr. Wiant was appointed to five terms on the Colorado Water Quality Control Commission and has served on the U.S. Environmental Protection Agency's National Drinking Water Advisory Council, the federal advisory committee for development of regulations for disinfection by-products in drinking water (Stages I and II), and he was the chair of the National Science Foundation's International Council of Public Health Consultants. Dr. Wiant is also the past president of both the National Environmental Health Association and the Colorado Public Health Association, and was a key participant in the negotiation of the cleanup (Record of Decision) at the Rocky Mountain Arsenal Superfund site. He has been appointed to and recognized for his service on a variety of other local, state, and national advisory committees and boards, and as a facilitator of solutions to community public health challenges. Dr. Wiant received his M.S. in health services administration,

M.P.H. from the University of Illinois, and Ph.D. in public policy with an emphasis in environmental policy from the University of Colorado.

STAFF

Elizabeth Barksdale Boyle, M.P.H., is a senior program officer in the National Academies of Sciences, Engineering, and Medicine's Health and Medicine Division's Board on Population Health and Public Health Practice. She previously served for several years as a program officer with the National Academies' Board on Environmental Studies and Toxicology. Formerly, she was an environmental health scientist at Westat, where she supported the U.S. Environmental Protection Agency, the National Institute of Child Health and Human Development, and the National Cancer Institute. Before her tenure at Westat, Ms. Boyle was a student epidemiologist at the Minnesota Department of Health and an industrial hygienist at a consulting firm in Cincinnati, Ohio. She is a fellow of the Bloomberg American Health Initiative at the Johns Hopkins Bloomberg School of Public Health, where she is pursuing a Dr.P.H. in environmental health. Ms. Boyle has an M.P.H. in environmental health from the University of Minnesota and a certificate in risk sciences and public policy from the Johns Hopkins Bloomberg School of Public Health, and is a certified industrial hygienist.

Marilee Shelton-Davenport, Ph.D., is a senior program officer at the National Academies of Sciences, Engineering, and Medicine, where she serves as the director of the innovative, cross-institutional Environmental Health Matters Initiative. Since 1999, Dr. Shelton-Davenport has worked to guide the country's best scientists and practitioners in providing authoritative advice to agencies and other organizations interested in biomedical research and regulatory issues. Specifically, she has experience developing and executing impactful activities that serve the National Institutes of Health, the U.S. Environmental Protection Agency, the U.S. Department of Defense, and the U.S. Food and Drug Administration. Her work with the National Academies' Board on Life Sciences and Board on Chemical Sciences and Technology has focused on emerging science related to the impact of environmental exposures on human health and defense. A leader of multidisciplinary and multisector teams, Dr. Shelton-Davenport is skilled at guiding experts in drilling down to the root of issues, generating strategies, and developing creative solutions. She has a B.S. in biochemistry from Clemson University and a Ph.D. in pharmacology from the University of North Carolina.

Kate Guyton, Ph.D., is a senior program officer with the Board on Environmental Studies and Toxicology within the Division on Earth and Life Studies at the National Academies of Sciences, Engineering, and Medicine. She has more than 20 years of experience applying her expert knowledge in mechanistic toxicology and carcinogenesis, and has been certified as a diplomate of the American Board of Toxicology since 1998. Her prior experience includes service as a senior toxicologist at the International Agency for Research on Cancer with the World Health Organization in Lyon, France (2014–2020). Previously, Dr. Guyton served as a toxicologist in the Office of Research and Development at the U.S. Environmental Protection Agency (2005–2014) and as the director of scientific affairs at CCS Associates (1998–2005), a woman-owned small business. She has authored more than 90 scientific articles in her areas of expertise with an overall h-index of 46. Dr. Guyton received her B.A. (cum laude) in biology from Johns Hopkins University, her Ph.D. in toxicological sciences from the Johns Hopkins Bloomberg School of Public Health, and her postdoctoral training at the National Institutes of Health.

Kaley Beins, M.P.H., is a program officer with the Board on Environmental Studies and Toxicology within the Division on Earth and Life Studies at the National Academies of Sciences, Engineering, and Medicine. Previously, she worked as a federal contractor for Abt Associates, where she supported the development of toxicological profiles for the Agency for Toxic Substances and Disease Registry, as well as green chemistry and product labeling programs for the U.S. Environmental Protection Agency. Ms. Beins focuses on the intersection of public health and toxicology, and has conducted research and led community engagement initiatives with nongovernmental organizations and local health departments, including service as a Fulbright Research Fellow. She is a board member for DC EcoWomen and the vice president of the Washington, DC, chapter of Sigma Xi Scientific Research Honor Society. Ms. Beins has a B.S. in environmental biology from Georgetown University and an M.P.H. in environmental health sciences from the University of Maryland.

Alexis Wojtowicz is an associate program officer who has supported the National Academies of Sciences, Engineering, and Medicine's Board on Population Health and Public Health Practice since 2016. She has also supported the Roundtable on Health Literacy, an action collaborative on preventing firearm violence, and consensus studies about Medicare payment and the public health effects of e-cigarettes. Before joining the National Academies, Ms. Wojtowicz conducted recruitment and intake at a culinary job training program in Washington, DC, and prior to that, coordinated an AmeriCorps VISTA program that placed summer associate members at anti-hunger nonprofit organizations across the United States. Ms. Wojtowicz has a B.A. in art history from the University of Maryland and is currently pursuing an M.P.H. at the Johns Hopkins Bloomberg School of Public Health, where she is a Bloomberg American Health Initiative fellow.

Alexandra McKay, M.A., is a senior program assistant in the Health and Medicine Division of the National Academies of Sciences, Engineering, and Medicine. During her graduate and undergraduate careers, she worked in museums and cultural heritage institutions, focusing on public education and assisting in database creation and exhibit curation. Ms. McKay also has experience working for the National Park Service as an interpretation ranger, concentrating on science education and public engagement. She received her M.A. in archaeological studies from Yale University.

COMMUNITY LIAISONS

Laurene Allen is a resident of Merrimack, New Hampshire, who co-founded the community group Merrimack Citizens for Clean Water after learning, in March 2016, that the water she and her family drank for decades was contaminated by PFAS. She has been an advocate for the needs of her PFAS-impacted community for the past 6 years, engaging local, state, and federal officials, and bringing together residents to work together on both a local and national level. After working to gather health data in her community, Ms. Allen co-authored an article in the journal *Environmental Health* titled "Making the Invisible Visible: Results of a Community-Led Health Survey following PFAS Contamination of Drinking Water in Merrimack, New Hampshire." Additionally, she co-founded and continues to serve in leadership of the National PFAS Contamination Coalition, composed of community members from across the nation working together to attain mutual federal needs.

Andrea Amico is the co-founder of the Testing for Pease community action group at the former Pease Air Force Base in Portsmouth, New Hampshire. In 2014, after learning her husband and two small children were impacted by drinking water highly contaminated by PFAS, she began advocating for her community and others, raising awareness and providing education, to achieve a common goal of reducing PFAS exposure to impacted communities. Ms. Amico testified at U.S. Senate hearings on PFAS in September 2018 and again in December 2021. She received the Citizen Excellence in Community Involvement Award from the U.S. Environmental Protection Agency Region 1 in November 2018. Ms. Amico attended the President's State of the Union Address as Senator Jeanne Shaheen's guest in February 2019 to raise awareness of PFAS contamination. She is the co-founder of the National PFAS Contamination Coalition, formed in June 2017 to bring PFAS advocates from around the country to organize for change at the federal level. Additionally, Ms. Amico helped organize the first two National PFAS Conferences held at Northeastern University in Boston, Massachusetts, in June 2017 and June 2019. She has given a TEDx talk on PFAS and presented at the Centers for Disease Control and Prevention's first Public Health Grand Rounds on PFAS in Atlanta, Georgia. Ms. Amico serves on the

Pease Community Assistance Panel with the Agency for Toxic Substances and Disease Registry and is community co-chair of the Pease Restoration Advisory Board with the U.S. Air Force. She is also a community co-chair on the City of Portsmouth Safe Water Advisory Group. Ms. Amico has a master's degree in occupational therapy.

Stel Bailey is the chief executive director of Fight For Zero and the co-facilitator of the National PFAS Contamination Coalition, is certified in wildlife monitoring, and is a content administrator behind several online publications. Her life took a dramatic turn in 2013 when she and her father, brother, and uncle, as well as the family dog, were diagnosed with cancer. Her family's case was so unique that a genetic counselor invited them to do genetic testing, which showed no mutation genes, indicating they did not have an increased risk of developing the disease. Determined to find answers, Ms. Bailey began to crowdsource information about other cancer cases in her hometown on Florida's space coast. As she spoke out about her family's case, she connected with many others affected by diseases in unusual ways. As a cancer survivor and military dependent, Ms. Bailey is passionate about helping families and veterans prevent disease by sharing her insights about harmful toxins, including PFAS, and their health effects. Her broad knowledge in coordinating community engagement projects has resulted in various opportunities to share environmental health information, such as contributing to local publications and speaking at events. She has led additional efforts focused on harmful contaminants, such as assisting in getting a state health assessment, collecting water samples for countywide projects, and developing partnerships that will help improve understanding of the burdens of PFAS contamination and related chronic diseases.

Kyla Bennett, Ph.D., is the director of science at Public Employees for Environmental Responsibility (PEER), a national nonprofit service organization for environmental and public health professionals, land managers, scientists, enforcement officers, and other civil servants dedicated to upholding environmental laws and values. Prior to working at PEER, Dr. Bennett was an environmental scientist and attorney for the U.S. Environmental Protection Agency (EPA) for 10 years. Her work at PEER includes the investigation of sources of PFAS contamination in artificial turf and pesticides, and of impacts on endangered species, such as the North Atlantic right whale, manatee, and Louisiana black bear. Dr. Bennett is currently working on exposing the effects of aerial pesticide spraying and investigating faulty risk assessments of pesticides and other chemicals at EPA. She has a Ph.D. in ecology and evolutionary biology from the University of Connecticut, and a law degree with a certificate in environmental and natural resources law from Lewis and Clark Law School.

Karen Blondel is an environmental organizer/advocate for the Fifth Avenue Committee. She has conducted field trainings and has a background in health and engineering. A native of Brooklyn, Ms. Blondel has led the Gowanus Neighborhood Coalition for Justice activist group to demand the city include the neighborhood's New York City Housing Authority (NYCHA) houses in its development plan. During her tenure with AmeriCorps, she coordinated a women's history event at the Red Hook Library; started a volunteer program to escort seniors on their errands; and launched a summer Safe Streets program, facilitating a shutdown of local roads for recreational use for local residents. In 2018, Ms. Blondel began a "Know Your Rights" workshop for Red Hook public housing residents to inform them on bylaws, requirements, and deadlines. In her current role as an environmental organizer, she educates public housing residents on environmental burdens within and near NYCHA housing developments that are often overlooked by the city.

Phil Brown, Ph.D., is a university distinguished professor of sociology and health science at Northeastern University, where he directs the Social Science Environmental Health Research Institute and co-directs its PFAS Lab, which has grants from the National Science Foundation to study social policy and activism concerning PFAS, and from the National Institute of Environmental Health Sciences (NIEHS) to study children's immune responses and community's responses to PFAS contamination, and to develop a nationwide report-back and information exchange. He directs an NIEHS T-32 training program, "Transdisciplinary Training at the Intersection of Environmental Health Science and Social Science," heads the Community Outreach and Translation Core of Northeastern's Children's Environmental Health Center (Center for Research on Early Childhood Exposure and Development in Puerto Rico, orCRECE), and is the research translation director and the community engagement core co-director of Northeastern's Superfund Research Program (Puerto Rico Testsite to Explore Contamination Threats, or PROTECT). Dr. Brown is a past member of the NIEHS Council. His books include *No Safe Place: Toxic Waste, Leukemia, and Community Action*; *Toxic Exposures: Contested Illnesses and the Environmental Health Movement*; and *Contested Illnesses: Citizens, Science and Health Social Movements*. Dr. Brown earned his Ph.D. in sociology from Brandeis University.

Alberto J. Caban-Martinez, D.O., Ph.D., M.P.H., is a public health board-certified faculty member in the Department of Public Health Sciences at the University of Miami with an educational background in medicine, epidemiology, and public health and more than 10 years of domestic and international practice and research experience. He is the assistant provost for research standards and an associate professor of public health sciences and physical medicine and rehabilitation at the University of Miami Miller School of Medicine; the deputy director of the Firefighter Cancer Initiative at the Sylvester Comprehensive Cancer Center; and the co-director and principal investigator of the national Federal Emergency Management Agency–funded Fire Fighter Cancer Cohort Study. He is a former fellow of the National Academies of Sciences, Engineering, and Medicine's Gulf Research Program, and in 2014, he was appointed to the Institute of Medicine's Committee on Gulf War and Health for 2 years to provide scientific expertise on occupational exposures and work-related health conditions. He also served on the National Occupational Research Agenda committee of the National Institute for Occupational Safety and Health (NIOSH), helping to set the national research agenda on worker health and safety. He served 4 years as a standing member of the NIOSH study section and 1 year as the chair. He has published more than 120 peer-reviewed publications and shared more than 215 scientific presentations on a wide range of occupational health and safety topics.

Cheryl Cail is a small business owner in Myrtle Beach, South Carolina, the vice chief of the Waccamaw Indian People, and the chair for SC Idle No More, a committee within the South Carolina Indian Affairs Commission. In December 2018, after her 20-year-old son was diagnosed with testicular cancer, Ms. Cail became aware of PFAS contamination in the groundwater at the former Myrtle Beach Air Force Base, and later learned of the extent of contamination from the use of aqueous film-forming foam at other U.S. Department of Defense sites throughout the state. She joined the National PFAS Contamination Coalition in 2019, and is currently working to raise awareness of the impact of PFAS contamination of both the environment and people in South Carolina. Ms. Cail received associate degrees from Horry-Georgetown Technical College in both legal studies and human services and received the Phi Theta Kappa Society's All-State Academic Team award.

Courtney Carignan, Ph.D., is an exposure scientist and an environmental epidemiologist at Michigan State University (MSU). Her research helps protect reproductive and child health by investigating exposure to contaminants in food and water, and consumer and personal care products. She currently helps lead biomonitoring studies (PFAS UNITEDD and PFAS REACH) funded by the U.S. Environmental Protection Agency and the National Institutes of Health, investigating PFAS exposure and immunotoxicity for communities impacted by drinking water contamination. Dr. Carignan is on the organizing committee for the National PFAS Conference and helps lead MSU's Center for PFAS Research. She received her Ph.D. in environmental health from the Boston University School of Public Health and completed postdoctoral training at Dartmouth College and the Harvard T.H. Chan School of Public Health.

Tracy Carluccio is the deputy director of the Delaware Riverkeeper Network (DRN), where she has worked as an environmental advocate since 1989. DRN is a nonprofit membership organization working throughout the entire length and breadth of the Delaware River Watershed to defend its outstanding values and restore them where needed. Ms. Carluccio works for the Watershed's protection in New Jersey, Pennsylvania, New York, and Delaware, addressing issues that include water quality; healthy habitats and communities; environmental regulation and policy; clean, efficient, and renewable energy; and biodiverse ecosystems. She has also worked on PFAS issues since 2005 and has commented and written extensively on community and regulatory matters related to these compounds in New Jersey, Pennsylvania, and Delaware. Ms. Carluccio serves on the New Jersey Department of Health Centers for Disease Control and Prevention Biomonitoring Advisory Committee on PFAS.

Jamie DeWitt, Ph.D., is an associate professor in the Department of Pharmacology & Toxicology of the Brody School of Medicine at East Carolina University. Her research focuses on the effects of environmental contaminants on the adult and developing immune systems, as well as on interactions between the immune and nervous systems. She is the principal investigator, co–principal investigator, and co-investigator on several funded studies concerning the immune effects of PFAS, especially those considered novel or understudied. Dr. DeWitt received her Ph.D.s in environmental science and neural science from Indiana University Bloomington and completed postdoctoral training in immunotoxicology at the U.S. Environmental Protection Agency under a cooperative training agreement with the University of North Carolina.

Emily Donovan is the co-founder of Clean Cape Fear. She is a tireless advocate for clean water, spending her free time educating the public on the dangers of PFAS and other toxins in drinking water and the environment. Ms. Donovan has testified before Congress twice regarding the influence of the DuPont/Chemours facility in Fayetteville, North Carolina, which is contaminating the downstream drinking water supply for 250,000 residents, including giving testimony during the first-ever congressional hearing on PFAS contamination. She participated in a *Washington Post Live* panel discussion with actor Mark Ruffalo and lawyer Rob Bilott. Most recently, Ms. Donovan worked to secure reverse osmosis filling stations in 49 public schools impacted by PFAS contamination in Brunswick and New Hanover counties. She regularly travels the country sharing her personal impact story, as well as those of her friends and neighbors. Ms. Donovan frequents Washington, DC, and Raleigh, North Carolina, pressuring lawmakers and regulators for swifter responses to the growing PFAS public health crisis.

Alan Ducatman, M.D., M.S., is a Mayo Clinic–trained internist and occupational physician, and a professor emeritus in the West Virginia University School of Public Health and School of Medicine. His clinical career has focused on medical screening related to environmental exposures, with research addressing environmental disease and disease prevention, including health communications to affected populations. He has designed community studies and actively published on population aspects of exposure to PFAS, and many of his more than 30 publications concerning PFAS are highly cited. As a clinical consultant to industry, labor, nonprofit organizations, and government organizations, Dr. Ducatman has also worked with a variety of community groups to provide information concerning PFAS clinical science. In addition, he has an active research program in quality assurance concerning clinician laboratory orders and interpretation, which informs his community service. His other public service has included serving as the chair of the external science advising committee to the National Center for Environmental Health and the Agency for Toxic Substances and Disease Registry of the Centers for Disease Control and Prevention, and as the chair of the Residency Review Committee in Preventive Medicine for the Accreditation Council for Graduate Medical Education.

Patrick Elder is the director of Military Poisons, an organization that works to draw attention to the role of the military in the environmental contamination through the use of PFAS in firefighting foams and other applications. Two hundred of his articles on PFAS have been published in several dozen publications, including *Global Geneva, Truthout, Consortium News, Common Dreams*, and *LA Progressive*. Mr. Elder seeks to draw public attention to food, especially seafood, as the primary pathway of human exposure to PFAS. He receives support from the Women's International League for Peace and Freedom (WILPF); WILPF US Earth Democracy; the Patagonia Foundation; the Peace Development Fund; and the Center for Health, Environment & Justice. Mr. Elder is presently working on campaigns in several New England states and Maryland, as well as in Germany and Japan, to influence policy makers to take steps to protect public health from the scourge of PFAS. He has toured the United States and Europe several times, speaking on contamination at military bases. Mr. Elder holds an M.A. in government from the University of Maryland and a B.A. in political science/education from St. Mary's College of Maryland.

Teresa Gerade is an advisory member of a Vermont-Canadian volunteer organization in the Northeast Kingdom known as DUMP (Don't Undermine Memphremagog's Purity), which came into being in June 2018 to appeal the application for a 51-acre expansion of Vermont's only permitted, privately owned landfill. The group's goal is to restore and protect the waters and watershed of Lake Memphremagog, an international lake located in the northeast corner of Vermont. The appeal was unsuccessful because of a lack of funding for legal representation. However, the group participated in a mediation session with the owners of the landfill and achieved a 4-year moratorium on the discharge of landfill leachate (the garbage juice that is created in landfills) into the Lake Memphremagog watershed. DUMP is now actively pursuing a state legislative designation of "Lake in Crisis" to garner funding and additional support for restoring and protecting the waters of Lake Memphremagog.

Hope Grosse is the co-founder of Buxmont Coalition for Safer Water and serves on the National PFAS Contamination Coalition, working to establish enforceable federal and state drinking water standards for this chemical class. She acts as the community liaison for outreach and leads social media efforts, updating residents about PFAS water contamination and related issues. Ms. Grosse has been featured on WHYY's *Radio Times, Live with Marty Moss-Coane,* discussing PFAS chemicals health and regulations. She also serves as a community representative on the Warminster Naval Airbase Technical Review Committee, which is charged with oversight of the environmental cleanup under the Comprehensive Environmental Response, Compensation, and Liability Act. Ms. Grosse grew up directly across the street from Warminster Naval Base and worked on the base after high school graduation. Her father died of cancer at age 52, and she was diagnosed with stage 4 cancer at age 25, which quickly spread to her lymph system. Other unexplained tumors have also been removed from her body over the years. After learning that the Warminster Naval Base was named a Superfund National Priority site, she began to get involved in environmental advocacy and is committed to bringing awareness of PFAS and other contaminants to others in her community.

Loreen Hackett has been advocating for families in Hoosick Falls, New York, since the discovery of severe contamination more than 6 years ago led to issuance of the first National Priorities List Federal Superfund Site declarations for PFOA. Through biomonitoring, her family exhibited some of the highest levels of perfluorooctanoic acid tested in the area. In June 2016, she created #PfoaProjectNY, which has gone worldwide, and continues to share information on PFAS. Ms. Hackett is the co-chair of the Hoosick Falls Community Action Working Group established for the superfund sites, serves on the Community Action Partnership Committee for the Centers for Disease Control and Prevention–awarded site study, and is on the Committee of the National PFAS Contamination Coalition. She has attended and submitted testimony in two congressional hearings in Washington, DC, and continues working with elected officials on bills regulating PFAS, as well as with various environmental organizations. She survived cancer that may have been linked to PFAS exposure.

Ayesha Khan worked at Vitamin Water in New York City for a decade before moving to the island of Nantucket, Massachusetts, where she worked for a local surgeon. Currently, she is a stay-at-home mom for her two young children. Her husband Nate, who works as a firefighter, was recently diagnosed with testicular cancer and is in remission. As Ms. Khan researched about her husband's cancer, she learned about PFAS and their connections to aqueous film-forming foam and firefighter gear. She co-founded Nantucket PFAS Action Group as an educational resource about the hazards of PFAS and as a space for local community members to openly discuss concerns, thoughts, and experiences. Ms. Khan holds a B.A. in applied mathematics and statistics from Boston University.

Rainer Lohmann, Ph.D., is a professor of oceanography and the director of the University of Rhode Island Superfund Research Center, where he and his group conduct research into the sources, transport, and bioaccumulation of anthropogenic pollutants, often relying on the use of passive samplers. In addition to PFAS, his research covers dioxins, polychlorinated biphenyls, legacy pesticides, and emerging contaminants. Dr. Lohmann is a scientific counselor for the U.S. Environmental Protection Agency's Sustainable and Healthy Communities Subcommittee and a speaker at the National Academies of Sciences, Engineering, and Medicine's Environmental Health Matters Initiative on PFAS. He was selected as an Alexander von Humboldt fellow, and as a Fulbright fellow for the 2021–2022 Arctic Initiative. Dr. Lohmann serves as the editor for *Environmental Toxicology and Chemistry* and is on the editorial boards of publications including *Environmental Science & Technology* and *Environmental Science & Technology Letters*.

Samraa Luqman works in social services with the federal government. A native of Dearborn, Michigan, she has experience working in public schools and serving in community hospitals and cultural centers, including the Arab Community Center for Economic and Social Services and the American Yemeni Women's Association, where she is the secretary and a board member. Ms. Luqman has observed a multitude of community members becoming ill with cancers, asthma, and other illnesses attributable to pollution, and has become involved in pursuing environmental justice, speaking at community meetings and town halls. In 2019, she joined the Environmental Health Research to Action Steering Committee, which educates high school students about the impacts of pollution, and the PFAS Alliance shortly thereafter. By 2020, Ms. Luqman also joined the Clean Air Council and presented to multiple audiences on the effects and existence of PFAS in both air and water. She has also worked with the Concerned Residents for South Dearborn, the University of Michigan's Environmental Interpretive Center, and Friends of the Rouge to raise awareness on the environmental issues of her community. Her most recent endeavors include providing input on the enactment and enforcement of a fugitive air dust ordinance; fighting for cumulative health impact studies; and partnering with organizations, universities, and other entities to add greenspace, trees, and rain gardens to the area. Ms. Luqman completed her bachelor's degree in psychology at the University of Michigan.

Beth Markesino is the founder of North Carolina Stop GenX in Our Water, advocating for regulations on GenX, PFAS, and perfluorooctane sulfonic acid in North Carolina. She is a participant in the first ever GenX human health study. Ms. Markesino has multiple endocrine issues associated with PFAS, including a thyroid tumor, an adrenal tumor, and placenta problems during her pregnancy that resulted in the death of her son. She actively advocates in her community for clean water and regulatory action addressing endocrine disruptor chemicals. With a grant from Hydroviv, she provided 120 filters for low-income residents in the Lower Cape Fear region. Additionally, she has lobbied against DuPont scientist Michael Dourson's appointment to the U.S. Environmental Protection Agency.

Aaron Maruzzo worked as a water lab analyst for a public utility company in the Northern Mariana Islands, serving the islands of Saipan, Tinian, and Rota. He is currently working toward an M.P.H. in environmental health science at the University of California (UC), Berkeley. In 2020, Mr. Maruzzo co-authored a report for the Berkeley Center for Green Chemistry to identify nontoxic alternatives to short-chain PFAS in molded fiber food packaging. His current interests are in community exposure to PFAS in drinking water, green technology, environmental epidemiology, and environmental justice, where he aims to leverage academic resources at UC Berkeley to benefit science and policy actions toward a PFAS-free future in the Western Pacific U.S. territories. Mr. Maruzzo received his B.A. in biology and comparative literature, with a concentration in public health, from Williams College, where he developed a strong interdisciplinary approach to diverse public health issues.

Tobyn McNaughton is a resident of Belmont, Michigan, who has been affected by PFAS. She is a former elementary teacher and now stays home with her two sons. Ms. McNaughton has become an activist and voice for people affected by PFAS.

Kristen Mello, M.S., is the director of Westfield Residents Advocating for Themselves (WRAFT), a community group formed in response to PFAS contamination of the drinking water in Westfield, Massachusetts, caused by the use of aqueous film-forming foam at the Barnes Air National Guard Base. With WRAFT, Ms. Mello has led efforts to get PFAS blood testing for Westfield residents and has served on the Massachusetts Department of Environmental Protection's PFAS MCL (Maximum Contaminant Level) Stakeholder Group. Her advocacy work on PFAS, in large part, led to her being elected a Westfield City Councilor At-Large in 2019 and 2021. Ms. Mello has a bachelor's degree in chemistry and a master's degree in analytical chemistry, specializing in chemometrics.

Pamela K. Miller is founder and Executive Director of Alaska Community Action on Toxics (ACAT). ACAT is an environmental health and justice research, policy, and advocacy organization. She brings more than 30 years of research, education, and advocacy experience to her present work. She serves as Principal Investigator for community-based participatory research projects funded by the National Institute of Environmental Health Sciences. Ms. Miller received a Meritorious Service Award from the University of Alaska and Alaska Conservation Foundation's Olaus Murie Award in recognition of her "long-term outstanding professional contributions to the conservation movement in Alaska." She holds a bachelor's degree in biology from Wittenberg University and a master's degree in environmental science from Miami University. Prior to coming to Alaska, she served as Ocean Issues Technical Coordinator for the Washington Department of Ecology and Director of a marine science education center at Nisqually Reach in southern Puget Sound. She received the Governor's Award for Environmental Excellence in Washington State.

Elizabeth Neary, M.D., is a pediatrician, as well as an educator and scholar in public and environmental health. Having practiced general pediatrics for 15 years, she has devoted herself to educating students, residents, legislators, and the general public about environmental health issues. Dr. Neary is the co-president of the Wisconsin Environmental Health Network, part of Physicians for Social Responsibility–Wisconsin. She is an adjunct assistant clinical professor of pediatrics at the University of Wisconsin–Madison School of Medicine and Public Health and the Wisconsin representative to the Region 3 Pediatric Environmental Health Specialty Unit. For the past 5 years, she has taught environmental health to pediatric residents. Dr. Neary has testified on the health effects of PFAS to the Wisconsin state legislature and to the Madison Water Utility Board.

Laura Olah is the executive director and the co-founder of Citizens for Safe Water Around Badger, a community-based group that organized in 1990 when the community learned that a plume of cancer-causing chemicals had poisoned private drinking water wells near Wisconsin's Badger Army Ammunition Plant. A member of the Sokaogon Ojibwa Community Mole Lake Band, Ms. Olah defines her role as part of a community "undefined by geography" that shares a deep concern for the human environment and the need to empower affected tribes and stakeholders in the decision-making process, ensuring powerful, long-term solutions to military and industrial toxins in our rural communities. She views the past 30 years of accomplishments not as hers alone but as shared with the many individuals who have and who continue to be at her side in the work for environmental justice.

Jacob Park, Ph.D., is an associate professor in the Castleton University College of Business. He specializes in innovation, entrepreneurship, and community health issues, with special focus and expertise in emerging and developing economies in Africa, Asia Pacific, and the Caribbean islands. Dr. Park is also a visiting professor at the University of Johannesburg (South Africa); a former Kevin Ruble fellow in conscious capitalism at the Rutgers University School of Management and Labor Relations; the Edmond J. Safra Network Fellow at Harvard University; and a research fellow in the Oxford University Smith School of Enterprise and the Environment. He has served as the coordinating lead author of the report for the United Nations (UN) Environment Programme's Global Environment Outlook (GEO-6); as lead author for the UN Millennium Ecosystem Assessment initiative; and as an expert reviewer for a number of Intergovernmental Panel on Climate Change publications, including the Sixth Assessment Report. Dr. Park is a former member of the Renewable Energy and Adaptation to Climate Technologies Investment Committee of the Africa Enterprise Challenge Fund, and the chair of the U.S. Sustainable and Responsible Investment Forum's International Working Group Steering Committee. He received his master's degree in city planning from the Massachusetts Institute of Technology and his Ph.D. in social sciences/public administration from the Erasmus University of Rotterdam.

Sue Phelan[1] was the co-founder and the volunteer director of GreenCAPE, a nonprofit community advocacy organization formed to educate Cape Cod residents about the hazards of pesticides and other harmful chemicals (including PFAS) being used above the Cape Cod's vulnerable sole source aquifer. The focus of GreenCAPE's work is the community's exposure to PFAS via the public drinking water system and other as-yet-unidentified sources. Its major concerns include that PFAS-exposed Hyannis residents have not been provided with blood testing and the lack of health-protective PFAS regulations nationwide. Ms. Phelan was a board member of the Alliance for a Healthy Tomorrow/Boston and Clean Water Action/Mass, as well as a member of the Cape Chapter of the Sierra Club, the Sierra Club Toxics Team, and the National PFAS Coalition. She served as a community project partner and a member of the Cape Cod Advisory Committee for the Sources, Transport, Exposure, and Effects of PFAS Research Center and was one of only two community representatives invited to be on the Massachusetts Department of Environmental Protection's PFAS MCL (Maximum Contaminant Level) Stakeholder Group. After working as a medical microbiologist/virologist, Ms. Phelan returned to the University of Connecticut College of Agriculture, graduated from the environmental horticulture program, and worked in research and development on insect pheromones and light-mediated changes in plant growth and development.

Andrea Rich is an active participant in fighting for clean water in Wisconsin. She works with S.O.H2O (Save Our Water) to bring the impacts of PFAS to light, and share relevant information with the community in Marinette. Marinette is the third-largest PFAS contamination site in the nation. Ms. Rich has been researching PFAS and its health impacts, as well as filtration methods, for several years. She has been an active participant in public hearings and Department of Natural Resources presentations, and has provided testimony to Senate and Assembly committees on multiple bills proposed to regulate PFAS. Ms. Rich was also instrumental in bringing into the spotlight the contamination of farm fields and dozens of contaminated wells in rural Marinette County, some of which may otherwise still be unidentified. Her current focus is trying to get testing for PFAS in the agricultural goods coming from the contaminated farm fields, and blood testing for the community to determine exposure levels, and possibly identify additional sources of exposure. She is an honors graduate of Lakeland College, with a double major in business administration and marketing.

---

[1] Deceased January 2022.

Dana Sargent is the executive director of the environmental nonprofit Cape Fear River Watch (CFRW), based in Wilmington, North Carolina, where she works to protect and improve the water quality of the Cape Fear River basin for all people through education, advocacy, and action. Her work on PFAS began in 2017 when her community learned that DuPont and then Chemours had polluted their drinking water supply for more than 40 years. Her organization sued the polluter and the state regulatory body, which culminated in a Consent Order, parts of which she, along with her CFRW partners and their legal counsel, maintain oversight, ensuring the polluter and the regulatory agency are upholding the requirements of the Order. In addition, CFRW partners with academia on several PFAS research grants including the GenX exposure study, along with several ecological studies and outreach grants. She works among a coalition of advocates in North Carolina pushing for strong PFAS legislation and regulatory action and her organization is one of six that has sued the U.S. Environmental Protection Agency after it refused to hold a PFAS polluter accountable for funding health and toxicity studies. Ms. Sargent has published two websites and numerous articles and op-eds on PFAS, including an op-ed in which she questions whether PFAS exposure caused the brain cancer that took her brother's life in December 2019. He had been exposed for decades in the line of duty as a Chicago firefighter and a former U.S. Marine. She holds an M.S. in environmental sciences and policy from Johns Hopkins University and a B.A. in journalism with a music minor from San Diego State University.

Laurel Schaider, Ph.D., M.S., is a senior scientist in environmental chemistry and engineering at the Silent Spring Institute, where she leads water quality research on PFAS and other contaminants of emerging concern. Her research focuses on characterizing PFAS exposures from drinking water, diet, and consumer products; understanding health effects associated with PFAS; investigating socioeconomic disparities in exposures to drinking water contaminants; and working with communities to develop research studies and resources to address their concerns. Dr. Schaider is the principal investigator for the PFAS-REACH (Research, Education, and Action for Community Health) study, a researcher–community partnership that is evaluating PFAS exposures and immune system effects in children in communities with PFAS water contamination, and developing an online resource center for PFAS-affected communities. She is also the principal investigator of the Massachusetts PFAS and Your Health Study, one of seven projects within the Centers for Disease Control and Prevention PFAS Multi-site Health Study. Dr. Schaider co-leads the Community Engagement Core for the STEEP (Sources, Transport, Exposure, and Effects of PFAS) Superfund Research Program at the University of Rhode Island, including a study to evaluate PFAS levels in private wells on Cape Cod and identify contamination sources. Before joining the Silent Spring Institute, she was a research associate at the Harvard T.H. Chan School of Public Health, where she currently holds an appointment as a visiting scientist. Dr. Schaider served as a member of the planning committee for the 2020 National Academies of Sciences, Engineering, and Medicine's Workshop on Federal Government Human Health PFAS Research. She earned her master's degree and Ph.D. in environmental engineering at the University of California, Berkeley, and a bachelor's degree in environmental engineering science from the Massachusetts Institute of Technology.

Linda Shosie, a proud Latina woman, is a mother, grandmother, wife, and the founder and organizer of the grassroots community organization Mothers for Safe Air & Safe Water Force in Tucson, Arizona. In her fight for the human right to clean and safe water she has dedicated her entire life to protecting public health. Ms. Shosie has become one of the leading voices nationally regarding PFAS exposures in Latino barrios. In 2007, she lost her child to a rare disease she believes was caused by exposures to numerous toxic chemicals in the drinking water in the Tucson South-Side barrio. More recently Ms. Shosie has begun initiating a community-led PFAS health study in her community to assess for PFAS in human blood. She is also part of the National PFAS Contamination Coalition.

Lenny Siegel is the executive director of the Center for Public Environmental Oversight, where he has been serving since 1994. He is one of the American environmental movement's leading experts on both military facility contamination and the vapor intrusion pathway. Mr. Siegel runs two Internet newsgroups for his organization: the Military Environmental Forum and the Brownfields Internet Forum, as well as activist discussion lists on trichloroethylene and PFAS. He has served on several Interstate Technology and Regulatory Council work teams on environmental remediation, as well as a dozen National Academies of Sciences, Engineering, and Medicine committees addressing military environmental issues. In July 2011, Mr. Siegel was awarded the U.S. Environmental Protection Agency's Superfund Citizen of the Year award. In 2018, he served as mayor of Mountain View, California.

Mike Watters is a community organizer in the Fayetteville, North Carolina, area. After learning his well was contaminated with more than 16 PFAS, Mr. Watters set up a community outreach group. He created a Facebook group with more than 2,700 members and actively provides information to the community, ensuring wells are tested in a rapidly growing area around the Fayetteville Works Facility in North Carolina. Mr. Watters provides input at all North Carolina Science Advisory committee meetings, leads teams in assisting in research with the NC PFAS Testing Network, and works actively with North Carolina State University research teams. Additionally, he engages with state and federal authorities to ensure that violations and spills are documented and action is taken; two notices of violation are directly tied to his actions. Mr. Watters also works with the North Carolina Department of Environmental Quality, having participated in the Granular Activated Carbon Pilot Test. He currently has a state PFAS air monitoring station on his property and assists the state in gathering information. While in the military, Mr. Watters was trained and certified as a U.S. Department of Defense hazardous materials handler. He has college degrees in information technology and firearms technology.

La'Meshia Whittington is a professor in the Division of Sociology at Meredith College. She is also the deputy director for Advance Carolina and the campaigns director for the North Carolina Black Alliance. She is the co-convener of the North Carolina Black and Brown Policy Network, a former national democracy campaigner for Friends of the Earth, the chair of the FRENC Fund Administration, a founding member of Democracy Green, a member of the Burke Women's Fund in Western North Carolina, and a former North Carolina spokesperson on fair courts for The Leadership Conference on Civil and Human Rights. A leader in intersectional democracy and environmental justice, Ms. Whittington was the co-author of North Carolina Senate Bill 673, prioritizing environmentally contaminated communities of color in voting rights, and the co-author in several Pro-Democracy North Star legislation bills. She is a member of the NC PFAS Testing Network, anchoring legislation on aqueous film-forming foam, and she is a convener of the Black Firefighters Fighting PFAS Collective. Ms. Whittington has created and co-convened national, regional, and statewide tours and workshops on environmental justice, focusing on chemical contaminants and dirty corporations. She works continually with the Brody School of Medicine at East Carolina University, the U.S. Department of Health and Human Services, and other government agencies to ground community needs and strategies in alleviating health disparities in Black and Brown communities. Ms. Whittington led the development of a statewide map to highlight the intersection of environmental justice contamination zones and the frequency in which they are located within gerrymandered Black majority voting districts. She is a petitioner in two active petitions to the U.S. Environmental Protection Agency. Ms. Whittington is an Afro-Indigenous woman from North Carolina, hailing from a former environmental justice settlement: The Kingdom of the Happy Land. She received her education at Western Piedmont Community College and Meredith College.

Alan Woolf, M.D., is a professor of pediatrics at Harvard Medical School, a board-certified pediatrician and medical toxicologist, and a senior attending physician at Boston Children's Hospital (BCH). He is the director of the Pediatric Environmental Health Center at BCH, directing its fellowship training program in pediatric environmental health and the Region 1 New England Pediatric Environmental Health Specialty Unit. Dr. Woolf is a member of the Executive Committee of the Council on Environmental Health of the

American Academy of Pediatrics and a past president of both the American Association of Poison Control Centers and the American Academy of Clinical Toxicology. He is the editor and author of two books: *The Children's Hospital Guide to Your Child's Health and Development* (Perseus Publishers) and *History of Modern Clinical Toxicology* (Academic Press) and has authored or co-authored more than 300 scientific publications.

Cathy Wusterbarth is a registered dietitian nutritionist, and the co-leader and founder of NOW (Need Our Water) in her hometown, Oscoda, Michigan. NOW was created in 2017 to give Oscoda-AuSable and surrounding communities a voice. Its mission is to be a reliable resource and catalyst for education and communication while advocating for long-term health and environmental welfare plans on behalf of those affected by water contamination from the former Wurtsmith Air Force Base in Oscoda, Michigan. Ms. Wusterbarth and other NOW members have consistently demanded that the State of Michigan and the Air Force address the PFAS contamination that affects her community, as well as more than 700 military installations. She was invited to the 2019 State of the Union by Congressman Dan Kildee to represent communities affected by PFAS contamination and support the newly created Congressional PFAS Task Force. In addition to testing and monitoring the health of Oscoda residents, Ms. Wusterbarth believes the federal government should conduct the necessary epidemiological studies to correlate health outcomes among veterans and their families.

Sandy Wynn-Stelt is a clinical psychologist in private practice. In 2017, she learned that her groundwater had been contaminated by PFAS from a landfill that had been used by Wolverine Worldwide to dispose of Scotchgard-laden tannery waste. Since then, Ms. Wynn-Stelt has become an advocate for preventing similar contamination in other communities. She has spoken with several state and local legislators and the Michigan attorney general, and has testified in Washington, DC, at hearings on PFAS contamination. Ms. Wynn-Stelt currently participates on the leadership team of the local Wolverine Community Advisory Committee, as well as on the Michigan PFAS Action Response Team. She also participates in the Michigan Department of Health and Human Services' Health Study as a stakeholder for both the state PFAS study and the Agency for Toxic Substances and Disease Registry's Multisite PFAS Study. Ms. Wynn-Stelt participates in training with U.S. Environmental Protection Agency (EPA) staff on effective communications with community members. In 2020, she was awarded the Citizen Excellence in Community Involvement Award by the EPA for her work in the community and state in advocating for stronger regulation of this class of chemicals.

Appendix B

Summary of the Committee's Town Halls

As noted in Chapter 1, the committee engaged with community members throughout the study process. A key component of the community engagement was the conduct of three town halls, held on April 7, May 6, and May 25, 2021. The town halls were conducted remotely, because of the COVID pandemic, and designated Eastern, Middle, and Western. In total, they featured 36 people affected by PFAS.[1]

This appendix summarizes the discussions at the town halls. It was prepared by Anna Ruth Robuck, Ph.D., as a factual summary of what occurred. The statements made are those of the rapporteur or individual meeting participants and do not necessarily represent the views of all town hall participants, the study committee, or the National Academies of Sciences, Engineering, and Medicine.

OVERVIEW

The town halls included presentations by invited community representatives detailing information about exposures, health effects, and health care needs; discussion sessions enabled exchange between committee members and participants. The presenters described frustrating and harrowing ordeals navigating PFAS exposure and related health issues and concerns. Several key themes were echoed by multiple presenters and in discussion sessions, including:

- an immediate need for accessible PFAS blood testing,
- continued assessment of PFAS health effects,
- the need for equitable action that best supports the people who are most vulnerable and disproportionately affected, and
- a continuing need for comprehensive exposure assessments.

Accessible Testing Multiple speakers highlighted the importance of readily available, affordable PFAS blood testing. Such testing would serve to establish baselines of exposure, provide agency to exposed communities and families, show respect for community concerns, and inform precautionary health care. They stressed that testing must be financially accessible and recognized by insurance coverage. Many recounted reluctance or refusal by health care providers to order such testing because of the inability to definitively relate blood concentrations to health effects. While acknowledging pervasive exposure to PFAS in the U.S. population, speakers suggested that exposed communities and vulnerable populations should be prioritized for PFAS blood testing, using equity as a guide to design testing protocols.

Continued Assessment of Health Effects Many speakers highlighted the lack of studies detailing health outcomes related to PFAS exposures, citing this as a glaring and troubling data gap, given the long list of health concerns and trends identified by and in exposed communities. Presenters pointed out that the lack of identified health effects often gets cited as a reason not to perform research or desired testing, leading to a circular situation in which health effects are not identified because of a lack of study, and studies are not conducted because of a lack of identified health effects.

---

[1] PFAS are chemicals known as perfluoroalkyl and polyfluoroalkyl substances; see Chapter 1.

Equitable Action Numerous presenters acknowledged the importance of considering systemic inequities when designing health care interventions or protocols related to PFAS exposure. They noted that minority communities often bear disproportionate burdens of exposure, magnified by inequitable sociocultural and economic frameworks. Health care workers seeking to mitigate and address PFAS exposure must consider such compounding effects of inequity, relying on community participation and input to guide intervention. Some speakers pointed out that children, pregnant women, transient populations (such as service members and migrant workers), and people without health insurance or lacking accessible health care need to be considered given the demonstrated burden of PFAS on these populations. Moreover, information about PFAS and appropriate health care options must be framed and presented in culturally appropriate ways across the continuum of both formal and informal health care settings.

Comprehensive Exposure Assessment Throughout the town halls, speakers provided background information about PFAS exposure scenarios in their community or region. These descriptions revealed vast uncertainty about the scope and scale of exposure. Speakers suggested that exposure vectors beyond drinking water should be explored, including air, cooking and washwater, fish and seafood, wildlife products, garden vegetables, and other foods and agricultural products. Speakers also stated that a larger number of PFAS should be included in monitoring efforts considering the ever-expanding number of compounds included in the PFAS class.

EASTERN TOWN HALL

Laurene Allen (Merrimack Citizens for Clean Water)

Laurene Allen shared perspectives from the extensively exposed community of Merrimack, New Hampshire. Residents of this area learned about their exposure to PFAS, in 2016, related to local industrial activities. PFAS have since been found regionally in air, soil, groundwater, and drinking water across an area of more than 65 miles containing five towns. The community is therefore aware of their past exposure and frustratingly continues to grapple with current exposure from ongoing PTFE[2] used in fabric and film coating by a local manufacturing facility. Allen stated the community finds the continuing exposure to unregulated PFAS troubling, particularly in the absence of appropriate health care that considers both the history and current extent of PFAS exposure.

Allen described the inaccessibility of blood testing for PFAS and health screenings for PFAS-related ailments, with feedback from physicians that such information is expensive to obtain and may be unhelpful or difficult to frame. However, Allen asserted that such testing is important and validating to the community, and a key "piece of the puzzle" to establish baseline information about the evolving understanding of the nature of exposure. Allen also stated that the community sees patterns of disease and illness related to source proximity that are not currently acknowledged or understood by health care practitioners. Given the documented link between PFAS and specific health effects, including immune function and endocrine health, blood testing is vital for matched, relevant health care based on exposure history.

Allen noted a link between COVID-19 cases and PFAS exposure in New Hampshire, with the further suggestion that information about patient and community PFAS exposure should be incorporated into COVID-19 responses, vaccination protocols, and other public health considerations. Allen also underscored the importance of integrating environmental health and exposure history into health records to increase the capacity to monitor for PFAS-related health effects over time. Allen concluded by emphasizing the importance of improved guidance and support for physicians as to how to best incorporate PFAS exposure into clinical care and patient risk reduction to ensure support for highly exposed communities desperate for appropriate care.

---

[2] Polytetrafluoroethylene, a polymer PFAS; see Chapter 1.

"It's been really difficult to get chronic PFAS exposure and risks posed from this chemical class acknowledged by health care providers." *Laurene Allen*

Teresa Gerade (Don't Undermine Memphremagog's Purity)

Teresa Gerade provided information about landfills as sources of PFAS contamination and her Vermont community's concern about an adjacent landfill. PFAS-enriched landfill leachate is often treated by nearby wastewater treatment plants. Wastewater treatment plants are typically not designed to remove PFAS, resulting in the generation of effluent and biosolids enriched in PFAS. PFAS are then reintroduced to the water cycle through effluent discharge or may enter plants grown in soils amended with sludge-derived biosolids.

Gerade focused on information about Vermont's landfill neighboring Lake Memphremagog, a lake spanning the border of the United States and Canada. The community surrounding Lake Memphremagog is particularly concerned about the local and regional PFAS exposure associated with the Vermont landfill and how it affects the health of both the lake and the community. Malignant melanoma has been found in a certain species of lake fish, raising questions about the health of fish populations. The surrounding community has related concerns about the safety of eating freshwater fish from the lake and crops grown in biosolid-amended soils. Gerade suggested the need for PFAS screening levels in human blood, like those in place for cholesterol or glucose, to allow for exposure monitoring over time and related risk mitigation.

Ayesha Khan (Nantucket PFAS Action Group)

Ayesha Khan provided insight about the experiences of the firefighting community and specifically the firefighting and AFFF-exposed[3] community on Nantucket, Massachusetts. Firefighters are routinely exposed to PFAS through both their gear and firefighting foams. Although the Centers for Disease Control and Prevention (CDC) recommends limiting exposure to PFAS, many firefighters are unaware of PFAS or their unique occupational exposure to the pollutants. As a result, firefighters often receive little or no training to mitigate the risk related to gear handling and foam use. This unawareness has led to previous practices that caused undue PFAS exposure, such as using AFFF to clean vehicles or allowing children to play in AFFF.

Khan emphasized the importance of medical monitoring and PFAS bloodwork for firefighters and other PFAS-exposed communities. Khan asserted that such access would establish exposure baselines, raise awareness, and empower people who are exposed to be proactive in reducing exposure and managing risk. Khan illustrated the limited and frustrating accessibility of bloodwork by recounting an experience with a physician, one who is aware of the PFAS crisis and lectures on PFAS exposures. During care for a child, the patient's parent requested PFAS bloodwork. The physician responded by trying to dissuade the parent from seeking the testing, saying if PFAS measurements in blood were carried out it would be difficult to frame the information, especially since the child was currently healthy. Khan also recounted that the doctor misquoted data from studies by the Agency for Toxic Substances and Disease Registry (ATSDR) that measured PFAS in residents of exposed communities throughout the United States, stating these studies "did not find much."

The physician went so far as to jokingly cite a study that found higher reading ability in children whose mothers were prenatally exposed to PFAS. Khan stated the doctor's response felt as though they were vastly minimizing exposure concerns to the detriment of informed health care for the patient and parent. Khan went on to assert the importance of improving CDC guidance about PFAS provided to clinicians, as the current guidance recommends limiting exposure with little or no information about how to do so in practice. Khan stated that the vagueness of the current guidance can result in situations in which clinicians become barriers to information or action, thereby imposing a burden on concerned

---

[3] AFFF are aqueous film-forming foams, used in firefighting, that often contain PFAS.

patients to figure out how to reduce exposure and advocate for themselves in a health care setting. Khan reiterated the importance of providing candid information about PFAS to exposed communities, as well as to health care providers, lawmakers, union leaders, and other decision makers in order to help exposed communities and families rectify their burden of contamination.

> "We are not scientists or doctors. We assume our government is keeping us safe, and when there is a possible link to adverse health risks, we are led to believe that issue is promptly resolved and not with a regrettable substitution."

> "My hope is that [the committee] will educate physicians to be honest with those of us who have been exposed." *Ayesha Khan*

Kristen Mello (Westfield Residents Advocating for Themselves [WRAFT])

Kristen Mello offered perspectives from a PFAS-impacted community as a representative of Westfield, Massachusetts. Mello provided context about her experience with PFAS exposure by explaining she learned about her community's AFFF-contaminated drinking water in 2016 and founded WRAFT in response. An ATSDR study of 459 Westfield residents subsequently found that 92 percent of the city's residents had serum concentrations of at least one PFAS that exceeded the national average.

In considering the committee's task regarding challenges related to PFAS encountered in health care settings, Mello cited the frustrating lack of clinical guidance for health care providers as a major challenge. She noted that the current paucity of clinical guidance contrasts sharply with the availability of scientific information describing exposure assessment of specific human populations, animal studies, and toxicological models, and the rigorous collation of such scientific information provided by the ATSDR toxicological profile for PFAS.

Mello strongly stressed the need to collect data about PFAS exposure and associated health effects in exposed communities, citing the inherent disrespect of the plight of exposed communities when testing efforts are denied or discouraged. Without paired assessment of PFAS exposure and associated health outcomes, Mello suggested, the true scope of adverse health effects associated with specific PFAS thresholds cannot be identified, further limiting the advancement of clinical guidance for health care practitioners.

Mello highlighted health conditions of concern observed in her own community, including allergies, autoimmune and immune disorders, asthma and pulmonary disease, colon diseases, reproductive cancers, menstrual and fertility issues in women, diabetes and metabolic disorders, thyroid disease, cholesterol and liver disease, osteoarthritis and osteoporosis, cognitive and developmental disorders, neurological issues, brain cancer, kidney diseases, and bladder cancer. Mello further argued for the benefits of data collection in exposed communities by highlighting the value of PFAS exposure and health assessment in a tailored public health response. For example, Mello said, blood testing could facilitate identification of communities more likely to be immunocompromised due to PFAS exposure when designing COVID-19 intervention or protocols.

Even in the absence of information framing PFAS exposure levels in relation to specific health outcomes, however, Mello said exposure assessment is critical for affected communities. Mello described the contamination occurring in exposed communities as an intentional crime, without the same protections afforded to victims of equally degrading crimes. Mello argued that the accessibility and implementation of such testing validates and respects the plight of those contaminated without their knowledge or consent, akin to respecting the rights of a crime victim.

Mello also commented on the ubiquity of PFAS exposure in the general population, pointing out that scientists have identified thousands of PFAS while only nine are being considered as part of the committee's study. She further pointed out exposure vectors relevant to the general public, including the air and consumer products.

Mello encouraged the committee to recommend amending guidance for health care practitioners to better support exposure reduction, ensure respect and support for those exposed, and enable equitable and voluntary data collection to assess PFAS levels and associated health effects. Additionally, Mello emphasized the importance of patient agency and fully informed consent during the care process. She also stressed the immediate need to use existing knowledge to revise clinician guidance and treatment plans for those currently exposed to PFAS, rather than waiting for further data. She also recommended the utility of future monitoring and a voluntary national database of exposure and health effects information so exposure and health data can be stored and further explored.

"So we were asked as community liaisons what challenges we have had with our medical providers in dealing with PFAS exposure, and the challenge is that there's just no helpful information."

"You don't have a problem getting an insurance assessor when your car is hit, you don't have a problem getting an insurance assessor when you have a tornado, but this slow-motion unfolding environmental and public health disaster ... is intentionally keeping the information from us so that we cannot take action." *Kristen Mello*

Tracy Carluccio (Delaware Riverkeeper Network)

Tracy Carluccio summarized perspectives as a community activist and advocate based in New Jersey. Carluccio contextualized her perspectives by describing the scope of ongoing PFAS contamination and action in New Jersey. She acknowledged the state's heavy burden of legacy and novel environmental contamination, including prolific PFAS contamination, caused by several companies such as DuPont, 3M, and Solvay. In response, the state has taken ground-breaking regulatory steps to address PFAS, and it was the first to adopt maximum contaminant levels (MCLs) for drinking water that were lower than the health advisory guidelines proposed by the U.S. Environmental Protection Agency (EPA).

Despite New Jersey's progressive stance on PFAS, however, Carluccio said that the narrative in the state is primarily driven by affected communities, and large information gaps still exist. Some blood testing has been carried out in specific localities, but there has been limited blood testing across the state or near specific industrial sites. The PFAS crisis in New Jersey cannot be fully understood until the public has access to data claimed as confidential business information by companies producing or using PFAS, she said. Additionally, Carluccio argued, thorough exposure assessment, health studies, and medical monitoring must be carried out to fully characterize the scope of PFAS contamination and related effects for New Jersey residents.

Carluccio provided two specific examples to underscore the importance of community access to information and monitoring in pursuit of the most appropriate health care. She described legacy and emerging PFAS contamination originating from Solvay, an industrial user of PFAS located in West Deptford. Industrial activities by the company were found to contaminate the drinking water of 50,000 residents of the surrounding area with legacy PFNA,[4] first identified in 2013. Since that time, the company has shifted to new PFAS to replace legacy compounds, while providing little information to the community about the occurrence and health implications of these new compounds. Carluccio also stated that the company has thus far refused to comply with a state directive requiring PFAS users and producers to provide information about production activities to the state, prompting a lawsuit.

The lack of data describing community exposure around Solvay has left the community in a precarious position, unable to appropriately steer their own health care and family choices. Carluccio also detailed PFAS exposure in communities adjacent to military bases in the state. She emphasized that company and agency recalcitrance to share information about legacy and novel PFAS use and contamination actively thwarts the needs of surrounding communities. She suggested that "ignorance is not bliss," as it may lead to misinformed health decisions; instead, informed community members gain

---

[4] Perfluorononanoic acid, a PFAS; see Chapter 1.

agency and can best decide whether they want to avoid tap water, install additional drinking water treatment, move out of a contaminated area, or consider exposure in making reproductive and family decisions.

Loreen Hackett (PFOA Project New York)

Loreen Hackett provided insight about her experiences as an organizer from Hoosick Falls, New York. In 2015 the town was found to be severely contaminated with PFAS, specifically PFOA,[5] and has since been designated as the first PFOA site on the EPA Superfund national priorities list in the United States. The town is now home to three federal sites, as well as several additional state Superfund site declarations, with more currently being investigated in the small community.

Hackett detailed her intimate experiences with PFOA contamination by citing the highly elevated levels of PFOA found in her own blood and the blood of her grandchildren, comparing these staggering figures to national averages that are hundreds of times lower. Hackett commented that the alarming levels found in her family's blood required guidance from health care professionals, yet thus far her family and community health care needs have been poorly met in a clinical setting. As an example, Hackett described a situation shortly after the community learned about exposure. The New York Department of Health organized a community meeting with a pediatrician who lacked PFAS expertise. The pediatrician responded to community questions about health concerns by stating "I don't believe any of your illnesses are caused by PFOA." Hackett described feeling mortified and frustrated at this response, given the community was familiar with existing research that decisively indicated the opposite.

Hackett also described results from a community health study that supported the community's concerns, listing cases of kidney cancer, testicular cancer, thyroid disease, pregnancy-induced hypertension, and ulcerative colitis in the community. Hackett stated the people had to do research for themselves, which was arduous and confusing at times. This fact-finding process took up valuable time that could have helped mitigate exposure-related effects in the community. Hackett indicated that state and local regulatory agencies are now taking PFAS contamination more seriously, with the state setting more protective drinking water MCLs and continued community blood testing.

Hackett stated that too few doctors are trained in the ramifications of environmental toxins and associated care protocols. As a result, Hackett described her habit of bringing research studies to appointments with specialists or unfamiliar doctors to illustrate relationships between PFAS exposure and health effects. She indicated that few listen, and she is reluctant or unwilling to follow up with those who do not listen, therefore missing out on further treatment and testing options, as well as any associated benefits.

Hackett also mentioned that Hoosick Falls residents often rely on the closest hospital, which is in Bennington, Vermont, which is itself a PFAS-contaminated community. Yet even there, Hackett stated, many health care providers are unaware of PFAS and associated health implications. She provided an example in which a provider suggested her breast cancer was genetic despite family history and medical history suggesting otherwise. She articulated that health care practitioners unfamiliar with the issue often made her feel intellectually inferior because of her lack of formal medical credentials, rather than acknowledging and validating her significant health concerns. She stated this dynamic continues to occur even with health care providers directly situated in currently contaminated communities.

Hackett also provided an example depicting the benefit of informed health care, by stating that a local general practitioner took time to learn about the medical ramifications of PFAS exposure; this awareness has resulted in concerted follow-up on PFAS-related health issues in the immediate community to keep pace with the emergence of health effects. Hackett suggested the implementation of continuing education credits to incentivize continued training on environmental health issues for health care practitioners.

---

[5] Perfluorooctanoic acid, a PFAS; see Chapter 1.

Hackett also stressed the need to establish new health care paradigms and protocols specifically tailored to exposed communities. For example, she questioned whether blood donation is safe in highly exposed communities like Hoosick Falls. Additionally, she suggested that exposed women should be informed by obstetrical/gynecological care providers that PFAS will be passed to babies in utero and in breast milk. Health care mantras like "breast milk is best" need to be thoroughly reevaluated in exposed communities, given that breastfeeding may double or triple PFAS levels in infants in comparison with their mothers' levels. Families in exposed communities cannot make informed reproductive choices or other family decisions without information tailored to their situation. Hackett relayed concerns from community members now expressing guilt at unknowingly poisoning their child over the course of pregnancy and breastfeeding. Hackett also suggested that health care norms, such as visiting times, need to be adjusted to better fit environmental concerns because health care discussion about PFAS require more attention than a short office visit.

Hackett emphasized the need for health care professionals to trust their patients and their observations and knowledge on PFAS issues. By trusting local community members, practitioners and scientists can better use community data and experiences in building effective health care for people who have been exposed. Hackett described her participation on the Community Advisory Panel with University of Albany for their CDC-awarded multisite study, collecting and sharing published health studies related to PFAS for the study website. Working in this context, Hackett stated, the study design has shifted to include more tests beyond liver, kidney, and thyroid function, as immune suppression, endocrine disruption, neurological effects, reproductive issues, and breast cancer become increasingly salient community concerns.

Hackett also detailed ongoing contamination concerns in her community due to continued stack emissions and exposure to unregulated, short-chain compounds designed to replace PFOA that studies show to be as toxic as long-chain, legacy PFAAs.[6] As a result of regrettable substitution, Hackett advised that bioaccumulation and total body burden of numerous PFAS has to be considered for those requesting medical direction through continued blood testing beyond the limited number of PFAS currently under scrutiny.

Patrick Elder (Military Poisons)

Patrick Elder articulated concerns and insights about the understudied role of PFAS exposure from food. Elder stated that he believes there is too much emphasis on PFAS levels in drinking water, with too limited a focus on PFAS exposure from food, particularly seafood. Elder contextualized this position by detailing his experiences testing surface water and seafood items near his home in southern Maryland, adjacent to the Naval Air Station Patuxent River Webster Field Annex. Elder's efforts resulted in the detection of significant PFAS concentrations in surface water and seafood items. Elder published the results in the local press, leading to concern and outrage in the community. He indicated that a subsequent public meeting with Navy officials resulted in an unsatisfactory exchange of information with the local community, as Navy officials reiterated that the chemicals in question were no longer in use and there was no medical treatment to reduce PFAS in the human body. The community sought increased testing on seafood items and water, expressing disagreement with the Navy's assertion that not enough is known about PFAS in seafood and the human body to justify immediate intervention.

Elder further highlighted the untenable data gaps for PFAS in food in the United States through a comparison: the European Food Safety Authority recommendations suggest up to 86 percent of PFAS exposure stems from food intake while the U.S. Food and Drug Administration (FDA) issued a statement suggesting there is no evidence that dietary choices should consider PFAS contamination. Overall, Elder emphasized the importance of better limiting PFAS exposure from food and seafood items in the United States and incorporating this vector of exposure when considering health effects and health studies.

---

[6] Perfluoroalkyl acids, a class of PFAS; see Chapter 1.

Hope Grosse (Buxmont Coalition for Safer Water)

Hope Grosse described her experiences in Bucks and Montgomery counties (BuxMont), Pennsylvania. Grosse described a lifetime of exposure due to drinking water contaminated by AFFF use at Warminster Naval Base and Willow Grove Naval Air Station.

Grosse reported that people in the BuxMont community have voiced concern regarding their PFAS exposure and a number of diseases and health conditions including thyroid cancer, non-Hodgkin's lymphoma, kidney cancer, testicular cancer, breast cancer, liver cancer, brain tumors, ovarian cancer, lung cancer, bladder cancer, melanoma, bone cancer, altered metabolism and obesity, fertility issues, birth defects, diabetes, cholesterol, high blood pressure, preeclampsia in pregnant women, decreased infant birthweight, autoimmune diseases, chronic inflammation, immune response, and alterations in liver enzyme levels. The implications of exposure are not limited to physical ailments: the community has also collectively experienced serious experiences with fear, anxiety, grief, emotional and physical stress, and a feeling of being forgotten.

Grosse said she personally wrestles with an unbelievable lack of trust, fear, and emotional scars accumulated related to the premature death of her father, her own early cancer diagnosis, and deaths of multiple cherished community members. Grosse further revealed that she struggles with shame and fear due to the feeling that she inadvertently poisoned children because she was unaware of her exposure. Grosse also described the financial stress imposed on exposed communities to pay for bottled water, increased public water rates, home filtration, diagnostic testing, medical fees, and loss of wages due to health issues and related loss of productivity.

Grosse stressed the importance of increased awareness and resources to better educate health care practitioners about the health effects of PFAS. She indicated that communities need to be able to trust that their caregivers and practitioners are knowledgeable on the issue and capable of advising action to address their concerns. She stated that the community feels they have no local medical resources or health care providers to answer their questions or advocate for them in the health care system. During her own health struggles, she said she felt belittled and embarrassed for asking questions about environmental health and its relation to her health maladies. Grosse recommended considering practical measures to improve the patient experience for exposed people, such as requiring questions about exposure history or concerns in health questionnaires and forms or designing better collaborations between insurance companies and physicians so practitioners can readily issue scripts for bloodwork and diagnostics that are covered by insurance.

Grosse stressed that such clinical guidance must particularly provide proper direction detailing how to best protect and treat children, given the particular risks of PFAS exposure during critical developmental periods. Mothers need awareness and guidance to test for PFAS in blood and breast milk and make effective parenting choices based on exposure results to best safeguard children in utero or during breastfeeding. Grosse mentioned she lacked such information and agency raising her own children and would have greatly valued the knowledge and associated opportunity to switch to filtered water or bottle-feed her children to reduce their early-life exposures.

Grosse also emphasized that blood testing remains imperative to characterize exposure and changes in PFAS blood levels over time. Grosse posited that the value of blood testing is not predicated on how well the derived information can be explained or compared to thresholds. Rather, documentation of the exposure can retrospectively serve communities, as health science progresses to more fully understand the health ramifications of exposure. Grosse also asserted that blood testing facilitates appropriate health planning, as exposed individuals have a right to preventive health measures and testing tailored to their exposure reality, such as titer testing to deduce PFAS-altered vaccine responses or tests for PSA and liver enzymes.

"When I would go to the doctor's and tell them about some of the exposures of over 50 chemicals that I was exposed to, the doctors would laugh and say no. Clearly they didn't have any information about environmental components [of disease]. They made me feel small, they made me feel stupid and embarrassed even just asking the question." *Hope Grosse*

Le'Meshia Whittington (North Carolina Black Alliance)

Le'Meshia Whittington discussed the perspectives and needs of Black and Brown communities in North Carolina. Whittington emphasized that Black and Brown communities face cumulative impacts in addition to environmental health concerns, resulting in disparate burdens and concerns for exposed minority communities well beyond drinking water. Whittington exemplified this reality as related to PFAS exposure in the Cape Fear River system in southeastern North Carolina. In 2017, the public learned about significant PFAS contamination in regional surface water and drinking water due to the industrial activities of Chemours in Fayetteville, North Carolina. Whittington highlighted that this news broke amid a history of climate-related disasters (e.g., hurricanes) that repeatedly damaged the regional water system over decades. As a result, with the occurrence of Hurricane Florence in 2018, regional communities were severely affected by floodwaters contaminated with coal ash, animal waste, and industrial wastes, as well as PFAS. However, the ramifications of PFAS-laden floodwaters were not addressed following Florence.

Whittington described the multifaceted and overlapping exposure vectors of concern for Black and Brown communities, highlighting that the interplay of these factors can cause undue cumulative exposure and subsequent harm. She highlighted that about 35 percent of North Carolinians are renters, leading to potential exposure from carpets, textiles, and building materials with no capacity to remove or replace these items. Whittington also pointed out 80 of the state's 100 counties are considered food deserts or contain food deserts, where residents face significant hurdles accessing fresh fruits, vegetables, and other sustainable foods. In these areas, people may unduly rely on packaged, processed, and fast foods prepared using PFAS-laden water, following storage or transport in food packaging materials that contain PFAS. This type of food sourcing leads to concentrated PFAS exposure. Whittington again highlighted the additive impact of many cumulative factors, noting many of these food desert areas in North Carolina are downstream from continued contamination from Chemours and are also routinely affected by climate-related flooding events.

Whittington also pointed out that occupational exposures require increased health monitoring, considering the cumulative risk imposed when different industrial activities collide in a given region. She stated that more information is needed describing how food service, essential, agricultural, and industrial workers may be exposed occupationally given inherent contact with potential PFAS sources in those jobs, as well as increased risk to those workers when their employment location is in a region rife with known contamination. For example, Whittington questioned how agricultural and abattoir workers who work in PFAS-contaminated zones may be routinely exposed through industrial air emissions and constant contact with PFAS-imbued agricultural products. Whittington suggested increased institutional study and support for workers through the Occupational Safety and Health Administration or other agencies, and the need for increased health care provisions for those closer in proximity and more consistently exposed to this group of chemicals.

Whittington also highlighted high-priority populations within wider Black and Brown communities. She identified pregnant women and children as two such priority populations given the demonstrated opportunity for maternal transmission of PFAS to their children. Whittington stressed the need for blood testing for expecting parents and children, and increased attention to fertility concerns. Whittington also highlighted veterans, people in law enforcement custody, people with disabilities, firefighters, residents of care and extended-stay facilities, and students as underserved priority populations. She stated that blood testing for PFAS should be made available and accessible to these and all demographic groups, with a critical need for exposure assessment over time.

Furthermore, Whittington stressed, it is important to have updated clinical guidance about PFAS blood testing, health concerns, and standards of care for health care professionals. This information must be disseminated among the wide range of practitioners, agencies, and institutions serving high-priority populations and wider Black and Brown communities. Whittington stressed that such guidance is also urgently needed to ensure health insurance claims related to PFAS exposure are viable and compensable, increasing access to exposure-informed health care.

> "This is what we are asking: our communities are willing to undergo testing, they will participate in studies, it just needs to be designed for our communities, alongside our communities."
> *Le'Meshia Whittington*

Mike Watters (Gray's Creek Residents United Against PFAS in Our Wells and Rivers)

Mike Watters provided perspective as a resident of a community affected by the industrial activities of Chemours in Cumberland and Bladen counties in North Carolina. Watters described concerted efforts to engage his community to ensure awareness and engagement. One such effort included a community health survey of 100 people that highlighted shared community health concerns related to PFAS exposure, including thyroid function and hyperthyroidism, weight gain, arthritis, asthma, autoimmune disorders, skin cancer, chronic inflammatory conditions, vitamin D deficiency, type 2 diabetes, prediabetes, hair loss, high blood pressure, high cholesterol, high triglycerides, irritable bowel syndrome, and itchy scalp. Watters emphasized that continued measurements of PFAS in blood are urgent to associate these health concerns with PFAS concentrations and to allow health studies to catch up with the state of exposure. Watters also stressed the value to the community of baseline exposure assessment and the importance of tracking exposure over time through continued blood testing given evolving exposure scenarios. This work is currently being done through the North Carolina State University GenX (PFAS) Exposure Study with participation from people in the community, as well as pets from their homes.

Watters also discussed understudied vectors of exposure, such as fish, air, rain, locally grown vegetables, cooking water, cleaning water, clothing washed in contaminated water, and showers. The contribution of these sources to PFAS exposure and associated health outcomes are poorly characterized. These understudied exposure sources therefore stand as sustained concerns for community members. Watters suggested the continued need for more monitoring to keep pace with the unique exposure scenario in North Carolina and to address these understudied routes of exposure, given growing evidence that air emissions of novel PFAS continue to affect a growing number of residents across the region.

Emily Donovan (Clean Cape Fear)

Emily Donovan provided insight about the community experience of residents in southeastern North Carolina subject to high levels of PFAS exposure. Donovan explained that regional communities have been overexposed to hundreds of novel and legacy PFAS for decades because of regional industrial activities and that contamination is continuing. Donovan noted that existing health guidance does not adequately address the amount or types of ongoing PFAS exposure in the state. She also described the frustrating inability of state and federal agencies to halt ongoing discharge of the complex PFAS mixture impacting the region. She noted the prevalence of PFAS occurrence in regional wildlife, sediment, rainwater, and consumer products, pointing out that communities in the region are concerned about PFAS exposure from these and other sources given the excessive PFAS exposure from their drinking water. Donovan said that clinical guidance tailored to ongoing exposure in the state must consider this and protect community members from routes of exposures beyond drinking water.

Donovan noted the community's limited capacity to obtain blood testing for the suite of PFAS found in North Carolina as a major and exasperating challenge. Analytical methods that screen for all relevant PFAS are uncommon, and those that do exist are expensive or possess detection limits that are too high to be useful. Donovan mentioned that she has heard commentary that the exposure profile experienced in southeast North Carolina is "too unique" to warrant development of rigorous testing; she countered this idea by highlighting the use of PFAS produced by Chemours at manufacturing facilities across the nation.

Donovan cited the GenX Exposure Study as an example of blood testing that occurred regionally, which found a range of novel and legacy PFAS in 400 local residents. However, she asserted that a larger proportion of the more than 300,000 people exposed should be included in further testing, citing the opportunity for the population to be used in larger health effects studies, given the long history of unique exposure across the area and documented health concerns shared across the population. She asserted this overexposed, large population is the "statistical power" required to identify understudied health effects so far missed by institutional investigation.

Donovan presented specific and heart-wrenching examples of personal contacts who are currently uncertain and worried about links between serious health conditions and their history of exposure. She detailed a litany of regional health concerns, including pediatric bone cancers, osteosarcoma and brain tumors, pediatric kidney cancers and diseases, bladder cancer, gallbladder dysfunction, testicular cancer, pancreatic cancer, liver cancer, nonalcoholic fatty liver disease, leukemia, blood cancers, colon cancer, thyroid cancer and dysfunction, autoimmune diseases, digestive issues, multiple sclerosis, skin disorders, infertility, premature births, developmental delays, learning disabilities, autism, breast cancer, and non-Hodgkin's lymphoma.

Donovan stated that the currently available data indicate increased risk of negative health effects, including birth defects, kidney disease, and increased cholesterol related to PFOA exposure. However, many PFAS found in southeast North Carolina were designed as replacements for PFOA, yet no such thresholds exist for these next-generation PFAS. Donovan stated that North Carolina communities and beyond need relevant clinical guidance that adequately identifies, protects, and monitors those who have high levels of diverse PFAS exposure, as current clinician guidance falls well short of these goals. Donovan specified that the regional community needs information about relationships between specific health outcomes and cumulative serum levels for total PFAS in their blood. Health care guidance must consider the complex mixture of exposures experienced by communities and go beyond clinical guidance for legacy PFAS. Donovan cited her current experience as an example, where she often finds herself informing her children's pediatrician about the most current research rather than vice versa; she described this as counterintuitive and discouraging.

Donovan exhorted the committee to investigate information gaps confounding or limiting health studies, suggesting that underreporting of relevant toxicity and health data by corporate producers contributes to damaging data gaps. Donovan stated the committee and the public should have full access to industry data related to production, toxicity, and health outcomes to fully explore the range of linkages between human health and complex PFAS exposures in this region.

"Sadly, it feels like guinea pigs are treated better, because at least their exposures are thoroughly studied for the betterment of humanity."

"At the moment I'm informing my pediatrician about the latest toxicity studies related to PFAS placental transfer and breast milk contamination. This is backwards and depressing." *Emily Donovan*

Cheryl Cail (South Carolina Indian Affairs Commission/SC Idle No More)

Cheryl Cail discussed her experiences as a Native American community member and a member of a PFAS-impacted community in Horry County, South Carolina.

The path to PFAS advocacy started for Cail when her 20-year-old son Trevor was diagnosed with testicular cancer. She shared a photo of her son's back showing bright red spots indicative of unknown but acute dermatitis. Cail explained they went to the doctor to identify a cause and solution; the local doctor was not aware of environmental health concerns and did not ask the right questions to quickly arrive at a diagnosis. As a result, Trevor's cancer progressed unchecked for 9 months.

Cail related that Trevor was working and attending school adjacent to Myrtle Beach Air Force Base, though at the time they did not realize the importance of this location and their regional exposure. Cail explained they then watched a movie entitled "The Devil We Know" about PFAS contamination in

West Virginia, and she started questioning whether PFAS were the cause of Trevor's testicular cancer given the demonstrated link between this type of cancer and PFOA exposure. The family also found a map detailing PFAS contaminated sites around the United States and realized their community's proximity to Myrtle Beach Air Force Base was problematic, as massive concentrations of PFAS had been found in the environment around the base despite remedial activities. Cail noted that right before the COVID-19 pandemic more state agencies began mobilizing to address the issue within her community and across the state, but this was only after years of inaction. She noted that she had seen this before, as institutions prioritize "wealth before health."

With more information about regional PFAS exposures, Cail and her family started trying to find answers from the medical community regarding how to get PFAS blood testing and how to get treatment for their exposure and related health effects. Cail mentioned that while her community is only beginning to understand the scope of PFAS contamination and related health effects, they have seen a concerning increase in cancers, autoimmune disorders, fertility issues, and birth-related defects. She stated that regional physicians were unaware of PFAS, its effects, and even the regional exposure event itself. Cail also described searching for PFAS blood testing options and the frustrating experience of trying to relay that information to a local physician. They found that regional testing was limited to people with occupational-related exposure, such as firefighters. Cail related that this left her son feeling defeated, but she saw a challenge that the committee and other PFAS-focused organizations could resolve.

Cail closed her statements by detailing specific asks from the committee, such as establishing PFAS as an environmental health issue and educating those working in the medical community. This education should involve establishing testing and treatment protocols for those in affected communities and those with diagnosed medical conditions linked to PFAS. She also petitioned the committee to seek expansive data collection to meaningfully assess the full scope of PFAS health effects in order to best design health-related services in exposed communities, including monitoring for children. Cail also relayed a statement from a local physician who has now treated thousands of exposed individuals in her community, who said that physicians need a protocol much like the protocol for lead poisoning; it is a nationwide issue, and so is PFAS poisoning.

Stel Bailey (Fight for Zero)

Stel Bailey offered perspectives as a cancer survivor and member of a military family from Florida's space coast. Bailey first recounted her personal journey with PFAS exposure to explain her path to advocacy. She explained that while her husband was deployed, multiple family members were diagnosed with various cancers in a span of a few months. Given this unusual frequency, she herself was told by multiple doctors that it was "impossible" for her to have cancer, and six different doctors provided care prior to reaching a cancer diagnosis despite continued breathing problems and swollen glands.

Within this harrowing process, Bailey was asked, "where did you grow up?" and subsequently made it her life's work to figure out how and why her background might be related to cancer outcomes. This endeavor took a major turn in 2018 with the release of a key report from the U.S. Department of Defense (DoD) explaining that the region around the facilities of the National Aeronautics and Space Administration, Cape Canaveral Air Force Base, and Patrick Air Force Base was highly contaminated with PFAS, with up to 4.3 million parts per trillion (ppt) of PFOS[7] and PFOA found in surrounding groundwater.

As a result of that report and prior community work investigating autoimmune and cancer cases in the surrounding population, Bailey connected with oncologist Julie Greenwalt. This physician went to high school near Patrick Air Force Base and was aware that dozens of high school classmates had since been diagnosed with cancer, starting at very young ages. Bailey stressed the importance of this advocate and resource for their community, stating "this support meant everything in our community." Bailey suggested that without Greenwalt's personal experience and concern, the surrounding community would

---

[7] Perfluorooctanesulfonic acid, a PFAS; see Chapter 1.

still be woefully unaware of health issues related to their substantial PFAS exposure. Since 2018, Dr. Greenwalt has helped the community document 54 cases of various types of cancer in individuals under age 40, which served to bring awareness and resources to the issue. This community health information also helped frame relationships between health outcomes and PFAS levels in drinking water after it was discovered that regional drinking water was also grossly contaminated with PFAS. Bailey chronicled that a large group of people in the community helped crowdsource community health information and address this issue, yet many have been lost since the 2018 DoD report revealing the scope of contamination, including a girl who passed away from brain cancer just 3 days after her 17th birthday.

Community efforts have more recently focused on mapping autoimmune and cancer cases surrounding Patrick Air Force Base. Using this crowdsourced data, the community was able to push forward their own health assessment in the county. The assessment found an increased risk of certain cancers, including urinary, bladder, leukemia, liver, lymphoma, breast, and testicular cancer. Other concerns captured by the health assessment include liver damage, increased risk of thyroid disease, and birth defects. Bailey also described continued routes of exposure that concern the community, such as consumption of regional fish and wildlife and produce irrigated with contaminated water. Bailey mentioned that drinking water and groundwater contamination remain problematic, noting the emergence of short-chain, understudied compounds such as PFBA[8] in these sources.

Bailey emphasized that her primary message to the committee centers on the fact that early detection of cancer saves lives and reduces health care costs. Since learning about the PFAS problem in tandem with regional medical resources encouraging care, Bailey provided multiple examples of early interventions across the community that served to detect or treat cancer or other health problems before severe disease. Bailey concluded by outlining specific community needs from health care providers, including physician education, medical monitoring, PFAS blood testing, preventive health screening and assessments, documentation on medical records of PFAS exposure and environmental health attributes, and recognized guidance that ensures insurance coverage. Bailey particularly emphasized the need to normalize the inclusion of such questions as "Have you been exposed to any environmental toxins or chemicals in your home or workplace?" and "Where do you live?" on medical forms and records.

> "We are doing physicians a big disservice by not providing them the help or information. Lacking this guidance is only harming people. We need action to save lives now." *Stel Bailey*

Several other people at the Eastern Town Hall provided public testimony, which is available on YouTube[9]:

Andrea Amico, Testing for Pease
Katie L. Bryant, Clean Haw River
Jovita Lee, Democracy Green
Beth Markesino, North Carolina Stop GenX in Our Water
Meg Seymour, National Center for Health Research
Yolanda Taylor, Advance Carolina
Sanja Whittington, Democracy Green

---

[8] Perfluorobutanoic acid is a PFAS; see Chapter 1.

[9] See https://www.youtube.com/watch?v=YrYSj9BPbEQ&list=PLGTMA6QkejfimvAGwR7o_7hP9nXbfF fcr&index=24 (accessed June 29, 2022).

MIDDLE TOWN HALL

Andi Rich (Save Our Water [S.O.H2O])

Andi Rich offered insight as a community member from Marinette, Wisconsin, home to the JCI Fire Technology Center. This site is considered the third-largest PFAS contamination site in the United States. Rich described the extent of contamination around the site, stating that PFAS levels in groundwater around the site reach 400,000 ppt, and the contamination plume has spread for miles around the site. Rich mentioned community concern about discharge of PFAS-laden effluent into regional surface waters, application of contaminated biosolids on agricultural lands, and potential air contamination related to continued onsite outdoor testing involving unknown PFAS.

Rich also described an ongoing class action lawsuit focused on PFOA and PFOS levels in well water, noting that the suit does not consider people exposed through other, more complex exposure pathways that affect the community. Rich indicated this leaves residents poorly informed and inadequately prepared to make decisions in the lawsuit, which in turn affects the fairness of payout claims. She said that lawsuit participants and community members have repeatedly requested blood testing and medical monitoring to further understand their exposure. These requests have been met with sustained resistance. Rich indicated that the community has been told that PFAS exposure does not equate to illness, that blood testing would not be useful, and that blood testing is not recommended. Rich pointed out that blood testing would be highly useful to ensure lawsuit payouts are based on internal exposure, as payouts are currently slated to be distributed arbitrarily.

Rich shared some personal health history to further highlight the utility of PFAS blood testing and care guidelines for practitioners. She related that a physician recommended bloodwork to check thyroid function, which overwhelmed and frustrated Rich given her awareness of the association between PFAS exposure and thyroid health. The doctor tried to comfort Rich, but Rich found out the doctor was visiting from Nashville and had no knowledge of local exposure issues. Rich stated that practices such as traveling practitioners makes blood testing, medical monitoring, and results-based guidelines that much more imperative to ensure environmental health factors are comprehensively and continuously considered by transient health care providers. Rich also stated that she has not yet sought the recommended thyroid bloodwork, due to logistical constraints, though she said she would be far more likely to prioritize follow-up if blood testing indicated PFAS exposure. Rich compared the utility of PFAS blood testing to the breast cancer gene (*BRCA*) test that detects a person's genetic proclivity for the disease. Each test serves as an indication of the increased potential for illness that can inform preventive care routines.

Rich went on to state that Marinette is an impoverished community where few can afford the cost of PFAS blood testing. The majority of community members have been denied covered testing through state agencies and insurance providers. Rich said that the committee's recommendations have the potential to improve care and community health outcomes by making sought-after care accessible and affordable, and she urged the committee to recommend blood testing and medical monitoring for residents in exposed communities, ensuring those exposed through complex pathways are included. Rich also emphasized that the committee should actively avoid "not doing anything because of lack of data, and not collecting data because of a lack of evidence of harm."

> "How can above average incidence of PFAS-related illness be identified in a community where the doctors aren't in town long enough to identify the trend?"

> "We need help putting a stop to the contamination, as the corporate polluters are far more powerful than our voices." *Andi Rich*

Laura Olah (Citizens for Safe Water Around Badger)

Laura Olah commented as a community member affected by the Badger Army Ammunition Plant in Wisconsin. Olah's remarks focused on the major lack of exposure information available to the public and the urgent need for medical monitoring. Olah stated that minimizing PFAS exposure leads to reduced health risks, but communities and physicians trying to accomplish this run into challenges trying to obtain PFAS exposure information in individuals and in the environment. She provided several examples illustrating the pervasive institutional secrecy that prevents transparency about exposure and risk.

She described a public meeting hosted by the U.S. Army to discuss groundwater contamination that has migrated offsite, contaminating rural drinking water wells and discharging into the Wisconsin River. Hydrogeological experts recommended testing drinking wells beyond the currently monitored area, including on the other side of the Wisconsin River. Army personnel argued against this, saying there would be no way to ensure PFAS in well water was derived from the base, and they were not certain which wells should be tested. Olah highlighted this as "ridiculous," pointing out that all wells should be tested given uncertainty about which wells are most at risk as there are no offsite groundwater data. Olah also described several frustrating experiences seeking existing information about PFAS in drinking water. For example, a request for information about PFAS in drinking water around Volk Air National Guard base required a formal Freedom of Information Act request, which has now been pending for 3 years. Another request for a written report describing PFAS contamination around a military base in Tennessee was also unmet, and is now 2 years old. Olah also highlighted that this lack of transparency was not limited to the military, as private corporations refer to PFAS products like AFFF formulations as proprietary mixtures or confidential business information and do not disclose PFAS content in their products. This lack of information makes it impossible to assess and prevent exposure and possible health risks.

Olah asserted that we cannot predict the potential implications and future benefits of medical testing today, but without these data, exposed communities face battles akin to Vietnam War veterans who are still fighting for presumptive care based on exposure to Agent Orange and other toxicants. PFAS blood testing helps baseline exposures and raises awareness so people can take steps in reducing exposures. Baseline testing could help answer health questions in the future and help secure health studies in communities at risk

"Given all these barriers [to PFAS exposure information], the public and the medical community cannot identify which patients are at greatest risk to and harm from PFAS; therefore, our care must be presumptive."

"Without medical monitoring data now, presumptive care will be out of reach for civilians and service members and children exposed to PFAS now and in 50 years from now." *Laura Olah*

Samraa Luqman (Concerned Residents for South Dearborn)

Samraa Luqman offered insights as an environmental advocate and a Yemeni-American community member from the South End in Dearborn, Michigan. Luqman framed her remarks by providing context about her community, describing the South End of Dearborn as a community that grew around Ford Motor Company. The continued need for cheap labor attracted immigrant populations over time, with a more recent influx of Middle Eastern immigrants. A high percentage of the population possess limited English skills, are considered low income, rely on some kind of food or government assistance, and rent their homes. The surrounding area is also home to various industrial activities, in addition to the Ford Motor Company, that use diverse types of environmental pollution. PFAS contamination has recently been identified in the area, resulting in the inclusion of the town on Michigan's PFAS Action Response Team website.

Luqman went on to highlight the importance of considering PFAS exposure beyond drinking water, including inhalation, dermal contact, and maternal offloading of PFAS from mother to child. These understudied exposure vectors are ongoing in the South End, along with continuous air quality issues, lead exposure, and other environmental health concerns. Luqman emphasized the importance of considering such cumulative multiple exposures, citing sustained community concern about deciphering causality or relationships between observed health effects and PFAS in what is clearly a complex exposure scenario. Luqman highlighted that PFAS blood testing can help clarify required follow-up care by identifying exposed individuals most prone to PFAS-related health effects. Luqman expressed particular concern about the relationship between PFAS exposure and immune impacts in the context of the ongoing COVID-19 pandemic. She stated that COVID-19 outcomes have been worse for her community compared with nonexposed communities. She also stated that a lack of PFAS exposure information for her community affects behavioral choices with real-world consequences. Luqman offered the example that someone from the South End exposed to PFAS may be more vulnerable to COVID-19 and require more diligent mask use compared with someone residing 10 miles away, but would never know this given the ongoing lack of exposure information.

Luqman also discussed the need for medical professionals to have awareness about environmental contamination and its potential implications, even decades after a person's leaving a contaminated area. She highlighted that health risks from environmental exposures must be assessed and considered by state agencies and health care practitioners; then this information must be provided to residents and patients to ensure full awareness of exposure. She provided an example of a community member who moved away from the South End area and is now dealing with cancer approximately 20 years later. Luqman stated that this person's doctor explained to them that despite moving away, the carcinogens and pollutants accrued while in the area do not go away. Luqman also underscored the need for health and exposure information to be communicated in culturally appropriate ways, acknowledging and overcoming language and sociocultural barriers.

Cathy Wusterbarth (Need Our Water)

Cathy Wusterbarth offered insights as a community member from Oscoda, Michigan. This community is adjacent to the Wurtsmith Air Force Base and is subject to major contamination of regional groundwater and surface water from AFFF use on the base. This widespread contamination has resulted in historical and ongoing PFAS exposure for service and community members, as well as the area's fish and wildlife.

Wusterbarth's remarks focused on several primary community needs: PFAS blood testing, improved guidance for health care practitioners, exposure assessment in the environment, and exposure mitigation. Wusterbarth highlighted the movie *No Defense* as a crucial example of the health effects of PFAS contamination. The film chronicles the serious lifelong health effects experienced by Mitchell Minor and his family, residents of Oscoda. Wusterbarth also introduced James Bussey, a service member in Oscoda. Wusterbarth explained that Bussey was too ill to present to the committee, but had asked her to provide his medical records for committee consideration. These records include a long list of ailments and physician recommendations based on his exposure to PFAS, which was documented through blood testing.

Wusterbarth emphasized that these real-world examples capture a crucial disconnect between exposure information and health outcomes, which can easily be rectified by improved access to PFAS blood testing. Wusterbarth declared testing should begin immediately to establish baseline exposure levels. Annual testing should occur thereafter, just as is done for other risk factor measurements, such as cholesterol. Wusterbath also identified that given the presence of PFAS in 98 percent of the U.S. population, everyone should have access to blood testing. However, priority should be given to communities and individuals with identified or hypothesized high-level PFAS exposure. Wusterbarth also suggested that testing should be implemented using simple, affordable labs, highlighting a recently opened laboratory in Oscoda as an example. Wusterbarth also recommended that all PFAS be included in blood testing, unless specific PFAS measurements are required to discern the role and liability of a specific polluter or product.

Wusterbarth further described a crisis of trust, stating that her community was experiencing the degradation of trust with the very institutions they believed would protect them, such as EPA, other government agencies, and the military. She detailed discouraging recent information from the DoD, noting that more than 700 military installations have now been recognized as PFAS hot spots, starting from just one in 2012. She also stated that health care providers have a duty to discern the causes of disease in their patients, but without guidance from the CDC or other medical authorities, practitioners are unaware of health risks associated with PFAS exposure to the detriment of patient care. Wusterbarth also emphasized the gravity of the committee's recommendations based on the town halls, stating that state agencies and communities are relying on the forthcoming guidance document to inform physicians and care. She shared a personal experience in which she provided PFAS information to her doctors, only to be told this information was too lengthy to review despite her history of cancer and immunological disease. Wusterbarth stressed the need for swift action by the committee to immediately curtail ongoing harm in exposed communities.

> "We've tested the fish; we've tested the deer; we've tested the groundwater, the waterways, and the foam. When are we going to test the people?"

> "The only risk [of testing] is to the polluters who do not want us to link them to our exposure."
> "This study is a result of the PFAS communities telling you changes are needed."
> *Cathy Wusterbarth*

Sandy Wynn-Stelt (Wolverine Community Advisory Committee, Michigan PFAS Action Response Team)

Sandy Wynn-Stelt lives in a PFAS-impacted community, Belmont, Michigan. Wynn-Stelt opened her remarks by sharing her own story of PFAS exposure. Wynn-Stelt and her husband moved to Belmont in 1992, seeking an idyllic and quiet home. They were not aware that the Christmas tree farm adjacent to the property was a dumping site for a major PFAS user, Wolverine Worldwide. In 2016 her husband's health rapidly deteriorated, and after only a few short weeks, he died of liver cancer, which Wynn-Stelt described as an unbelievable loss. Shortly thereafter, the Michigan Department of Environment, Great Lakes, and Energy tested her drinking water, and Wynn-Stelt learned the well water she and her husband had drank for more than 20 years was contaminated with PFAS at levels up to 80,000 ppt. On learning about her exposure, Wynn-Stelt sought blood testing. She described this process as challenging. She ultimately paid $800 for blood testing through a commercial laboratory in California. Her bloodwork indicated alarmingly high levels of PFHxS, PFOA, and PFOS, well above levels seen In blood studies as part of the National Health and Nutrition Examination Survey (NHANES) and in some other exposed communities.

Wynn-Stelt followed up with her community doctor regarding her bloodwork results; she described this physician as "very proactive and engaged." She stated that her doctor reviewed the ATSDR physician and patient guidance about PFAS that was available at the time. They concluded the guidance was not helpful and instead agreed on a plan to monitor those things listed as potential risks of PFAS exposure. As a result, when Wynn-Stelt experienced breathing problems several years later, this physician recommended thyroid testing based on their knowledge of Wynn-Stelt's PFAS exposure, although this diagnosis would normally be an unlikely candidate. This testing resulted in the identification of thyroid cancer. Wynn-Stelt credits this diagnosis and quick action to prior PFAS blood testing, adding this vital information to her medical history for consideration by her physician.

Wynn-Stelt provided an additional example of a neighbor exposed to PFAS from the same groundwater source. This neighbor's child received PFAS blood testing and was found to have elevated levels of PFOS. This information was provided to the child's pediatrician, who then monitored vaccine response given the child's exposure history. The child was found to have reduced vaccine responses and required some boosters not normally required. Wynn-Stelt reported similar examples of adverse health effects across the community, with cases of thyroid disease, cancer, kidney disease, liver disease, and cholesterol issues in children and adults. Wynn-Stelt also highlighted the mental and emotional toll of exposure and the lack of discussion of this in exposed communities and in care guidelines.

Wynn-Stelt also took time to address often-cited statements used to deny community access to PFAS blood testing. Wynn-Stelt countered the assertion "that it is not ethical to test for something we can't treat" by explaining knowledge of environmental health and exposure is just as medically relevant as asking any patient for a history of diabetes or heart disease. All these pieces of information are required for patients and physicians to monitor potential problems and reduce risk. Wynn-Stelt also addressed the concern that patients may panic given hard-to-frame PFAS information: she countered that exposed patients are not fragile, and knowledge is power. Wynn-Stelt labeled the assertion that PFAS levels in blood cannot be definitively linked to health effects, and therefore should not be monitored as circular logic, stating if testing is not occurring at various scales, links cannot be identified, further and erroneously justifying a lack of testing.

Wynn-Stelt closed her remarks by stating explicitly her community's needs, including accessible PFAS blood testing for people in various exposed communities. She also urged expansive testing for the entire class of PFAS, rather than a few targeted chemicals. Wynn-Stelt also requested research about possible mechanisms to lower PFAS body burdens in exposed individuals and further research to better understand how PFAS levels may impact health care choices, such as blood or organ donation.

Tom Johnson (Clean Water Action)

Tom Johnson shared his perspectives with the committee as an environmental advocate and organizer working with exposed communities across Minnesota and beyond. Johnson explained that much of his work involves public education, seeking to inform people about toxic chemicals overall. When conducting educational activities in the East Metro region of Minnesota, Johnson indicated that PFAS are a central topic of interest, as this area is rife with PFAS producers and users whose activities have massively contaminated the regional environment. This contamination resulted in an $850 million legacy settlement. Johnson suggested that communities are increasingly aware that, although this settlement sum will cover long-term drinking water treatment for all area residents, there will be little or no money left for regional remediation.

Johnson attested to high levels of frustration from exposed communities in the region, with unanswered questions about the likelihood of current or future health effects for families and their children. Despite the lack of citable research, the community possesses large amounts of anecdotal and experiential evidence observed over decades that suggests links to health effects and informs the community concerns. Johnson described the experience of a mother of a low-birthweight baby, who is now concerned about a similar outcome with her second pregnancy. Her doctor has no ability to comment on the likelihood of this outcome given vast uncertainties surrounding her exposure and risk due to a lack of needed exposure assessment and health effects research. Johnson also discussed the importance of accessible and expansive biomonitoring for exposed communities and more broadly. In the case of the East Metro contaminated area, biomonitoring was available for some people in this community, but it is not available for all Minnesotans, leaving data gaps for those who have moved from the area and for other exposed communities. Johnson also highlighted the utility of testing to assess the efficacy of interventions over time and the ability of testing to illuminate links between exposure and understudied health effects.

Vicki Quint (Foam Exposure Committee)

Vicki Quint is a firefighter advocate through her work as a co-chair with the Foam Exposure Committee. Following a massive tire fire in Watertown, Wisconsin, and her husband's death due to cancer, Quint learned about significant exposure to PFAS in the fire service. Firefighters are exposed to

PFAS through AFFF use and the use of PFAS-containing protective gear. Quint provided further details about PFAS in AFFF and current activities, highlighting the current availability of fluorine-free alternatives to AFFF. She noted that there are several sites in Wisconsin contaminated with PFAS as a result of the use or product of AFFF.

Quint followed this information about firefighter exposure and AFFF use by describing the plight of exposed firefighters seeking PFAS blood testing. Military firefighters are now eligible for PFAS blood testing, but retired military firefighters and civilians are not. Quint related concerns from the fire service community regarding the ramifications of blood testing results and options to lower PFAS levels in blood. Quint emphasized the need for all fire departments to discontinue use of all fluorine-containing AFFFs, as there are no regulations requiring municipal fire departments to use AFFF.

Art Schaap (fourth-generation dairy farmer)

Art Schaap described his experience as a dairy farmer in New Mexico whose farm and family were unknowingly subjected to severe PFAS contamination due to military AFFF use.

Schaap stated he has been on his farm for more than 30 years and considered investments in his farm as his version of a 401(k). When he was approached about PFAS water testing, Schaap stated he readily agreed without prior knowledge about PFAS. Those tests revealed levels of PFAS up to 30,000 ppt in drinking well water from his property. These findings led Schaap to contact the New Mexico Department of Agriculture asking for PFAS testing in milk. The department informed Schaap that if PFAS were found in his milk, Schaap would need to dump the load. He insisted on the testing despite that risk, stating he did not want to distribute products with PFAS to customers around the nation. Schaap stated that PFAS testing in milk from his cows subsequently revealed PFAS concentrations ranging between approximately 800 and 2,500 ppt; this testing has been ongoing for more than 2.5 years.

The contamination has resulted in the devastation of his livelihood, as the cows can only be minorly rehabilitated, pose an economic burden, and reflect a major investment that will be devastating to lose without any possible profit. Schaap has limited options available to get rid of the contaminated animals in any profitable way. Dairy, beef, and rendering industries do not want PFAS-contaminated animals. Schaap described his efforts to filter water for his cows, only to find that it takes years for cows to eliminate PFAS from their body (also described by John Kern, below). PFAS have also been found in the soil on Schaap's farm, further exposing his herd beyond drinking water. This untenable situation has unfolded tragically for Schaap and has resulted in the stranding of at least 4,000 cows, the deaths of 1,200, and the dumping of 1,500 loads of milk.

Schaap enumerated a number of institutional failures that have left him with few resources to rectify his situation. He said that multiple government agencies continue to eschew responsibility for the situation and other PFAS exposure concerns, related to DoD pressure and the lack of actionable EPA standards. Schaap pointed out that if FDA provided standards for PFAS in food products, under the Comprehensive Environmental Response, Compensation, and Liability Act (Superfund), contamination on his farm would require action and remediation. He also mentioned the lack of discussion and accountability about this topic across the dairy industry at large.

Schaap concluded his remarks by recounting health effects he has observed over time in cattle, including decreased lactation, premature births, dwarf calves, reduced pregnancy rates, poor body condition, and increased mortality. He also described health concerns from his family, including high cholesterol, hypertension, kidney damage, kidney stones, diabetes concerns, and infertility.

John Kern (RuttenKern Policy Group)

John Kern offered comments to the committee from the perspective of a litigator and environmental advocate residing in New Mexico. Kern explained that in addition to concerns over drinking water, they also focus on PFAS exposure in the water relied on agriculture and a lack of associated safety standards needed to protect the food chain. Kern's testimony dovetailed with the testimony of the previous presenter, Art Schaap.

Kern framed his perspectives by providing information about Cannon Air Force Base near the city of Clovis. AFFF contamination at the military based resulted in groundwater contamination spreading 5–6 miles from the base. The region is home to multiple dairies and cheese plants. Water from this source is provided to livestock, which are then the source for milk, cheese, and meat. This water is also used to cultivate silage vegetation and other crops. PFAS-contaminated groundwater has resulted in PFAS contamination of regional livestock, leading to sundry economic and health questions and concerns unrelated to drinking water exposure.

Kern displayed data from a regional dairy with high levels of PFAS in groundwater; these data show a decrease in lactation in dairy cows over time. Kern also described increasing mortality and birth defects over time in cattle from the same farm. Additionally, Kern presented data about PFAS in milk, stating that the FDA cites 400 ppt as an acceptable level of PFAS in milk. Measurements in milk from the Highland Dairy Farm varied between approximately 900 and 4,600 ppt from November 2018 to August 2020; the data show that dairy milk slowly reflected reduced PFAS exposure. A shorter-term study conducted by the Food Safety Inspection Service showed increased variability and a faster rate of PFAS elimination in dairy milk from exposed cattle compared with data from a longer-term study conducted on the same farm. Kern pointed out these rates have serious implications regarding how quickly cattle are considered rehabilitated from PFAS exposure and allowed for market purchase or dairy use. The slow elimination rates in the longer-term study also pose severe economic ramifications for farmers, given the average lifespan of cattle is around 6 years.

Kern concluded his remarks by describing a frustrating current impasse with the DoD regarding standards. Federal legislation compelled the DoD to clean up agricultural waters in 2018, yet the DoD responded that they would take no action until a federal agency tasked with ensuring food safety (e.g., the FDA or the U.S. Department of Agriculture) set standards for PFAS in food products. Kern emphasized the need for agencies to set standards to allow effective action on PFAS at multiple scales.

Two other people at the Middle Town Hall provided public testimony, which is available on YouTube[10]:

> Pam Ladds, Don't Undermine Memphremagog's Purity
> Beth Markesino, North Carolina Stop Genx in Our Water

WESTERN TOWN HALL

Liz Rosenbaum (Fountain Valley Clean Water Coalition)

Liz Rosenbaum provided testimony as an AFFF-exposed community member in El Paso County, Colorado. Rosenbaum described hearing about regional drinking water contamination in a group setting based on *The New York Times* reporting. Following the initial news of exposure, the community focused on learning about PFAS and government agencies that could assist in an appropriate response. The community response also entailed building partnerships with water districts and assuaging community anger toward these entities, who learned about the contamination at the same time as the community with no prior knowledge of the issue. Since 2018, Rosenbaum stated, the community has been focused on navigating the regulatory and legislative system to reap meaningful action as soon as possible, turning their anger into action through state legislation. This has involved building connections to EPA, state and county health departments, and elected officials at city and state levels. Rosenbaum reported that, as a result of these organizing efforts, legislation has been adopted increasing fines for polluting corporations.

---

[10] See https://www.youtube.com/watch?v=mGeABDDTCuI&list=PLGTMA6Qkejfg1GIgAPeqkMCDncdx 04qHN&index=16.

Beyond detailing the exposure reality for her community, Rosenbaum described priority groups that require increased attention in further PFAS studies. She stated that military families move frequently and may be exposed at multiple sites; care must be provided to military families, as well as service members, from all branches. Rosenbaum also suggested that service members require assistance from the U.S. Department of Veterans Affairs (VA) like that provided to Vietnam veterans. Rosenbaum also indicated the need for equity in exposure assessment and mitigation for Black, Brown, and Indigenous families, as well as rural families, mentioning that rural families in her county are not being offered drinking water at the same nondetect level offered to households in PFAS-exposed municipalities. Rosenbaum stated that a recent ATSDR community-level exposure assessment revealed that this clean municipal water facilitated a decrease in PFHxS[11] levels in the blood of municipal residents.

Rosenbaum also pointed out the limited number of health care options in El Paso County, stating there are no hospitals in the southern half of the county. This lack of health care facilities exacerbates contamination issues because there is no capacity to monitor health outcomes, as families seek health care in the northern part of the county where doctors may not be able to identify localized health outcome patterns. Rosenbaum emphasized the immediate need for blood testing to establish a baseline that serves as evidence of contamination and an indicator of potential health effects. Rosenbaum concluded by emphasizing the need for continued patience and engagement between health care practitioners and researchers on this issue to find shared vocabulary. She explained that communities are living through frightening and unfamiliar exposures but may not have the appropriate technical vocabulary to describe their concerns or ailments.

> "Half the battle for the community has involved learning how to ask the right questions to get what we need to have clean and safe communities for our working families." *Liz Rosenbaum*

Martha Dina Argüello (Physicians for Social Responsibility)

Martha Dina Argüello provided perspectives as a Latina and an environmental advocate in California with experience connecting health care providers and communities. Argüello posited that guidance from the committee must include provisions for increased biomonitoring as many communities facing exposure are as yet unaware of the problem. She also discussed the distinct legacy of industrial contamination in Los Angeles. This history has resulted in community exposures to multiple pollutants, which means that approaches that focus on one chemical at a time do not address the lived reality of complex contamination. Argüello pointed out that these factors have led to distrust of water quality, particularly in the Latino community. She highlighted that Latino communities often pay more for water per capita than for fuel as a result of their reliance on bottled water. Comprehensive testing and proper health education for communities and practitioners is key to rebuild trust in water quality and address the comprehensive reality of complex environmental exposures.

Argüello also explained that communities often face the downplaying of anxieties and risks surrounding exposure; this reflects a lack of training that leads to an inability by clinicians to validate the lived experiences of exposed communities. This dynamic further compromises exposed communities. Argüello emphasized that it is essential for clinical guidance about PFAS exposure to include instruction about how to help concerned patients minimize exposure rather than brush off environmental health concerns.

Argüello cited the dynamics of PFAS regulation in California as an example illustrating how to avoid this paradigm of patient treatment. Argüello highlighted that helpful legislative efforts have been under way, yet early warning systems detecting PFAS in drinking water and through biomonitoring studies have not been appropriately heeded. She specifically flagged the response of a physician who is a member of the state legislature, who stated that exposure should not be quantified for patients as the worry is worse than the exposure. Argüello stated this exemplifies how physicians are trained to respond

---

[11] Perfluorohexanesulfonic acid, a PFAS; see Chapter 1.

to environmental health concerns, and yet this lack of training and understanding of exposure often leads to minimizing people's experience with exposures.

Argüello also urged that clinical guidance should take an anticipatory approach to ensure health care practitioners are provided some education and literacy surrounding the PFAS issue and complex exposure scenarios: health care practitioners should be capable of some interpretation of water quality results and be able to comment on options for filtration. Argüello also specified that clinical guidance must provide intervention options accessible to different socioeconomic and cultural backgrounds. This stratification of intervention is critical to ensure appropriate guidance is provided for all types of exposed communities as many cannot "buy their way out of being exposed." Argüello also highlighted tools that allow tailored exposure assessment and risk reduction strategies, citing how doctors are now writing prescriptions for new carpeting or other household materials to aid rental tenants in substandard housing. Argüello also touted the use of a geospatial tool that allows physicians to assess cumulative exposure risk based on patient location overlaid with information about multiple ambient environmental health exposures.

"When scientists fear speaking truth to power, we know that truth dies."

"We need physicians to step forward because we have regulatory agencies that are actually not preventing exposure."

"We can't change this broken system without the partnership of science and physicians."
Martha Dina Argüello

Mark A. Favors (Fountain Valley Clean Water Coalition)

Mark A. Favors provided commentary as an Army veteran and community member from Colorado Springs, Colorado. Favors and his family learned of serious PFAS contamination affecting their region from activities at Peterson Air Force Base and other regional military bases, prompting Favors to assume an ardent advocacy role seeking safe water. Favors pointed out that while important state legislation regulating PFAS has been passed in Colorado (also discussed by Liz Rosenbaum, above), DoD is exempt from most state legislation. This exemption stands despite the fact the Air Force admitted to dumping AFFF into regional water resources three times per year for multiple decades.

Favors detailed that sustained contamination from AFFF exposure has led to high levels of PFHxS in the blood of people in the community: measured levels were the highest in the country except for individuals who directly manufactured the chemical. With this exposure in mind, Favors chronicled harrowing details about his family's struggle with health problems. For example, there have been many cases of kidney disease and cancer in family members of many ages in the contaminated zone, some requiring kidney transplants that were further complicated by development of cancer in the donated kidney. Favors stated that the abundance of serious kidney ailments in his family in Colorado Springs is particularly striking considering that no family members who live outside the contamination zone have encountered these issues.

Favors explained that these health issues assumed urgent relevance for his family in 2016 when the Air Force disclosed dumping PFAS into drinking water sources, given the known links between PFOA and kidney issues found at other locations. Favors reported that DoD also revealed the detection of PFAS in some community drinking water sources at concentrations up to 8,000 ppt, well above the EPA's health advisory limit of 70 ppt. Favors indicated that despite these levels, the state of Colorado has not provided expansive PFAS blood testing to all residents, even knowing the inherent value of blood testing as an indicator of exposure and potential health effects. Favors emphasized the particular importance of transparent information and access to testing when considering the transitory lifestyles of military members and other community residents. He pointed out that he did not find out about the contamination in his hometown until 2018 while visiting his mother in the region. Favors provided an additional example of a cousin who previously lived in the contaminated zone and has since moved, only to find unexplained liver tumors in one of their young children years later.

Favors also stressed the need for transparency and education in the health care community and shared his experience as a health professional in endoscopy. He stated that despite evidence linking PFAS to irritable bowel disease, gastroenterologists in his department were unaware of PFAS. Favors highlighted that such transparency and access to exposure information is key for both practitioners and community members, especially for exposed children and their families who are concerned about developing health problems in the future.

> "I have family members ... buried at Fort Logan National Cemetery after surviving combat tours in Vietnam, Korea, and Afghanistan.... They're now buried in the National Cemetery from cancer after the military admitted dumping this toxic chemical into their water." *Mark A. Favors*

Andria Ventura (Clean Water Action)

Andria Ventura provided context about past and ongoing actions addressing PFAS in California from the vantage point of an environmental advocate. Ventura opened her remarks by describing her continued struggle engaging with the medical community about environmental health concerns as a long-time advocate, describing it as difficult for clinicians to make the connection between exposure and health outcomes. Ventura hypothesized that this stems from reluctance or avoidance by clinicians to assume an advocacy role. As a patient and a resident of a PFAS-affected city, she also stated she has never been asked about toxic exposure or environmental health background by her doctors or other medical caregivers, despite dealing with several chronic concerns over decades. She had to actively broach this topic with providers, likening this experience to the testimony of other presenters to the committee. However, it has been clear that doctors and nurses are not trained to take toxic exposure into consideration.

Ventura went on to provide abundant information about the PFAS problem overall. She highlighted the multifaceted ways humans can be exposed to PFAS, including through drinking water, surface waters, consumption of wildlife, or consumption of agricultural products exposed to PFAS in soil or water. She also reiterated the importance of considering PFAS as a class, stating thousands of PFAS have been identified and all are considered persistent, accumulative, mobile, and hazardous to some degree. Ventura also showed that multiple PFAAs have been linked to health effects that affect immunity, development and reproduction, fat and metabolism, liver function, endocrine function, and blood systems. She stressed the potential for additive and synergistic effects related to exposure to multiple PFAS, flagging this as a required point of awareness for health care practitioners. Ventura also emphasized that lack of research on novel PFAS should not hinder consideration of their health effects, stating that evidence is mounting that newer, short-chain PFAS have negative health effects. Ventura also stated that many novel PFAS degrade or transform into PFAAs.

Ventura went on to elaborate about specific actions and concerns in California, describing the recent position of the state to implement phased drinking water monitoring and site investigation. She added the caveat that this phased approach has failed to assess small water systems and private wells: this is a key data gap considering health care practitioners need detailed information about exposure to adequately consider environmental health concerns during care. Despite data collection limitations, data thus far indicate catastrophic PFAS contamination problems across the state that have so far not been addressed by sluggish regulatory efforts. Ventura also stressed key data gaps, such as uncertainties about PFAS levels in surface waters, PFAS in fish and wildlife consumed by humans, and implications for crops grown in PFAS-laden biosolids or irrigation water.

Ventura concluded her remarks by delineating key exposure assessment and health care needs in California. She emphasized the need for expanded understanding of the scope of the problem, including all PFAS in assessment efforts, as well as improved understanding about PFAS health effects. She suggested such fact-finding efforts should include expanded water monitoring; PFAS monitoring in diverse environmental media; assessment of diverse exposure vectors; identification of those most at risk of harm from exposure; expanded health studies, including mixture exposure scenarios; and communication of health risks in culturally appropriate ways.

Linda Shosie (Mothers for Safe Air & Safe Water)

Linda Shosie provided perspective as a Latina and exposed community member in the South Side neighborhood of Tucson, Arizona. Shosie provided information to the committee collected as part of community-based exposure and health assessment efforts, highlighted egregious environmental justice issues, and explained her own path to advocacy. Shosie explained that her community is affected by PFAS contamination from military sites and airport activities, resulting in PFAS concentrations in drinking water up to 13,000 ppt. This high level of PFAS exposure is plaguing a majority-Latino community across a 3-mile contamination plume; Shosie provided a number of maps providing geospatial context about the extent of contamination. She described leading health assessments within the community, which revealed high rates of cancer and immune system disease around the Tucson International Airport and Morris Air Force Base. These community-derived results led the county health director to conduct a more in-depth epidemiological study in 2017. Shosie stated this study indicated "significant invasive cancer incidence rates" compared with people living in other areas around Tucson.

Shosie expressed frustration, sadness, and anger at the lack of action addressing PFAS contamination in her community and at other Superfund sites, saying, "environmental justice provisions continue to fail meanwhile thousands of contaminated sites remain unclean for more than 40 years." She also stated that those sites that have been remediated are in White or upper-class neighborhoods, while sites like the one in her low-income community remain unresolved to the detriment of community health and trust in governing institutions. Shosie also shared emotional details about losing her daughter, prompting her own path to find answers about environmental health issues in her community.

> "We cannot rely on state, CDC, and other local government officials who continue to turn their backs on the people who are affected."

> "Many people asked me why I got involved in the fight for … clean water, and demand government transparency and accountability. I got involved because I witnessed the death of my daughter out of this devastation I knew that I needed to find out why my daughter got so sick. *Linda Shosie*

Aaron Maruzzo (University of California, Berkeley, School of Public Health)

Aaron Maruzzo commented as a voice representing PFAS-contaminated communities in Western U.S. territories like Saipan in the Northern Mariana Islands. Maruzzo was born in Saipan and returned to work in the territory as a water quality analyst. Maruzzo framed the PFAS contamination problem in Saipan by leveraging data collected as part of the EPA's Third Unregulated Contaminant Monitoring Rule program. These program data capture PFAS concentrations in select public water systems from around the United States. Maruzzo comparison revealed major contamination in Saipan, with an average concentration of 1,700 ppt in territory drinking water, well above EPA health advisory limits and the levels observed in other U.S. states and territories. The highest detectable concentration of PFOS was also found in Saipan. Maruzzo said these data have been hiding in plain sight for many years and illustrate an ongoing environmental injustice.

Maruzzo went further to explain that PFAS contamination across the small island is unevenly distributed, with the highest levels found in water resources adjacent to the only airport found on the island, home to firefighting training facilities that used AFFF. Sixteen villages along the south and southwest margins of the island were disproportionally exposed to PFAS from this drinking water source, including the village where Maruzzo grew up. The southern portion of the island tends to include more non-White, noncitizens who may be easily missed in health data collection efforts as a result of frequent immigration and emigration. Health effects in this population may be also missed because of reluctance to seek medical care due to costs and accessibility. Despite these data collection challenges, preliminary reports suggest deep impacts on morbidity and mortality in the region, including heart disease, cancer, abnormal birth outcomes, hypertension, obesity, and cholesterol issues.

Maruzzo indicated that environmental justice issues are a key concern for his community, as "toxic exposures to hazardous chemicals continue to be disproportionately placed on communities of color and the poor working class." Maruzzo also stressed the continued systemic exclusion of U.S. territories like Saipan from continued dialogue about issues like PFAS.

With all this context, Maruzzo explained that the community needs further information about PFAS exposure routes specific to a small island setting. He indicated the need for further information about PFAS in the water of neighboring islands in the Northern Mariana Islands and the levels in private water systems, bottled water, and wastewater. Maruzzo flagged information gaps surrounding AFFF, questioning when AFFF was first used on the island and the composition of AFFFs over time. Maruzzo also questioned how military testing, imperialism, and globalization affect the contamination crisis today. Additionally, Maruzzo pointed out a lack of information describing PFAS in many consumer products, such as textiles and food packaging. Maruzzo indicated data gaps surrounding PFAS in fish and highlighted this as a problem, considering the cultural and economic importance of fishing on the island.

Following discussion of the unique exposure scenario ongoing in Saipan, Maruzzo offered comments, recommendations, and questions for the committee to consider while drafting clinical guidance. He explained that monitoring in humans can help answer questions about the importance of all these exposure gaps, while establishing an important baseline perspective. Maruzzo also argued for the importance of biomonitoring for optimal health care, as it allows families to make informed health care and lifestyle choices while empowering individuals to take action in their communities and environments. Biomonitoring studies also help constrain health effects, even as the candidate list of health effects explicitly related to PFAS requires further study. Maruzzo suggested that despite many unknowns plaguing our understanding of health effects related to PFAS, biomonitoring should be prioritized as it enables a precautionary approach to allow identification of unknown PFAS and health effects. Maruzzo mentioned the highly persistent nature of PFAS, pointing out that it is unadvisable to allow highly persistent chemicals to remain in our body only to find out about health effects later. Maruzzo also said this characteristic should be emphasized to physicians to help them understand the complexity of PFAS given typically low awareness about the issue in health care settings and in communities.

Maruzzo echoed other participants by pointing out that highly exposed and vulnerable populations should be prioritized for PFAS blood testing, using equity as a guide to design testing protocols. He further suggested that inclusion in these categories should be constrained by occupation, location, and biosocial vulnerabilities. Maruzzo stressed the importance of considering who is in the 95th or 99th percentile of exposure, while questioning who is missing from the dataset, to maximize understanding of the issue while minimizing exposure harms. Maruzzo raised the issue that there is also a moral component to blood testing that must be considered by the committee, asking how physicians will be prepared to educate exposed community members who do not already know about issues related to PFAS exposure. He asked whether an established protocol, cost considerations, or health care access would drive such decisions. Maruzzo also discussed the need to develop guidelines describing how to care for transient populations like migrants and military personnel. He asked the committee to consider how testing programs and rigorous exposure assessment should account for high loss of follow-up and discontinuity of care for these populations. Maruzzo also highlighted the value of culturally appropriate guidance and communication, stating that general scripts about risk reduction are "useful for a broad audience but there should be a mechanism to specify what's known into a local context."

"If there's one takeaway today, I think it's important to consider the implications of what happens when you don't listen to the voices at the margins."

"The absence of evidence or the absence of consensus doesn't mean a PFAS compound is safe."

"One of the most PFAS-polluted places in the United States is a U.S. territory, and this is a serious environmental injustice."

"Risk communication should prioritize not solely action but meaningful action which is guided through the lens of equity." *Aaron Maruzzo*

Pamela Miller (Alaska Community Action on Toxics)

Pamela Miller spoke as a community member and environmental health advocate in Alaska. Miller provided abundant context about the unique exposure and community dynamics present in Alaska, while offering clear recommendations to the committee about Alaskan health needs and concerns. Miller explained that Alaska's strategic military importance has resulted in the establishment of multiple military installations across Alaska that have used AFFF. Miller also explained that Alaska is a very aviation-dependent state, with AFFF-using airports situated directly adjacent to communities and their water resources. Miller stated that PFAS have been found in at least 100 different sites across 30 locations spanning the entirety of the state, with multiple communities exposed to unsafe amounts of numerous PFAS. A total of 11 current and former military installations are currently under investigation for PFAS contamination, with results to date indicating the environmental occurrence of PFAS above health advisory levels. Despite the abundance of known sites, and the likelihood of as-yet-unidentified sites, Miller explained that there is no cohesive state plan to measure PFAS in the environment, fish, and wildlife, or in people. Miller mentioned an overall failure of regulatory efforts to guide the state's response to PFAS contamination, despite the support of attempted legislative efforts by associations of health care professionals, such as the Alaska Nurses Association. Only a handful of the 33 communities relying on water likely contaminated by AFFF or other PFAS sources have been able to access drinking water testing, contributing to widespread unawareness of the problem across the state.

Miller stated that limited assessment of drinking water and other environmental factors is matched by a lack of health assessments probing the effects of PFAS in exposed communities; only two health assessments have been conducted. One of these studies found positive correlations between the PFAS found in drinking water and those compounds found in serum, indicating an influence of drinking water contamination on human body burdens. Miller highlighted the particular plight of Alaskan remote regions as a hemispheric sink for persistent organic pollutants like PFAS. Remote polar regions receive undue burdens of mobile pollutants as a result of global distillation processes that transport pollutants to remote areas, including PFAS. Marine mammals and fish from polar regions therefore contain some of the highest burdens of persistent organic contaminants in the world because of these transport mechanisms and the bioaccumulative nature of PFAS. She shared the results of a community-based study in an Indigenous community on the island of Sivuqaq (St. Lawrence Island) in the Northern Bering Sea. This study tested the blood of community members reliant on traditional diets incorporating polar fish and marine mammals. The assessment found 13 PFAS in the blood of 85 people, as well as correlated concentrations of select PFAS to thyroid disruption.

Miller urged the need for biomonitoring in Alaskan communities with known or suspected PFAS contamination, including remote communities exposed through water sources, traditional foods, firefighting workers, and other exposed workers. Miller stated that human biomonitoring should be paired with assessments investigating exposure from water, dust, produce, fish, and wild game. These data are vital to inform health care providers and the wider community about exposures, possible associations with adverse health effects, appropriate risk reduction interventions, and relevant health care options. These data are also vital to inform policy.

Miller provided examples that emphasized the need to ensure that clinical guidance about PFAS is made accessible to diverse health care providers through various communication and training avenues. She explained that in Alaska health care needs are often addressed by community health aides as many communities lack doctors or nurses. Health aides in Alaska require the same clinical guidance and education afforded physicians, given their central role in meeting community health needs. Current, clear synthesis of scientific information, as well as an ability to execute biomonitoring and medical monitoring, is needed for physicians, nurses, and community health aides to allow these health care providers to inform and protect their patients from PFAS. Miller suggested that provider education should occur through professional organizations, as well as continuing education credits offered through public health, medical, and community training programs tailored to different provider types, including the Indian Health Service and regional health care providers.

> "PFAS are contaminating groundwater, surface waters, fish, wild game, garden produce, and people throughout Alaska."

> "The burden of proof should not be on our communities, and this must change so that laws reflect current scientific understanding and are protective of public health."

> "Health care professionals must be informed to become even more effective advocates for their patients and for ending nonessential uses of PFAS." *Pamela Miller*

Randy Krause (Port of Seattle Fire Department)

Randy Krause, the port fire chief, provided commentary as a veteran firefighter and fire chief who previously used AFFF. Krause detailed his career as a firefighter, spanning experience with the DoD, private industry, and a public-serving fire department. Krause explained that fire training activities with the DoD involved regular training with military-grade AFFF or "mil-spec" foams. Training scenarios were enacted and AFFF was sprayed abundantly onto the training props and into the wider environment. Krause also related that, in 1985, these foams were used routinely to clean floors and wash trucks, and were thrown on other firefighters during training exercises.

When he moved to private industry, Krause found a similar approach to training, where an open pit was used to stage fire scenarios, and AFFF was thrown on training fires. Krause clarified that at the time firefighters were assured these mil-spec foams were safe, biodegradable, and did not pose a risk to the environment, which has seen been learned to be incorrect. Krause became fire chief at Seattle International Airport in 2010. While the department had at one time used a similar open pit training set-up with mil-spec foams for fire training activities, Krause indicated this approach was not in practice when he arrived. The department refrained from use of fluorine-containing AFFFs for training due to state bans.

Krause emphasized that the safety of his team is a top priority as fire chief and described an opportunity to contribute to the Firefighter Cancer Cohort Study. This study is a multicity, long-term national research effort focused on assessing cancer in firefighters across the nation. Participation in this study provided Krause and several other department members access to PFAS blood testing. Krause shared his blood testing data in graphical and chart format detailing concentrations of isomers of PFOA, isomers of PFOS, PFHxS,[12] PFDeA,[13] PFNA, PFUA,[14] and Me-PFOSA-AcOH.[15] His results varied in proximity to provided benchmarks, but showed levels of linear PFOS and PFHxS close to or surpassing the nationwide 95th percentile. Krause explained he does not know what these results mean at this time, but discussion with Dr. Jeff Burgess from the University of Arizona suggests high levels of PFOS and PFHxS are commonly elevated in other firefighters. Krause also highlighted that the reporting techniques used by the Firefighter Cancer Cohort Study were particularly helpful, providing his exposure data, average amounts in firefighters from his department, the 50th and 95th percentiles based on NHANES data, and the range of all amounts measured in the firefighters in his department.

---

[12] Perfluorohexanesulfonic acid or perfluorohexane sulfonic acid, PFAS; see Chapter 1.

[13] Perfluorodecanoic acid, a PFAS; see Chapter 1.

[14] Perfluoroundecanoic acid, a PFAS; see Chapter 1.

[15] 2-N-methyl-perfluorooctane sulfonamido acetate, a PFAS; see Chapter 1.

Jean Mendoza (Friends of Toppenish Creek)

Jean Mendoza spoke as a community member from the Lower Yakima Valley in Washington. Mendoza stated her community and organization have concerns about PFAS because of the prolific application of PFAS-imbued biosolids in agricultural fields across the area. Washington state statutes require biosolid application to the fullest extent possible. This raises concern as the area is highly reliant on agricultural activities and is home to large dairy cow populations; no information has yet investigated crop and livestock safety in response to biosolid-driven PFAS exposure in this area.

Mendoza conveyed discontent and frustration about blatant disregard of environmental health concerns in her community. Mendoza cited efforts by state agencies to measure PFAS in some areas across the state, but noted that PFAS measurements in central Washington and across the Yakima watershed seem to be intentionally omitted. Mendoza also pointed out that Yakima County ranks poorly in the state with respect to health outcomes and health factors, reflecting poor environmental health in an area home to tribal communities and majority Latino populations. In light of these findings, Mendoza stated that the state seems to be blatantly neglecting the exposure of people of the Yakima Valley.

Mendoza listed community health issues, including asthma, myocardial infarction, and low birthweights, noting the region is home to multiple complex exposures including PFAS, air pollution, nitrates, and pesticides that are associated with a number of adverse health outcomes. Mendoza raised the point that it is challenging for Yakima Valley residents to attribute health effects to any one pollutant, given the cumulative exposure to so many pollutants in the region. Mendoza also stated that concern about social maladies often trumps concern about PFAS in the region, given limited awareness of the problem in the community and inaction on the issue by the Yakima Health District. Mendoza moreover detailed that some elected officials take an "ignorance is bliss" approach and opt to avoid investigating the regional PFAS problem to avoid taking action on the issue.

Mendoza enumerated several challenges observed in her community that should be considered by the committee when formulating clinical guidance about PFAS for her community and beyond. She noted that many patients do not understand public health and risk assessment, and clinicians in the area receive no support from local health districts about PFAS. She also highlighted that many of the most exposed are poor households that live paycheck to paycheck, with limited capacity to worry or plan for long-term illness. She also cited the intangible nature of the PFAS problem, with no ability for folks to see, taste, or smell the issue. Mendoza offered recommendations as well, asking the committee to consider educating and informing clinicians about PFAS and other environmental health risks through professional organizations, to support biomonitoring studies, to recommend PFAS testing in fish, and to encourage a moratorium on biosolid application.

> "We pay lip service to scientific evidence over here, but very often in Yakima Valley science is suppressed." *Jean Mendoza*

Rebecca Patterson (Vietnam Veterans of America)

Rebecca Patterson presented commentary as a Navy veteran and veteran advocate, highlighting the importance of PFAS blood testing for veteran's health care. Patterson explained that one of the largest sources of PFAS exposure for service members and military communities is the use of AFFF. These firefighting products have been used since 1970 to fight petroleum fires; legacy AFFFs contain PFAAs like PFOS and PFOA. While these are no longer in service, the DoD continues to use AFFF formulations containing PFAS despite the availability of fluorine-free alternatives. Patterson pointed out that use of these firefighting foams readily introduced PFAS into the environment and the water cycle and has led to widespread environmental contamination now documented at hundreds of military bases around the United States.

Patterson stated that her presentation was intended to educate the committee regarding how blood testing could help PFAS-exposed veterans gain access to VA health care. To accomplish this, Patterson shared information about the Veterans Health Administration (VHA). The VHA is tasked with providing care to eligible veterans, though not all veterans access health care through the VHA. This means community health care practitioners may be serving veterans without knowledge of their service history and related exposure; this necessitates that health care providers ask patients about their service history. Patterson also explained that after basic eligibility criteria are met, service members qualify for VHA care based on several factors, including service-connected disability, income, and exposure to toxicants and environmental hazards.

Service-connected disability is of particular importance to the veteran community; this designation typically requires specific evidence substantiating the connection between the given disease or injury and military service. Since medical concerns can arise years after service, it can be difficult for a veteran to connect an ailment to military service or exposure incurred through service. Disability approval results in tiered compensation and priority access to VHA care. With this context, Patterson concluded that access to health care and disability compensation can have a tremendous impact on a veteran's quality of life. PFAS blood testing can provide evidence of exposure that enables veterans to access vital health care. Patterson also asserted that PFAS blood testing can lead to more informed health care, allowing service members and veterans to screen for and potentially prevent health conditions specifically associated with PFAS exposure.

Bucky Bailey (son of former DuPont Washington Works plant employee)

William "Bucky" Bailey III provided perspective as a community member from Parkersburg, West Virginia. Bailey was born with multiple birth defects, including only one nostril, a keyhole eyelid, a serrated eyelid, and breathing difficulties. The family had no idea what caused the defects, given the lack of similar issues in Bailey's siblings. However, Bailey's mother worked as a full-time employee at the DuPont Washington Works plant in Parkersburg, where she controlled the production of PFOA in a confined area. Upon returning to work from maternity leave, Bailey's mother found that other pregnant women were removed from the Teflon production process. She also discovered that studies had been previously conducted by 3M that showed the same birth defects in laboratory animals exposed to PFAS. Despite these lines of evidence, DuPont denied that Bailey's birth defects were a result of his mother's occupational exposure, and the Bailey family found litigation impossible to pursue given DuPont's stature in the community.

Years later, Bailey met Rob Bilott, the lawyer who uncovered DuPont's malfeasance and pursued settlements for exposed residents of Parkersburg. Bailey described feeling relief finding out about links between PFOA exposure and health effects, following years of surgeries and underlying uncertainty regarding the cause of his deformities. Yet Bailey explained that joy following these revelations was also met with disheartenment and discouragement, knowing that the contamination that likely caused his deformities was entirely out of his control and had caused other health problems and untimely deaths in his family and wider community.

Bailey indicated that the C-8 Health Project did not find concrete links between his specific deformities and PFOA exposure, despite admission by DuPont scientists that the compound can cause birth defects. Bailey was also told his children would have a 50 percent chance of inheriting his health issues, which Bailey described as a tough and deeply troubling finding considering his marriage and his love of children. Bailey stated he struggled deeply with the decision to have kids, not wishing to put his children through what he went through as a child. Bailey and his wife ultimately decided to have children after wrestling with the question for over 10 years, and he reported he is the happy father to a healthy son and daughter. Bailey pointed out that knowledge of his contamination and related health risks delayed their family's decision to have children, which ultimately occurred after his father passed away. This timing deprived his children a relationship with their grandfather and deprived Bailey's father the opportunity to meet his grandkids. Bailey stated that his concerns now center on potential health effects like kidney and testicular cancer that he may encounter in the future given his significant PFOA contamination.

"It was joy for me to learn some of the things the scientific study found out."

"I hope we can all acknowledge that we need to move in the same direction at the same time and not point fingers and not fight and not quarrel but find out what we can do to stop this from happening because it is going to cost us our lives." *Bucky Bailey*

One other person at the Western Town Hall provided public testimony, which is available on YouTube[16]: Gina Solomon.

---

[16] See https://www.youtube.com/watch?v=WghqL6urt6w&list=PLGTMA6QkejfjVRybfjFf-vRkIQnEh Ob0a&index=22.

Appendix C

Public Meeting Agendas

COMMITTEE ON THE GUIDANCE ON PFAS TESTING AND HEALTH OUTCOMES

The Keck Center, 500 Fifth Street, NW
Washington, DC 20001

FEBRUARY 4, 2021

VIRTUAL MEETING

1:00	Purpose of Open Session and Introduction of Committee Members **Bruce N. Calonge**, *Committee Chair*
1:15–2:00	ATSDR Perspectives on Study Scope, Background, and Objectives **Patrick Breysse**, *Director, National Center for Environmental Health/Agency for Toxic Substances and Disease Registry*
2:00–2:45	NIEHS Perspectives on Study Scope, Background, and Objectives **Brian R. Berridge**, *Scientific Director, Division of the National Toxicology Program, Associate Director, National Toxicology Program*
2:45–3:35	Committee Discusses the Statement of Task with the Sponsor
3:35–3:40	Break
3:40–3:45	Instructions for Public Comment Session **Bruce N. Calonge**
3:45–4:45	Opportunity for Public Comment on Committee's Charge (must preregister, 1 person per organization, 3 minutes each)
4:45	ADJOURN

APRIL 7, 2021
EASTERN COMMUNITIES TOWN HALL (ATSDR REGIONS I–IV)
VIRTUAL MEETING

2:00–2:10	Welcome and Introductions **Bruce N. Calonge**, *Committee Chair*

SESSION A—Community Perspectives from ATSDR Region I

2:10–3:10	Community Perspectives from ATSDR Region I **Alan Woolf**, *Moderator*, Harvard Medical School/Boston Children's Hospital

2:10–2:20	**Laurene Allen**, Merrimack Citizens for Clean Water
2:20–2:30	**Teresa Gerade**, Don't Undermine Memphremagog's Purity (DUMP)
2:30–2:40	**Ayesha Khan**, Nantucket PFAS Action Group
2:40–2:50	**Kristen Mello**, Westfield Residents Advocating For Themselves (WRAFT)
2:50–3:10	Discussion

SESSION B—Community Perspectives from ATSDR Region II

3:10–3:45	Community Perspectives from ATSDR Region II **Laurel Schaider**, *Moderator*, Silent Spring Institute
3:10–3:20	**Tracy Carluccio**, Delaware Riverkeeper Network
3:20–3:30	**Loreen Hackett**, PFOA Project New York
3:30–3:45	Panel Discussion
3:45–3:55	Break

SESSION C—Community Perspectives from ATSDR Region III

3:55–4:30	Community Perspectives from ATSDR Region III **Maida Galvez**, *Moderator*, Icahn School of Medicine at Mount Sinai
3:55–4:05	**Patrick Elder**, Military Poisons
4:05–4:15	**Hope Grosse**, Buxmont Coalition for Safe Water
4:15–4:30	Panel Discussion

SESSION D—Community Perspectives from ATSDR Region IV

4:30–5:40	Community Perspectives from ATSDR Region IV **Linda Birnbaum**, *Moderator*, National Institute of Environmental Health Sciences, National Toxicology Program (retired)
4:30–4:40	**La'Meshia Whittington**, North Carolina Black Alliance
4:40–4:50	**Mike Watters**, Grays Creek Residents United Against PFAS in our Wells & Rivers
4:50–5:00	**Emily Donovan**, Clean Cape Fear
5:00–5:10	**Cheryl Sievers-Cail**, South Carolina Indian Affairs Commission/SC Idle No More
5:10–5:20	**Stel Bailey**, Fight For Zero
5:20–5:45	Panel Discussion

5:40–5:45   Break

SESSION F—OPEN COMMENT PERIOD

5:45–6:45   Public Comments

6:45   ADJOURN

MAY 6, 2021
MIDDLE COMMUNITIES TOWN HALL (ATSDR REGIONS V–VII)
VIRTUAL MEETING

3:00–3:10   Welcome and Introductions
Bruce N. Calonge, *Committee Chair*

SESSION A—Community Perspectives from ATSDR Region V

3:10–3:55   Community Perspectives from ATSDR Region V
Phil Brown, *Moderator*, Northeastern University
Andi Rich, Save Our Water (S.O.H20)
Laura Olah, Citizens for Safe Water Around Badger (CSWAB)
Samraa Luqman, Concerned Residents for South Dearborn
Cathy Wusterbarth, Need Our Water (NOW)

3:55–4:15   Panel Discussion with Session B Speakers

SESSION B—Community Perspectives from ATSDR Region V (continued)

4:15–4:50   Community Perspectives from ATSDR Region V
Courtney Carignan, *Moderator*, Michigan State University
Sandy Wynn-Stelt, Belmont, Michigan, resident
Tom Johnson, Clean Water Action
Vicki Quint, Foam Exposure Committee/Code PFAS

4:50–5:10   Panel Discussion

5:10–5:20   Break

SESSION C—Community Perspectives from ATSDR Region VI and VII

5:20–5:55   Community Perspectives from ATSDR Regions VI and VII
Alan Ducatman, *Moderator*, West Virginia University
Art Schaap, Highland Dairy, Cedar Rapids, Iowa Resident
John Kern, Clean Water Partnership

5:55–6:15   Panel Discussion

6:15–6:20   Break

6:20–7:00   Public Comments

7:00   ADJOURN

MAY 25, 2021
WESTERN COMMUNITIES TOWN HALL (ATSDR REGIONS VIII–X)
VIRTUAL MEETING

1:00–1:10   Welcome and Introductions
              Bruce N. Calonge, *Committee Chair*

SESSION A—Community Perspectives from ATSDR Region VIII and IX

1:10–2:00   Community Perspectives from ATSDR Region VIII and IX
              Elizabeth Neary, *Moderator*, Wisconsin Environmental Health Network
              Liz Rosenbaum, Fountain Valley Clean Water Coalition
              Martha Dina Argüello, Physicians for Social Responsibility, Los Angeles
              Mark A. Favors, Army Veteran

1:40–2:00   Panel Discussion

SESSION B—Community Perspectives from ATSDR Region IX

2:00–2:30   Community Perspectives from ATSDR Region IX
              Lenny Siegel, *Moderator*, Center for Public Environmental Oversight
              Andria Ventura, Clean Water Action Fund
              Linda Shosie, Environmental Justice Task Force, Tucson, Arizona
              Aaron Maruzzo, University of California, Berkeley, School of Public Health

2:30–2:50   Panel Discussion

2:50–3:00   Break

SESSION C—Community Perspectives from ATSDR Region X

3:00–3:30   Community Perspectives from ATSDR Region X
              Anna Reade, *Moderator*, Natural Resources Defense Council
              Pamela Miller, Alaska Community Action on Toxics
              Randy Krause, Port of Seattle/Washington State Association of Fire Chiefs
              Jean Mendoza, Friends of Toppenish Creek

3:30–3:50   Panel Discussion

SESSION D—Additional Community Perspectives

3:50–4:20   Additional Community Perspectives
              Celeste Anne Monforton, *Moderator*, Texas State University
              Rebecca Patterson, Vietnam Veterans of America
              Bucky Bailey, Son of former DuPont Washington Works plant employee

4:20–4:40   Panel Discussion with Session E Speakers

4:40–4:45   Break

4:45–5:00   Public Comments

5:00   ADJOURN

JULY 13–14, 2021
INFORMATION-GATHERING SESSION, MEETING 5
VIRTUAL MEETING
Thursday, July 13, 2021

1:30–1:40   Welcome and Introductions
                 Bruce N. Calonge, *Committee Chair*

SESSION A—Patient Perspectives on PFAS Testing and Health Outcomes

1:40–1:55   Patient Perspectives on PFAS Testing and Health Outcomes
                 Andrea Amico, Testing for Pease

SESSION B—Human Exposure Sources

1:55–2:25   Current Knowledge About the Contribution of PFAS Exposure Sources to
                 Human Exposure
                 Elsie M. Sunderland, Harvard University

2:25–2:55   Panel Reflection and Q&A from Committee
                 Chris Wiant, *Moderator*, Committee Member
                 Bruce H. Alexander, Colorado State University
                 Thomas F. Webster, Boston University School of Public Health
                 Laurel Schaider, Silent Spring Institute
                 Elsie M. Sunderland, Harvard University

2:55–3:05   Break

SESSION C—Human Exposure Reduction

3:05–3:35   Clinical Principles for Advising Patients to Reduce Exposure
                 Sheela Sathyanarayana, University of Washington

3:35–4:05   Panel Reflection and Q&A from Committee
                 Brian Linde, *Moderator*, Committee Member
                 Judy LaKind, LaKind Associates
                 Andrea Amico, Testing for Pease
                 Sheela Sathyanarayana, University of Washington

4:05–4:35   Open Comment Period

4:35   ADJOURN

HEALTH EFFECTS
Wednesday, July 14, 2021

1:00–1:10   Welcome, Purpose of Open Session
Bruce N. Calonge, Committee Chair

SESSION D—Overview of Putative Health Effects

1:00–2:10   Epidemiology: **David Savitz**, Brown University
Toxicology: **Jamie DeWitt**, East Carolina University

2:10–2:40   Panel Reflection and Q&A from Committee
Jane Hoppin, *Moderator*, Committee Member
Linda Birnbaum, National Institute of Environmental Health Sciences, National Toxicology Program (retired)
Joseph M. Braun, Brown University
Matthew Longnecker, Ramboll
David Savitz, Brown University
Jamie Dewitt, East Carolina University

2:40–2:55   Break

SESSION E—Evidence Synthesis and Its Application

2:55–3:25   Methods for Evidence Synthesis
Jonathan Samet, Colorado School of Public Health

3:25–3:55   Making Useful Recommendations
Rebecca L. Morgan, McMaster University

3:55–4:25   Panel Reflection and Q&A from Committee
Bruce N. Calonge, *Moderator*, Committee Member
Ellen Chang, Exponent
Nicholas Chartres, University of California, San Francisco
Holger Schünemann, McMaster University
Rebecca L. Morgan, McMaster University
Jonathan Samet, Colorado School of Public Health

4:25–4:55   Open Comment Period

4:55   ADJOURN

AUGUST 11–12, 2021
INFORMATION-GATHERING SESSION, MEETING 6
VIRTUAL MEETING
Wednesday, August 11, 2021

1:25–1:35   Welcome and Introductions
Bruce N. Calonge, *Committee Chair*

SESSION A—Frameworks for Making Decisions on Clinical Evaluation and Biomonitoring

1:35–2:05   Principles for Making Decisions on Clinical Evaluation and Biomonitoring
David Resnik, National Institute of Environmental Health Sciences

2:05–2:35   Application of Decision-Making Framework in the C-8 Medical Monitoring Panel
Dean Baker, University of California, Irvine

2:35–3:05   Panel Reflection and Q&A from Committee
Kevin Elliott, *Moderator*
Courtney Carignan, Michigan State University
Ayesha Khan, Nantucket PFAS Action Group
Jeffrey Brent, University of Colorado
Dean Baker, University of California, Irvine
David Resnik, National Institute of Environmental Health Sciences

SESSION B—Clinician Perspective on Advising Patients in PFAS-Exposed Communities

3:05–3:25   Clinician Perspective on Advising Patients in PFAS-Exposed Communities
Alan Ducatman, West Virginia University

3:25–3:55   Panel Discussion with Clinicians Who Have Advised Patients in PFAS-Exposed Communities
Laura Anderko, *Moderator*
Katie Huffling, Alliance of Nurses for Healthy Environments
Stewart Reed, University of California, Los Angeles
Maida Galvez, Icahn School of Medicine at Mount Sinai Hospital
Alan Ducatman, West Virginia University

3:55–4:25   Open Comment Period

4:25   ADJOURN

Thursday, August 12, 2021

1:00–1:10   Welcome, Purpose of Open Session
Bruce N. Calonge, *Committee Chair*

**SESSION C—Clinical Principles for Communicating Biomonitoring
Results and Relationship with Clinical Care**

1:10–1:40   Biomonitoring Results: Communication
                 Julia Brody, Silent Spring Institute

1:40–1:55   PFAS Clinical Guidance
                 Phil Brown, Northeastern University

1:55–2:25   Panel Reflection and Q&A from Committee
                 Erin Haynes, *Moderator*, Committee member
                 Marc A. Nascarella, Massachusetts Department of Public Health
                 Jessica Nelson, Minnesota Department of Health
                 Gary Ginsberg, New York Department of Health
                 Julia Brody, Silent Spring Institute
                 Phil Brown, Northeastern University

2:25–2:55   Open Comment Period

2:55   ADJOURN

Appendix D

Evidence Review: Methods and Approach

This appendix describes the approach and methods that the committee used to address the portions of the Statement of Task that asked for "an objective and authoritative review of current evidence regarding human health effects of those PFAS [per- and polyfluoroalkyl substances] being monitored in the CDC's [Centers for Disease Control and Prevention's] National Report on Human Exposure to Environmental Chemicals" (see Box 1-3 in Chapter 1). The Statement of Task specifically asked the committee to:

> Assess the strength of evidence for the spectrum of putative health effects suggested by human studies (including immune response, lipid metabolism, kidney function, thyroid disease, liver disease, glycemic parameters and diabetes, cancer, and fetal and child development) to establish a basis for prioritized clinical surveillance or monitoring of PFAS health effects. This assessment should characterize the likelihood of those health effects occurring (qualitative probability) given real world human exposures and identify the human populations at most risk (consider life stage, health status, exposure level). Data/evidence gaps that contribute to uncertainty about health effects of most concern should be annotated.

The committee decided that this portion of the Statement of Task required three different determinations:

- qualitative categories that describe the strength of evidence of PFAS putative health effects that can be used to prioritize clinical surveillance or monitoring,
- identification of the human populations at most risk (considering life stage, health status, exposure level) from PFAS exposure, and
- a scoping review that maps the data or evidence gaps that contribute to uncertainty about health effects of most concern.

To produce these outputs, the committee developed a multistage process. The first stage was to catalog what is known about PFAS and their health effects. The committee identified all authoritative reviews of the PFAS identified by the CDC (see Table 1-1 in Chapter 1) and all human health outcomes. The second stage was to identify any recent, high-quality systematic reviews between PFAS and any human health outcome; as noted below, the results of this stage were uninformative for the committee's goals. The third stage was to review the published research articles describing the association between exposure to PFAS and human health outcomes, based on the authoritative and systematic reviews. The committee's review approach improved efficiency while minimizing the risk of excluding scientific findings that would inform the committee's recommendations.

The committee was charged with assessing

> the strength of evidence for the spectrum of putative health effects suggested by human studies (including immune response, lipid metabolism, kidney function, thyroid disease, liver disease, glycemic parameters and diabetes, cancer, and fetal and

child development) to establish a basis for prioritized clinical surveillance or monitoring of PFAS health effects.

However, the committee did not restrict its evaluation to only the listed putative health effects.

The next section of the appendix covers the committee's analysis of the authoritative reviews; the following section covers its original literature review, and the final section covers the committee strength-of-evidence determination.

AUTHORITATIVE REVIEWS

The committee defined authoritative reviews to be reviews produced by government agencies or other bodies that publish strength-of-evidence determinations through a process that includes peer review. The committee focused on national or international organizations or agencies that influence other organizations. The following organizations met these criteria for authoritative reviews:

- C-8 Science Panel Reports
- European Food Safety Authority (EFSA)
- International Agency for Research on Cancer (IARC)
- Organisation for Economic Co-operation and Development (OECD)
- National Toxicology Program (NTP)
- Agency for Toxic Substances and Disease Registry (ATSDR)
- U.S. Environmental Protection Agency (EPA)

Table D-1 summarizes the authoritative reviews found by the committee. Among the authoritative reviews, the ATSDR's *Toxicological Profile for Perfluoroalkyls* included the greatest number of PFAS included in the committee's Statement of Task (MeFOSAA not included) and was the most recent (literature search conducted in September 2018). The other authoritative reviews were older and included chemicals that were also included in the ATSDR's *Toxicological Profile for Perfluoroalkyls*. Therefore, the ATSDR's *Toxicological Profile for Perfluoroalkyls* was used by the committee as the basis for the next stages of the review process.

TABLE D-1 Authoritative Reviews Found by the Committee

Review	PFAS Chemicals Covered in Review	Health Endpoints Covered in Review	Date of Last Literature Search by the Organization
ATSDR *Toxicological Profile for Perfluoroalkyls*	PFBA, PFHxA, PFHpA, PFOA, PFNA, PFDA, PFUnA, PFDoDA, PFBS, PFHxS, PFOS, FOSA	Not limited	September 2018
EFSA *Risk to Human Health Related to the Presence of Perfluoroalkyl Substances in Food*	PFOA, PFNA, PFHxS, PFOS	Fertility and pregnancy outcomes, development effects, neurotoxic outcomes, immune outcomes, endocrine effects, metabolic effects, kidney function, cardiovascular disease and mortality, bone mineral density	March 2013
EPA Health Effects Document (PFOA)	PFOA	Serum lipids, cardiovascular disease, liver disease, kidney disease, diabetes, developmental toxicity, thyroid effects, immunotoxicity, cancer: kidney and testicular cancer, neurotoxicity, steroid hormones	2015
EPA Health Effects Document (PFOS)	PFOS	Serum lipids, cardiovascular disease, liver disease, kidney disease, diabetes,	2015

		developmental toxicity, thyroid effects, immunotoxicity, cancer: testicular and kidney, neurotoxicity, steroid hormones	
IARC Monograph	PFOA	Cancer	June 2014
NTP Monograph	PFOA, PFOS	Immunotoxicity	May 2016
OECD Synthesis Paper	PFOA, PFOS, PFHxS, PFNA, PFDA, PFBS, PFBA, PFHxA, PTFE, PVDF, PFBE	Developmental toxicity, hypocholesteremia, ulcerative colitis, thyroid diseases, testicular cancer, kidney cancer, preeclampsia	Not presented, published 2013
C-8 Science Panel Probable Link Reports[a]	PFOA	Heart disease, kidney disease, liver disease, osteoarthritis, Parkinson's disease, autoimmune disease, infectious disease, neurodevelopmental disorders in children, respiratory disease, evaluation of stroke, thyroid disease, cancer, diabetes, birth defects, pregnancy-induced hypertension, miscarriage and stillbirth, preterm birth and low birthweight	Last report published in 2012

[a] See http://www.c8sciencepanel.org/prob_link.html (accessed July 1, 2022).
NOTE: ATSDR = Agency for Toxic Substances and Disease Registry; EFSA = European Food Safety Authority; EPA = U.S. Environmental Protection Agency; IARC = International Agency for Research on Cancer; NTP = National Toxicology Program; OECD = Organisation for Economic Co-operation and Development.

The committee's review did not assess the quality of the authoritative reviews, but it notes several areas where the ATSDR's *Toxicological Profile for Perfluoroalkyls* could be strengthened. First, the toxicological profile does not provide a detailed description of the evidence identification methods and does not document decisions as to why specific studies may have been excluded. Second, the study quality assessment does not appear to follow a standard approach, and in some cases it is difficult to identify the study designs that were included in the review. Third, the process to assess the strength of the evidence is not always clear.

REVIEW OF SYSTEMATIC REVIEWS

The committee's review of systematic reviews consisted of the following steps: literature search, screening of abstracts, full text review of studies identified in the abstract screening, evaluation of a final set of relevant studies, evidence assessment, and synthesis.

Literature Search

Systematic reviews were identified through searches of the medical and scientific literature on three databases: Embase Update, Medline, and Scopus. These three searchable databases index biological, chemical, medical, and toxicological publications. Search terms included full and abbreviated chemical names, common and manufacturer trade names, Chemical Abstracts Service (CAS) numbers, and MeSH[1] descriptors for each of the PFAS species of interest. Systematic reviews were included regardless of when they were published or where they were conducted. Systematic reviews were considered if they reviewed human studies, were classified as review papers, and were published in English. The databases were searched on June 28, 2021.

---

[1] MeSH descriptors are sets of terms naming descriptors in a hierarchical structure that permits searching at various levels of specificity.

Ovid Embase Update Search Terms

1	("335-76-2" or "335-67-1" or "375-92-8" or "375-95-1" or "355-46-4" or "1763-23-1" or "2058-94-8").rn.	3673
2	Limit 1 to (human and English language and "review")	154
3	exp perfluorooctanesulfonic acid/ or exp perfluorohexanesulfonic acid/ or exp perfluorononanoic acid/ or exp perfluorooctanesulfonic acid/ or exp perfluorodecanoic acid/ or exp perfluorooctanoic acid/ or exp perfluoroundecanoic acid/ or ("Methyl-perfluorooctane sulfonamide" or "Methylperfluorooctane sulfonamidoacetic acid" or "Perfluorodecanoic acid" or "Perfluoroheptanesulfonic acid" or "perfluorohexane sulfonic acid" or "Perfluorohexanesulfonic acid" or "Perfluorononanoic acid" or "Perfluorooctane sulfonic acid" or "Perfluorooctanesulfonic acid" or "Perfluorooctanoic acid" or "perfluoroundecanoic acid").mp. or ("MeFOSAA" or "PFHxS" or "n-PFOA" or "Sb-PFOA" or "PFOA" or "PFDA" or "PFUnDA" or "n-PFOS" or "Sm-PFOS" or "PFOS" or "PFNA" or "Perfluorinated chemical" or "perfluorinated compound" or "perfluorinated chemicals" or "perfluorinated compounds").mp.	7814
4	Limit 3 to (human and English language and "review")	352

MEDLINE Search Terms

1	("335-76-2" or "335-67-1" or "375-92-8" or "375-95-1" or "355-46-4" or "1763-23-1" or "2058-94-8").rn.	331
2	Limit 1 to (English language and "review articles" and humans)	1
3	Limit 1 to (English language humans and "review" or "scientific integrity review" or "systematic review")	1 (*same article*)
4	("Methyl-perfluorooctane sulfonamide" or "Methylperfluorooctane sulfonamidoacetic acid" or "Perfluorodecanoic acid" or "Perfluoroheptanesulfonic acid" or "perfluorohexane sulfonic acid" or "Perfluorohexanesulfonic acid" or "Perfluorononanoic acid" or "Perfluorooctane sulfonic acid" or "Perfluorooctanesulfonic acid" or "Perfluorooctanoic acid" or "perfluoroundecanoic acid").mp.	4159
5	Limit 4 to (English language and "review articles" and humans)	77
6	("MeFOSAA" or "PFHxS" or "n-PFOA" or "Sb-PFOA" or "PFOA" or "PFDA" or "PFUnDA" or "n-PFOS" or "Sm-PFOS" or "PFOS" or "PFNA" or "Perfluorinated chemical" or "perfluorinated compound" or "perfluorinated chemicals" or "perfluorinated compounds").mp.	5852
7	Limit 6 to (English language and "review articles" and humans)	192
8	Limit 4 to (English language humans and "review" or "scientific integrity review" or "systematic review")	81
9	Limit 6 to (English language humans and "review" or "scientific integrity review" or "systematic review")	203
10	("Methyl-perfluorooctane sulfonamide" or "Methylperfluorooctane sulfonamidoacetic acid" or "Perfluorodecanoic acid" or "Perfluoroheptanesulfonic acid" or "perfluorohexane sulfonic acid" or "Perfluorohexanesulfonic acid" or "Perfluorononanoic acid" or "Perfluorooctane sulfonic acid" or "Perfluorooctanesulfonic acid" or "Perfluorooctanoic acid" or "perfluoroundecanoic acid").rn.	2896
11	Limit 10 to (English language and "review articles" and humans)	49
12	Limit 10 to (English language humans and "review" or "scientific integrity review" or "systematic review")	51

Scopus Search Terms

(((TITLE-ABS-KEY (*"MeFOSAA"* OR *"PFHxS"* OR *"n-PFOA"* OR *"Sb-PFOA"* OR *"PFOA"* OR *"PFDA"* OR *"PFUnDA"* OR *"n-PFOS"* OR *"Sm-PFOS"* OR *"PFOS"* OR *"PFNA"* OR *"Perfluorinated chemical*"* OR *"perfluorinated compound*"*)) OR (TITLE-ABS-KEY (*"Methyl-perfluorooctane sulfonamide"* OR *"Methylperfluorooctane sulfonamidoacetic acid"* OR *"Perfluorodecanoic acid"* OR *"Perfluoroheptanesulfonic acid"* OR *"perfluorohexane sulfonic acid"*

OR *"Perfluorohexanesulfonic acid"* OR *"Perfluorononanoic acid"* OR *"Perfluorooctane sulfonic acid"* OR *"Perfluorooctanesulfonic acid"* OR *"Perfluorooctanoic acid"* OR *"perfluoroundecanoic acid"*)) OR (CHEMNAME (*"Methyl-perfluorooctane sulfonamide"* OR *"Methylperfluorooctane sulfonamidoacetic acid"* OR *"Perfluorodecanoic acid"* OR *"Perfluoroheptanesulfonic acid"* OR *"perfluorohexane sulfonic acid"* OR *"Perfluorohexanesulfonic acid"* OR *"Perfluorononanoic acid"* OR *"Perfluorooctane sulfonic acid"* OR *"Perfluorooctanesulfonic acid"* OR *"Perfluorooctanoic acid"* OR *"perfluoroundecanoic acid"*)) OR (CASREGNUMBER (*"335-76-2"* OR *"335-67-1"* OR *"375-92-8"* OR *"375-95-1"* OR *"355-46-4"* OR *"1763-23-1"* OR *"2058-94-8"*))) AND (INDEXTERMS (*human**) OR TITLE-ABS-KEY (*human**))) AND NOT INDEX (*medline*) AND (LIMIT-TO (DOCTYPE , *"ar"*)) AND (LIMIT-TO (LANGUAGE , *"English"*))

Screening of Abstracts

The literature search identified 639 potentially relevant systematic reviews. The publications were imported into PICO Portal, a web-based tool for collaborative citation screening for systematic reviews.[2] After importing to PICO Portal, 119 articles were identified as duplicates, leaving 520 for title and abstract screening. The review used the following population exposure comparison and outcome (PECO) statement:

> **Population:** Systematic reviews of health effects of PFAS in humans
> **Exposure:** PFAS species measured in the CDC's *National Report on Human Exposure to Environmental Chemicals* (see Chapter 1, Table 1-3)
> **Comparison:** Any comparison groups, including internal controls
> **Outcome:** Any human health outcome

The inclusion and exclusion criteria related to the PECO statement were as follows:

> **Inclusion Criteria:** Includes human evidence; includes the PFAS species measured in the CDC's *National Report on Human Exposure to Environmental Chemicals*; assesses the evidence for an association of PFAS and a health outcome in humans; and has a methods section
> **Exclusion Criteria:** Did not review health effects of PFAS; reviewed only animal or mechanistic studies; reviewed chemicals other than the PFAS included in the Statement of Task; reviewed generic classes of chemicals such as "endocrine disruptors" or "persistent organic pollutants;" or reviewed the ecological effects of PFAS or PFAS exposure pathways

Title and abstract screening was completed by two screeners. Disagreements were resolved by an adjudicator who helped facilitate a consensus decision. Fifty-four articles were included for full-text review.

Full Text Review

For the full text review, articles were excluded because they did not include human studies (n = 2), did not include relevant PFAS (n = 5), did not have a methods section (n = 18), or did not evaluate the association of PFAS exposure with a human health outcome (n = 3) (see Figure D-1). Thus, the full text review covered 26 articles.

---

[2] See https://picoportal.net (accessed July 1, 2022).

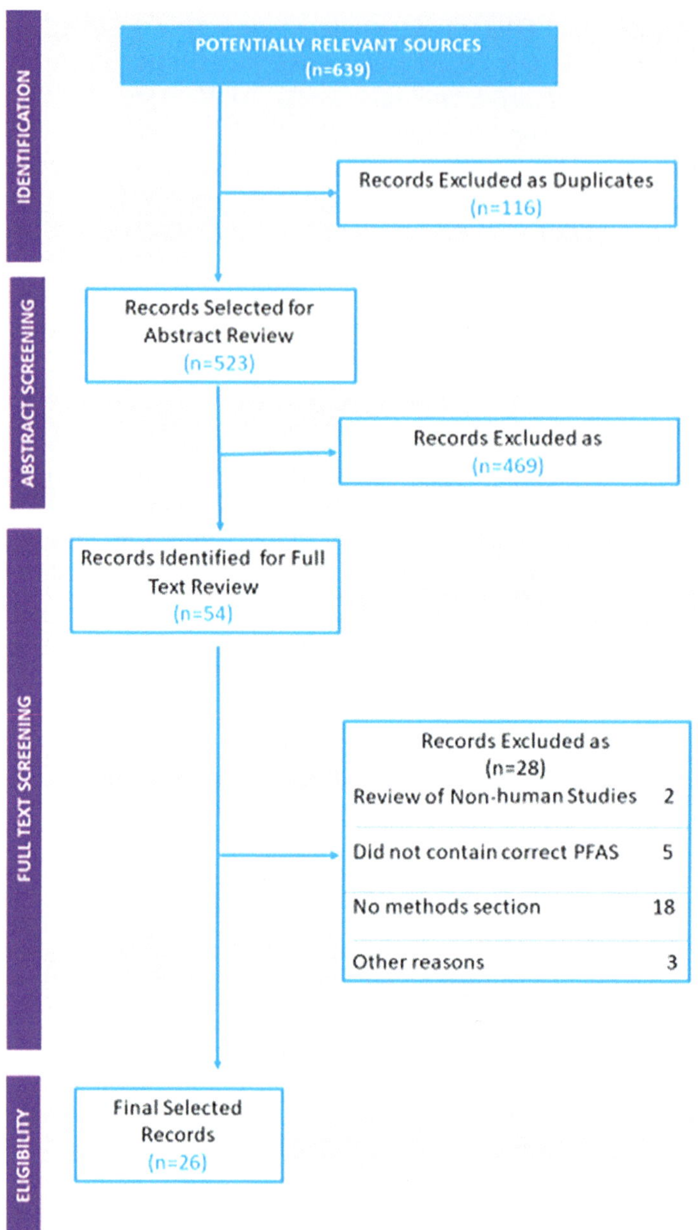

FIGURE D-1 Preferred reporting items for systematic reviews and meta-analyses (PRISMA) diagram for the committee's review of reviews on health effects of PFAS.

Evaluation and Evidence Assessment

The quality of the 26 included systematic reviews was evaluated using the AMSTAR-2 (A MeaSurement Tool to Assess systematic Reviews) (Shea et al., 2017), which has been used by several other committees of the National Academies of Sciences, Engineering, and Medicine (NASEM, 2019, 2021). AMSTAR-2 includes several critical appraisal domains (see Box D-1). The quality assessment was conducted by a staff member and was confirmed by a committee member. The committee conducted a critical appraisal of the systematic reviews because systematic reviews can be subject to a range of biases.

BOX D-1
AMSTAR-2 Critical Domains and Overall Confidence in the Results

Critical Domains
- Protocol registered before commencement of the review (item 2)
- Adequacy of the literature search (item 4)
- Justification for excluding individual studies (item 7)
- Risk of bias from individual studies being included in the review (item 9)
- Appropriateness of meta-analytical methods (item 11)
- Consideration of risk of bias when interpreting the results of the review (item 13)
- Assessment of presence and likely impact of publication bias (item 15)

Rating: Overall Confidence in the Results of the Review
- *High:* No or one noncritical weakness: the systematic review provides an accurate and comprehensive summary of the results of the available studies that address the question of interest.
- *Moderate:* More than one noncritical weakness: the systematic review has more than one weakness but no critical flaws. It may provide an accurate summary of the results of the available studies that were included in the review.
- *Low:* One critical flaw with or without noncritical weaknesses: the review has a critical flaw and may not provide an accurate and comprehensive summary of the available studies that address the question of interest.
- *Critically Low:* More than one critical flaw with or without noncritical weaknesses: the review has more than one critical flaw and should not be relied on to provide an accurate and comprehensive summary of the available studies.

Synthesis: Results

Of the 26 systematic reviews identified by the committee, 9 included studies published after the ATSDR's *Toxicological Profile for Perfluoroalkyls* (Bartell and Vieira, 2021; Boesen et al., 2020; Deji et al., 2021; Dzierlenga et al., 2020; Ferrari et al., 2019; Luo et al., 2020; Petersen et al., 2020; Steenland and Winquist, 2020; Xie et al., 2020; Zare Jeddi et al., 2021). Although all were of moderate quality, some reviews covered the same original data and papers, and the committee found it challenging to synthesize across them. As a result, the systematic reviews were used as sources for reference in the committee's determination of the biologic plausibility between PFAS and a health effect, but they were not formally included as part of the final strength-of-evidence determination.

ORIGINAL LITERATURE REVIEW

The original literature review consisted of the following steps: literature search, screening of abstracts, full text review of studies identified in the abstract screening, evidence mapping and evaluation, and data abstraction.

Literature Search

As was done for the review of reviews, the literature for the original literature review was identified through searches of the medical and scientific literature on three databases: Ovid Embase, Ovid Medline, and Scopus. These three searchable databases index biological, chemical, medical, and toxicological publications. Search terms included full and abbreviated chemical names, common and manufacturer trade names, the CAS numbers, and MeSH[3] descriptors for each of the PFAS species of interest. There were no time constraints or geographic constraints included in the search. The only constraints were human studies and English language only. The databases were searched on March 30–31, 2021.

Ovid Embase Search Terms[4]

1	("335-76-2" or "335-67-1" or "375-92-8" or "375-95-1" or "355-46-4" or "1763-23-1" or "2058-94-8").rn.	3547
2	exp perfluorooctanesulfonic acid/	3114
3	exp perfluorohexanesulfonic acid/	625
4	exp perfluorononanoic acid/	890
5	exp perfluorooctanesulfonic acid/	3114
6	exp perfluorodecanoic acid/	713
7	exp perfluorooctanoic acid/	3493
8	exp perfluoroundecanoic acid/	386
9	("Methyl-perfluorooctane sulfonamide" or "Methylperfluorooctane sulfonamidoacetic acid" or "Perfluorodecanoic acid" or "Perfluoroheptanesulfonic acid" or "perfluorohexane sulfonic acid" or "Perfluorohexanesulfonic acid" or "Perfluorononanoic acid" or "Perfluorooctane sulfonic acid" or "Perfluorooctanesulfonic acid" or "Perfluorooctanoic acid" or "perfluoroundecanoic acid").mp.	5503
10	("MeFOSAA" or "PFHxS" or "n-PFOA" or "Sb-PFOA" or "PFOA" or "PFDA" or "PFUnDA" or "n-PFOS" or "Sm-PFOS" or "PFOS" or "PFNA" or "Perfluorinated chemical" or "perfluorinated compound" or "perfluorinated chemicals" or "perfluorinated compounds").mp.	6579
11	1 or 2 or 3 or 4 or 5 or 6 or 7 or 8 or 9 or 10	7610
12	(exp animal/ or nonhuman/) not exp human/	6205240
13	11 not 12	5435
14	Limit 13 to english language	5232
15	Limit 14 to "pubmed/medline"	540
16	14 not 15	4692
17	Limit 16 to article	3517

Scopus Search Terms[5]

(((TITLE-ABS-KEY (*"MEFOSAA"* OR *"PFHXS"* OR *"N-PFOA"* OR *"SB-PFOA"* OR *"PFOA"* OR *"PFDA"* OR *"PFUNDA"* OR *"N-PFOS"* OR *"SM-PFOS"* OR *"PFOS"* OR *"PFNA"* OR *"PERFLUORINATED CHEMICAL*"* OR *"PERFLUORINATED COMPOUND*"*)) OR (TITLE-ABS-KEY (*"METHYL-PERFLUOROOCTANE SULFONAMIDE"* OR *"METHYLPERFLUOROOCTANE SULFONAMIDOACETIC ACID"* OR *"PERFLUORODECANOIC ACID"* OR *"PERFLUOROHEPTANESULFONIC ACID"* OR *"PERFLUOROHEXANE SULFONIC ACID"* OR *"PERFLUOROHEXANESULFONIC ACID"* OR *"PERFLUORONONANOIC ACID"* OR *"PERFLUOROOCTANE SULFONIC ACID"* OR *"PERFLUOROOCTANESULFONIC ACID"* OR *"PERFLUOROOCTANOIC ACID"* OR *"PERFLUOROUNDECANOIC ACID"*)) OR (CHEMNAME (*"METHYL-PERFLUOROOCTANE SULFONAMIDE"* OR *"METHYLPERFLUOROOCTANE SULFONAMIDOACETIC ACID"* OR *"PERFLUORODECANOIC ACID"* OR *"PERFLUOROHEPTANESULFONIC ACID"* OR *"PERFLUOROHEXANE SULFONIC ACID"* OR *"PERFLUOROHEXANESULFONIC ACID"* OR *"PERFLUORONONANOIC ACID"* OR

---

[3] MeSH descriptors are sets of terms naming descriptors in a hierarchical structure that permits searching at various levels of specificity.

[4] Not all chemical names mapped to a heading.

[5] The search was limited to articles and English language.

"PERFLUOROOCTANE SULFONIC ACID" OR *"PERFLUOROOCTANESULFONIC ACID"* OR *"PERFLUOROOCTANOIC ACID"* OR *"PERFLUOROUNDECANOIC ACID"*)) OR (CASREGNUMBER (*"335-76-2"* OR *"335-67-1"* OR *"375-92-8"* OR *"375-95-1"* OR *"355-46-4"* OR *"1763-23-1"* OR *"2058-94-8"*))) AND (INDEXTERMS (*HUMAN**) OR TITLE-ABS-KEY (*HUMAN**))) AND NOT INDEX (*MEDLINE*) AND (LIMIT-TO (DOCTYPE , *"AR"*)) AND (LIMIT-TO (LANGUAGE , *"ENGLISH"*))

Ovid Medline Search Terms

1	("335-76-2" or "335-67-1" or "375-92-8" or "375-95-1" or "355-46-4" or "1763-23-1" or "2058-94-8").rn.	330
2	("Methyl-perfluorooctane sulfonamide" or "Methylperfluorooctane sulfonamidoacetic acid" or "Perfluorodecanoic acid" or "Perfluoroheptanesulfonic acid" or "perfluorohexane sulfonic acid" or "Perfluorohexanesulfonic acid" or "Perfluorononanoic acid" or "Perfluorooctane sulfonic acid" or "Perfluorooctanesulfonic acid" or "Perfluorooctanoic acid" or "perfluoroundecanoic acid").mp.	4050
3	("MeFOSAA" or "PFHxS" or "n-PFOA" or "Sb-PFOA" or "PFOA" or "PFDA" or "PFUnDA" or "n-PFOS" or "Sm-PFOS" or "PFOS" or "PFNA" or "Perfluorinated chemical" or "perfluorinated compound" or "perfluorinated chemicals" or "perfluorinated compounds").mp.	5700
4	("Methyl-perfluorooctane sulfonamide" or "Methylperfluorooctane sulfonamidoacetic acid" or "Perfluorodecanoic acid" or "Perfluoroheptanesulfonic acid" or "perfluorohexane sulfonic acid" or "Perfluorohexanesulfonic acid" or "Perfluorononanoic acid" or "Perfluorooctane sulfonic acid" or "Perfluorooctanesulfonic acid" or "Perfluorooctanoic acid" or "perfluoroundecanoic acid").rn.	2837
5	1 or 2 or 3 or 4	6194
6	Animals/ not (Animals/ and Humans/)	4772259
7	5 not 6	4664
8	Journal Article/	30061095
9	7 and 8	4548
10	Limit 9 to English language	4379

Abstract Screening

The literature search identified 5,172 potentially relevant studies. The studies were imported into PICO Portal, a web-based tool for collaborative citation screening for systematic reviews.[6] After importing the studies to PICO Portal, 112 articles were identified as duplicates, leaving 5,060 articles to be screened. The titles and abstracts were screened for relevance to the research questions for the review. The review used the following PECO statement:

> **Population:** Studies of health effects of PFAS in humans
> **Exposure:** PFAS species measured in the CDC's *National Report on Human Exposure to Environmental Chemicals*
> **Comparison:** Any comparison groups, including internal comparisons
> **Outcome:** Any health outcome measured in humans

The inclusion and exclusion criteria related to the PECO statement were as follows:

> **Inclusion Criteria:** Is an epidemiologic or human study; includes a quantitative measure of the PFAS species measured in the CDC's *National Report on Human Exposure to Environmental Chemicals*; assesses the evidence for an association of PFAS and a health outcome in humans; and English language only

---

[6] See https://picoportal.net (accessed July 1, 2022).

Exclusion Criteria: Did not review health effects of PFAS; reviewed only animal or mechanistic studies; was about chemicals other than the PFAS included in the Statement of Task; did not include a quantitative measure; or was on the ecological effects of PFAS or PFAS exposure pathways

The literature search identified 5,172 potentially relevant studies. After removal of duplicates (112 articles), 5,060 articles were subject to title and abstract screening by two independent reviewers.

Full Text Review

For the full text review, 4,434 of the articles identified in the literature search were excluded because the titles and abstracts did not meet the inclusion criteria, so 626 articles were subject to full text review. During that review, additional articles were excluded if they were published before 2018 or listed in the references to the ATSDR's *Toxicological Profile for Perfluoroalkyls* (n = 320); were cross-sectional in design (n = 160); were not published in English (n = 1); did not provide risk estimates associated with PFAS exposure (n = 3); or were not studies in humans (n = 3) (see Figure D-2). Cross-sectional studies were largely excluded because this study design measures exposure and disease at the same time so cannot determine cause and effect. Thus, the full text review covered 139 articles.

Evidence Mapping and Evaluation

The committee then categorized the 139 articles according to the human health outcomes studied. The committee mapped the evidence with the goal of determining evidence gaps and to inform strategies for the evidence evaluation and evidence synthesis (see Figure D-3).

The committee focused on those endpoints for which additional review might change the committee's understanding of the association between PFAS exposure and health outcomes.

The committee conducted a narrative evaluation of the study quality and considered factors that may contribute to the study's risk of bias (see Box D-2) using a tool adapted from the Navigation Guide (Woodruff and Sutton, 2014). Bias is a systematic error that leads to study results that differ from the actual results. Bias can lead to an observed effect when one does not exist or to no observed effect when there is a true effect. Risk of bias is the appropriate term, as a study may be unbiased despite a methodological flaw (Higgins et al., 2019). The risk-of-bias assessment in a systematic review is based on the quality of the individual component studies (Eick et al., 2020).

A trained reviewer from ICF International, the EPA, the National Academies, or Johns Hopkins University abstracted the critical domain information from each study that the committee used to support its judgment determinations regarding a study's risk of bias, and the committee made the final risk of bias judgments for each study.[7] Each paper was given an overall assessment of its risk of bias (low, probably low, probably high, or high risk of bias).

---

[7] For some studies included in the committee's review, the data had been previously abstracted by ICF or the EPA to support the EPA's ongoing assessments of PFAS; newer evidence was abstracted by ICF or consultants at Johns Hopkins.

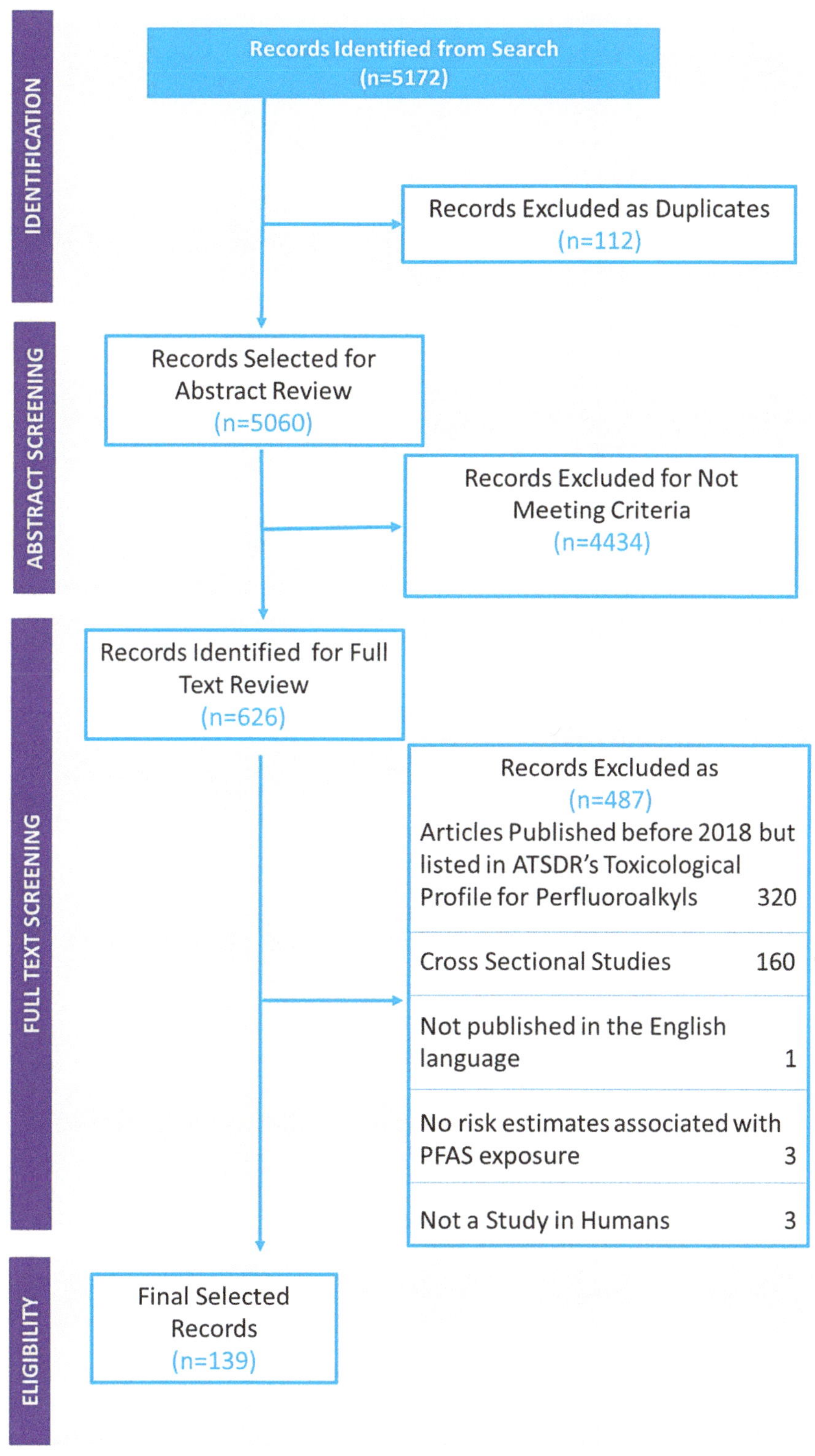

FIGURE D-2 Preferred reporting items for systematic reviews and meta-analyses (PRISMA) diagram for the committee's reviews on the health effects of PFAS.

Health Effect Category	Chemical								Grand Total
	MeFOSAA	PFDA	PFHxS	PFNA	PFOA	PFOS	PFUnDA	sumPFAS	
Bodyweight	4	4	10	10	13	12	2	0	13
Cancer	2	3	5	4	7	7	4	0	7
Cardiometabolic	7	9	21	20	19	20	6	3	21
Developmental	6	25	36	38	44	46	16	4	47
Endocrine	3	8	12	12	11	12	6	2	13
Hepatic	1	2	3	2	4	3	1	0	4
Immunological	0	8	13	13	14	14	10	2	14
Musculoskeletal	0	1	2	2	2	2	1	1	2
Neurological	1	5	9	10	10	11	3	2	11
Other	0	3	3	3	4	4	2	0	4
Renal	3	1	3	3	4	3	0	1	4
Reproductive	3	16	25	24	25	27	11	1	29
Respiratory	0	6	9	10	10	10	8	0	10
Grand Total	20	72	118	118	130	133	56	13	139

FIGURE D-3 Evidence map describing the number of studies found by PFAS for each health outcome category.

BOX D-2
Critical Domains Used by the Committee to Assess Risk of Bias

Exposure Measures: Assay information, quality control measures, repeat measures, validation studies

Outcome Measures: Sources of the effect measure, blinding to exposure status or level, methods of measurement or classification, incident versus prevalent disease, evidence from validation studies

Participant Selection: Study design, where and when was the study conducted, and who was included? Recruitment process, exclusion and inclusion criteria, type of controls, total eligible, comparison between participants and nonparticipants (or followed and not followed), final analysis group. Does the study include potential vulnerable/susceptible groups or life stages?

Potential Confounding: Potential confounders and outcome; degree of exposure to the confounder in the population

Analysis: Extent (and, if applicable, treatment) of missing data for exposure, outcome, and confounders, approach to modeling, classification of exposure and outcome variables (continuous versus categorical), testing of assumptions, sample size for specific analyses, and relevant sensitivity analyses

Selective Reporting: Are results presented with adequate detail for all of the endpoints of interest? Are results presented for the full sample as well as for specified subgroups? Were stratified analyses (effect modification) motivated by a specific hypothesis?

Sensitivity: What exposure range is spanned in this study? What are the ages of participants (e.g., not too young in studies of pubertal development)? What is the length of follow-up (for outcomes with long latency periods)? Choice of referent group and the level of exposure contrast between groups (i.e., the extent to which the "unexposed group" is truly unexposed, and the prevalence of exposure in the group designated as "exposed")

Source of Funding: Description of the disclosed sources of funding for the paper

---

Data Abstraction

Data abstraction was completed by a trained scientist from ICF, the EPA, the National Academies, or Johns Hopkins.[8] The data abstraction form included the following:

Reference: Author year and DOI number if relevant
Chemical: Acronym of specific PFAS
Endpoint: Name of the specific outcome for the risk estimate
Subpopulation: Description of the specific subpopulation for the risk estimate
N: Sample size that informs risk estimate
Exposure Levels: That apply to the risk estimate
Comparison: Description of type of risk estimate, such as "SMR for bladder cancer in the high exposure group compared to no workplace exposure" or "change in ln (TSH) per standard unit increase in serum PFOA."
Risk/Effect Estimate: Reported number
Lower Confidence Interval: Reported number
Upper Confidence Interval: Reported number

---

[8] For some studies included in the committee's review, the data had been previously abstracted by ICF or the EPA to support the EPA's ongoing assessments of PFAS; newer evidence was abstracted by ICF or consultants at Johns Hopkins University.

Effect estimates from the individual studies included in the review were extracted into a database and uploaded to a public website (Tableau Public) to allow for visualizations, such as evidence maps and forest plots.[9] The effect estimates in the Tableau represent those from the model most adjusted for confounders.

STRENGTH-OF-EVIDENCE DETERMINATION

To assess the strength of evidence regarding the potential for PFAS to cause a particular health effect, the committee then integrated the evidence reviewed in the ATSDR's *Toxicological Profile for Perfluoroalkyls* and other authoritative reviews with the evidence from the original literature review from the epidemiologic studies.

The synthesis of available data was guided by a framework based on the Hill considerations (Hill, 1965), which help to determine whether associations are causal (see Box D-3). The committee did not consider the Hill considerations to be a heuristic for assessing causation in isolation, that is, as a checklist where each item must be met to establish causality. Rather, the committee considered them as a list of possible considerations meant to generate thoughtful discourse by the committee to help inform its determinations for the strength of evidence (Fedak, 2015; NASEM, 2018).

BOX D-3
Bradford Hill Considerations

Strength: A small association does not mean that there is not a causal effect, though the larger the association, the more likely that it is causal.

Consistency: Consistent findings observed by different people in different places with different samples strengthens the likelihood of an effect.

Specificity: Causation is likely if there is a very specific population, at a specific site, with a specific disease with no other likely explanation.

Temporality: The effect has to occur after the cause.

Biological Gradient: Greater exposure should generally lead to greater incidence of the effect. In other cases, greater exposure leads to lower incidence.

Plausibility: A plausible mechanism between cause and effect is helpful.

Coherence: Coherence between epidemiological and laboratory findings increases the likelihood of an effect but "lack of such [laboratory] evidence cannot nullify the epidemiological effect on associations" (Hill, 1965, p. 298).

Experiment: Occasionally it is possible to appeal to experimental evidence.

Analogy: The effect of similar factors may be considered.

SOURCE: Hill, 1965.

---

[9] The committee's public Tableau is available at https://public.tableau.com/app/profile/nationalacademies/viz/NASEMPFASEvidenceMaps/PFASEvidenceMap (accessed July 1, 2022). The information may be viewed as an evidence map or as a forest plot. Within forest plots, filters can be accessed using the "toggle filters" function in order to restrict the view to data on specific health effect categories and other factors (such as reference, chemical, study design, study population).

Animal and Mechanistic Studies

The Statement of Task called for the committee to "assess the strength of evidence for the spectrum of putative health effects suggested by human studies." The committee focused on more recent epidemiological literature in line with its Statement of Task. However, the committee recognizes the importance of toxicologic evidence in making strength-of-evidence conclusions and relied on the toxicological data included in the ATSDR's *Toxicological Profile for Perfluoroalkyls*, other authoritative reviews, and systematic reviews that provide integrative conclusions based on multiple lines of evidence. In synthesizing evidence in this manner, the committee acknowledges that animal evidence greatly improves the interpretation of the human studies.

An observed association between PFAS exposure and a health effect does not necessarily mean that the exposure is the cause of that outcome. Toxicologic evidence, whether it supports or conflicts with evidence from epidemiologic studies, provides insights about biologic processes and informs how an observed association might be interpreted. The degree of biologic plausibility itself influences whether the committee perceives positive findings to be indicative of a pattern or the product of statistical fluctuations. Ultimately, the results of the toxicology studies should be consistent with what is known about the human disease process if they are to support a conclusion that the development of the disease was influenced by an exposure (NASEM, 2018).

Categories of Association

Sufficient Evidence of an Association

For effects in this category, a positive association between PFAS and the outcome must be observed in studies in which chance, bias, and confounding can be ruled out with reasonable confidence. For example, the committee might regard as sufficient evidence of an association evidence from several small studies that is unlikely to be due to confounding or to otherwise be biased and that shows an association that is consistent in magnitude and direction. Experimental data supporting biologic plausibility strengthen the evidence of an association but are not a prerequisite, nor are they sufficient to establish an association without corresponding epidemiologic findings.

Limited or Suggestive Evidence of an Association

In this category, the evidence must suggest an association between exposure to PFAS and the outcome in studies of humans, but the evidence can be limited by an inability to rule out chance, bias, or confounding with confidence. One high-quality study may indicate a positive association, but the results of other studies of lower quality may be inconsistent.

Inadequate or Insufficient Evidence to Determine an Association

If there was not enough reliable scientific data to categorize the potential association with a health effect as "sufficient evidence of an association," "limited or suggestive evidence of an association," or on the other end of the spectrum, "limited or suggestive evidence of no association," the health outcome was placed in the category of "inadequate or insufficient evidence to determine an association" by default. In this category, the available human studies may have inconsistent findings or be of insufficient quality, validity, consistency, or statistical power to support a conclusion regarding the presence of an association. Such studies may have failed to control for confounding factors or may have had inadequate assessment of exposure.

Limited or Suggestive Evidence of No Association

A conclusion of "no association" is inevitably limited to the conditions, exposures, and observation periods covered by the available studies, and the possibility of a small increase in risk related to the magnitude of exposure studied can never be excluded. However, a change in classification from inadequate or insufficient evidence of an association to limited or suggestive evidence of *no* association would require new studies that corrected for the methodologic problems of previous studies and that had samples large enough to limit the possible study results attributable to chance.

REFERENCES

Bartell, S. M., and V. M. Vieira. 2021. Critical review on PFOA, kidney cancer, and testicular cancer. *Journal of the Air and Waste Management Association* 71(6):663–679. https://doi.org/10.1080/10962247.2021.1909668.

Boesen, S. A. H., M. Long, M. Wielsoe, V. Mustieles, M. F. Fernandez, and E. C. Bonefeld-Jorgensen. 2020. Exposure to perflouroalkyl acids and foetal and maternal thyroid status: A review. *Environmental Health* 19(1):107. https://doi.org/10.1186/s12940-020-00647-1.

Deji, Z., P. Liu, X. Wang, X. Zhang, Y. Luo, and Z. Huang. 2021. Association between maternal exposure to perfluoroalkyl and polyfluoroalkyl substances and risks of adverse pregnancy outcomes: A systematic review and meta-analysis. *Science of the Total Environment* 783:146984. https://doi.org/10.1016/j.scitotenv.2021.146984.

Dzierlenga, M. W., L. Crawford, and M. P. Longnecker. 2020. Birth weight and perfluorooctane sulfonic acid: A random-effects meta-regression analysis. *Environmental Epidemiology* 4(3):e095. https;//doi.org/10.1097/EE9.0000000000000095.

Eick, S. M., D. E. Goin, N. Chartres, J. Lam, and T. J. Woodruff. 2020. Assessing risk of bias in human environmental epidemiology studies using three tools: Different conclusions from different tools. *Systematic Reviews* 9(1):249. https://doi.org/10.1186/s13643-020-01490-8.

Fedak, K. M., A. Bernal, Z. A. Capshaw, and S. Gross. 2015. Applying the Bradford Hill criteria in the 21st century: How data integration has changed causal inference in molecular epidemiology. *Emerging Themes in Epidemiology* 12:14. https://doi.org/10.1186/s12982-015-0037-4.

Ferrari, F., A. Orlando, Z. Ricci, and C. Ronco. 2019. Persistent pollutants: Focus on perfluorinated compounds and kidney. *Current Opinion Critical Care* 25(6):539–549. https://doi.org/10.1097/MCC.0000000000000658.

Higgins, J. P. T., J. Thomas, J. Chandler, M. Cumpston, T. Li, M. J. Page, and V. A. Welch. 2019. *Cochrane Handbook for Systematic Reviews of Interventions*. Hoboken, NJ: John Wiley & Sons.

Hill, A. B. 1965. The environment and disease: Association or causation? *Proceedings of the Royal Society of Medicine* 58(5):295–300.

Luo, Y., Z. Deji, and Z. Huang. 2020. Exposure to perfluoroalkyl substances and allergic outcomes in children: A systematic review and meta-analysis. *Environmental Research* 191:110145. https://doi.org/10.1016/j.envres.2020.110145.

NASEM (National Academies of Sciences, Engineering, and Medicine). 2018. *Advances in Causal Understanding for Human Health Risk-Based Decision-Making: Proceedings of a Workshop—in Brief*. Washington, DC: The National Academies Press.

NASEM. 2019. *Review of DOD's Approach to Deriving an Occupational Exposure Level for Trichloroethylene*. Washington, DC: The National Academies Press.

NASEM. 2021. *The Use of Systematic Review in EPA's Toxic Substances Control Act Risk Evaluations*. Washington, DC: The National Academies Press.

Petersen, K. U., J. R. Larsen, L. Deen, E. M. Flachs, K. K. Haervig, S. D. Hull, J. P. E. Bonde, and S. S. Tottenborg. 2020. Per- and polyfluoroalkyl substances and male reproductive health: A systematic review of the epidemiological evidence. *Journal of Toxicology Environmental Health B Critical Reviews* 23(6):276–291. https://doi.org/10.1080/10937404.2020.1798315.

Shea, B. J., B. C. Reeves, G. Wells, M. Thuku, C. Hamel, J. Moran, D. Moher, P. Tugwell, V. Welch, E. Kristjansson, and D. A. Henry. 2017. AMSTAR 2: A critical appraisal tool for systematic reviews that include randomised or non-randomised studies of healthcare interventions, or both. *British Medical Journal* 358:j4008. https://doi.org/10.1136/bmj.j4008.

Steenland, K., and A. Winquist. 2021. PFAS and cancer, a scoping review of the epidemiologic evidence. *Environmental Research* 194:110690. https://doi.org/10.1016/j.envres.2020.110690.

Woodruff, T. J., and P. Sutton. 2014. The Navigation Guide systematic review methodology: A rigorous and transparent method for translating environmental health science into better health outcomes. *Environmental Health Perspectives* 122(10):1007–1014. https://doi.org/10.1289/ehp.1307175.

Xie, W., W. Zhong, B. M. R. Appenzeller, J. Zhang, M. Junaid, and N. Xu. 2020. Nexus between perfluoroalkyl compounds (PFCs) and human thyroid dysfunction: A systematic review evidenced from laboratory investigations and epidemiological studies. *Critical Reviews in Environmental Science and Technology* 1–46. doi: 10.1080/10643389.2020.1795052.

Zare Jeddi, M., R. Soltanmohammadi, G. Barbieri, A. S. C. Fabricio, G. Pitter, T. Dalla Zuanna, and C. Canova. 2021. To which extent are per- and poly-fluorinated substances associated to metabolic syndrome? *Reviews in Environmental Health* 37(2):211–228. https://doi.org/10.1515/reveh-2020-0144.

Appendix E

White Paper: Review of the PFAS Personal Intervention Literature

Prepared for:

Elizabeth B. Boyle
Senior Program Officer
Board on Environmental Studies and Toxicology
Board on Population Health and Public Health Practice
National Academies of Sciences, Engineering, and Medicine

and

Committee on the Guidance on PFAS Testing and Health Outcomes
Washington, DC, USA

Prepared by:

Judy S. LaKind, Ph.D.
LaKind Associates, LLC
Catonsville, MD, USA

Josh Naiman, B.A.
LaKind Associates, LLC
Philadelphia, PA, USA

Contents

Acronyms and Abbreviations

AC	activated carbon
ATDSR	Agency for Toxic Substances and Disease Registry
CDC	Centers for Disease Control and Prevention
CI	confidence interval
EPA	U.S. Environmental Protection Agency
Et-PFOSA-AcOH	2-(n-ethyl-perfluorooctane sulfonamido) acetic acid
EtFOSAA	2-(n-ethyl-perfluorooctane sulfonamido) acetic acid
FDA	U.S. Food and Drug Administration
GAC	granular activated carbon
H2PFDA	2H,2H-perfluorodecanoic acid
H4PFOS	1H,1H,2H,2H-perfluorooctane sulfonic acid
H4PFUnDA	2H,2H,3H,3H-perfluoroundecanoic acid
L-PFHpS	sodium perfluoro-1 heptanesulfonate
LOD	limit of detection
LOQ	limit of quantitation
MDH	Minnesota Department of Health
Me-PFOSA-AcOH	2-(n-methyl-perfluorooctane sulfonamido) acetic acid
MeFOSAA	2-(n-methyl-perfluorooctane sulfonamido) acetic acid
MRL	Minimum Reporting Level
n-EtPFOSAA	n-ethyl-perfluoro-1 octanesulfonamido acetic acid.
n-MePFOSAA	n-methylperfluoro-1 octanesulfonamido acetic acid
n-PFOA	n-perfluorooctanoic acid
n-PFOS	n-perfluorooctane sulfonic acid
NEtFOSE	N-ethyl perfluorooctane sulfonamidoethanol
NHANES	National Health and Nutrition Examination Survey
NIEHS	National Institute of Environmental Health Sciences
PBDE	polybrominated diphenyl ether
PCB	polychlorinated biphenyl
PFAA	perfluoroalkyl acid
PFAS	per- and polyfluoroalkyl substances
PFBA	perfluorobutanoate
PFBS	perfluorobutane sulfonic acid
PFC	perfluorinated compound
PFCA	perfluoroalkyl carboxylic acid
PFDA	perfluorodecanoic acid
PFDcA	perfluorodecanoate
PFDoA (PFDoDA)	perfluorododecanoic acid
PFDeA	perfluorodecanoic acid
PFDS	perfluorodecane sulfonate

PFHpA	perfluoroheptanoic acid
PFHpS	perfluoroheptane sulfonate
PFHxA	perfluorohexanoate
PFHxS	perfluorohexane sulfonic acid
PFNA	perfluorononanoic acid
PFOA	perfluorooctanoic acid
PFOS	perfluorooctane sulfonic acid
PFOSA or FOSA	perfluorooctane sulfonamide
PFPeA	perfluoropentanoate
PFTeA	perfluorotetradecanoate
PFTeDA	perfluorotetradecanoic acid
PFTrA	perfluorotridecanoate
PFTrDA	perfluorotridecanoic acid
PFUA	perfluoroundecanoate
PFUnA	perfluoroundecanoic acid
PFUnDA	perfluoroundecanoic acid
POE	point of entry
POTW	publicly owned treatment work
POU	point of use
RO	reverse osmosis
Sb-PFOA	branched perfluorooctanoic acid
SD	standard deviation
Sm-PFOS	perfluoromethylheptane sulfonic acid
UCMR	Unregulated Contaminant Monitoring Rule
ww	wet weight

Abstract

This white paper provides an overview of the published literature on whether personal behavior modifications can demonstrably reduce exposure to per- and polyfluoroalkyl substances (PFAS) (e.g., by showing decreases in serum levels). The reviewed studies are presented by exposure source. The preponderance of the identified literature relates to diet and drinking water. Literature on interventions for other exposure sources, such as dust and consumer products, is more limited. Breastfeeding is an important potential source of exposure for infants; the effect of lactation on mothers' PFAS levels is unclear. For communities with high levels of PFAS in drinking water, interventions related to tap water filtration showed some efficacy in reducing PFAS levels in the water. It is possible that an intervention may reduce PFAS levels in a particular medium, but if this medium is not a major source of overall exposure, then that intervention may not contribute significantly to reduction in human exposures. Overall, the intervention literature is sparse and has many limitations. Thus, the committee may have to rely on assumptions and other bodies of evidence to make recommendations to individuals and communities about exposure reduction.

E-1

Introduction

Per- and polyfluoroalkyl substances (PFAS) are anthropogenic chemicals that have been produced and utilized globally since the 1940s.[1] PFAS have garnered attention for several reasons, including their ubiquitous presence in the environment (Ahrens and Bundschuh, 2014; von der Trenck et al., 2018) and in humans (Calafat et al., 2019; Göckener et al., 2020; Health Canada, 2019; Kannan et al., 2004), and because—as their epithet "forever chemicals" suggests—many of these chemicals are persistent both in the environment and in humans, with half-lives estimated to be several years (Li et al., 2018; Myers et al., 2012). Exposure to PFAS has been linked with such health endpoints as reduced immune response, lipid metabolism, and kidney function; thyroid disease; liver disease; glycemic parameters and diabetes; cancer; and impaired fetal and child development (ATSDR, 2020).

Activities to limit the production and use of exposure to PFAS compounds include regulatory limits, voluntary reductions in manufacture (Butenhoff et al., 2006) and use in products, cleanup of contaminated sites, and modifications to publicly owned treatment works (POTWs) to reduce PFAS in drinking water. However, as evidenced by studies of measurements of PFAS in serum from nationally representative populations in the United States (Calafat et al., 2019), as well as serum measurements in communities near sites with known contamination (Herrick et al., 2017), exposure to PFAS is ongoing. It has been well documented that PFAS are present in numerous media and products, including drinking water; breast milk; other foods and food packaging material; cosmetics; and household products, including carpets, stain- and water-repellent fabrics, nonstick products, polishes, waxes, paints, and cleaning products (D'Hollander et al., 2010; EFSA, 2020; Eichler and Little, 2020; Fromme et al., 2009; Sajid and Ilyas, 2017; Sunderland et al., 2019).[2]

Communities impacted by PFAS exposure would like advice on how they can prevent its potential health effects. To help clinicians respond to patient concerns about PFAS exposure, the Agency for Toxic Substances and Disease Registry (ATSDR) has published *PFAS: An Overview of the Science and Guidance for Clinicians on Per- and Polyfluoroalkyl Substances* (referred to hereafter as the ATSDR PFAS Clinical Guidance) (ATSDR, 2019). This guidance summarizes general information about PFAS and PFAS health studies and suggests answers to example patient questions. Some people living in PFAS-impacted communities have voiced frustration that the clinical guidance lacks clear recommendations to their physicians about what people can do to protect their health, which prompted the ATSDR and the National Institute of Environmental Health Sciences (NIEHS) to request that the National Academies of Sciences, Engineering, and Medicine convene a committee to provide advice for clinicians about PFAS testing, such as when to test, whom to test, how to test, what to test for, and the risks of testing. The committee is also charged with developing principles clinicians can use to advise patients on exposure reduction.[3] The committee commissioned this white paper to determine whether evidence exists that supports the effectiveness of these types of behavior changes. The literature review in this white paper is intended to help the National Academies committee evaluate possible evidence-based recommendations for improving the ATSDR's PFAS Clinical Guidance. This white paper is not intended to be a comprehensive review of human exposure to PFAS; the National Academies have other sources for that information.

Various organizations have provided suggestions for personal actions to lower individual PFAS exposure. These include avoiding contaminated water or fish and selecting personal care products that do

---

[1] See https://www.epa.gov/pfas/basic-information-pfas (accessed May 12, 2021).

[2] See also https://www.epa.gov/pfas/basic-information-pfas (accessed May 12, 2021).

[3] See the committee's full Statement of Task at https://www.nationalacademies.org/our-work/guidance-on-pfas-testing-and-health-outcomes (accessed May 19, 2021).

not contain PFAS and related compounds (ATSDR, 2020; EWG, 2016; Loria, 2019; ODH, 2020). Reducing intake of PFAS should reduce exposure, but people may not necessarily know whether their foods, beverages, or products contain PFAS. This review addresses the following question: Based on current research, are there interventions or *personal* changes that individuals can make to *effectively* reduce their PFAS exposure? This question includes two key concepts. First, "personal" indicates that the focus is on research related to media and products that people may be able to control partly or wholly (see Figure E-1) as opposed to activities that occur on a larger scale, such as contaminated site cleanup, changes in occupational exposures, or modifications to publicly owned water treatment plants. Second, "effectively" refers to changes in personal behavior that can result in measurable or substantial reductions in exposures.

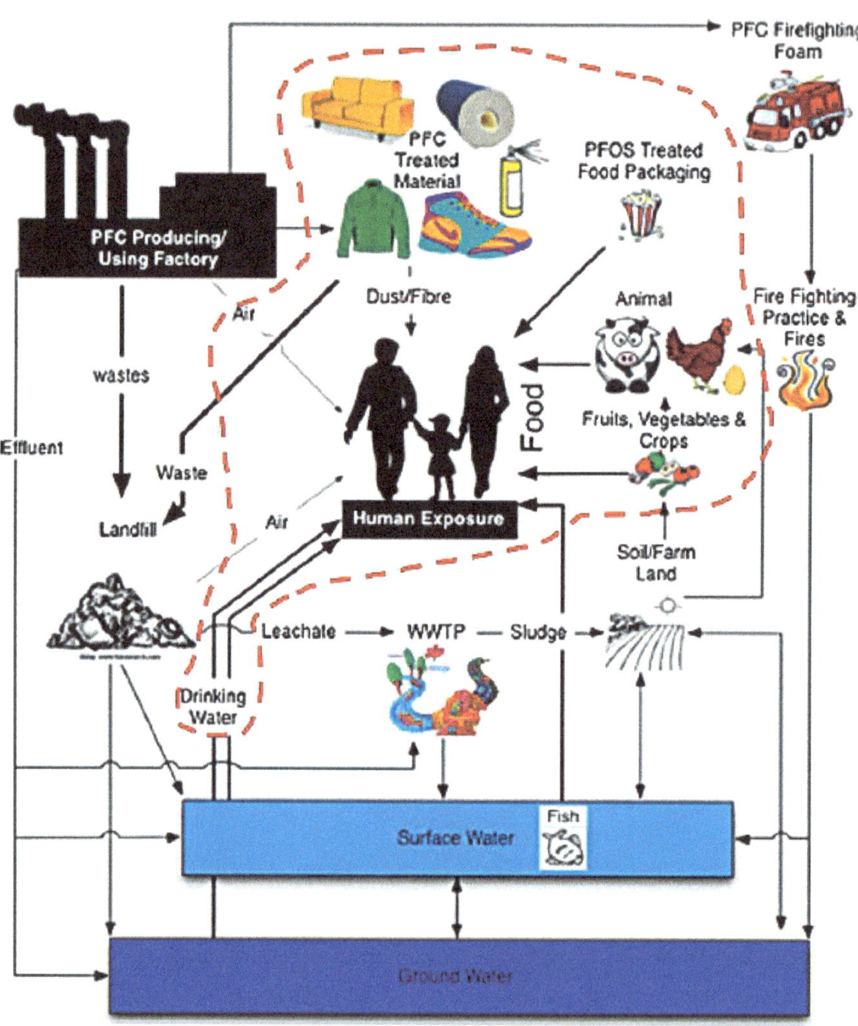

FIGURE E-1 Human PFAS exposure pathways. The area inside the dashed red line denotes pathways for which individual actions may lower PFAS exposures.
NOTE: PFC = per-/polyfluorinated chemical; WWTP = wastewater treatment plant.
SOURCE: Adapted from Oliaei et al., 2013. CC BY 4.0.

To answer the above question, it is important to consider several factors related to reducing human PFAS exposure. First, PFAS must be present in the media or products of interest. While this seems like a straightforward notion, information on PFAS concentrations in local media (e.g., local drinking water) and specific products (e.g., cookware) may in fact be sparse or unavailable. Second, there must be a complete exposure pathway between the medium or product and humans. In other words, if PFAS are present in a medium or product but there is no human contact, then exposure will not occur. Third, an exposure pathway may be complete but not contribute substantially to overall human exposure. In this case, a reduction in the PFAS source may not result in a meaningful reduction in human PFAS exposure. Fourth, actions to reduce PFAS exposure from one pathway may result in exposure to PFAS or other chemicals from a new pathway (e.g., an action to remove exposure to one food item with known PFAS levels may result in exposure to another food item that has not yet been analyzed for PFAS or other chemicals). Finally, PFAS tend to have long physiological half-lives; thus, interventions or changes in behavior may not produce near-term changes in internal PFAS levels (i.e., serum levels).

Taking these issues into consideration, this white paper explores the following three questions:

1. *Is there research that links specific interventions or changes to reductions in human exposures?* This paper focuses on studies seeking to establish that an intervention or behavioral change produces a quantifiable reduction in human exposure as evidenced by measurements of PFAS in the media of interest or directly in humans. As noted above, studies relying on biomonitoring to assess the efficacy of an intervention must consider the long half-lives of many PFAS and must be of appropriate duration to enable observation of postintervention decreases in serum levels.

2. *Can information from exposure assessments that estimate human intakes from multiple pathways of exposure and exposure routes be used as the basis for individual or community recommendations?* In this type of study, measurements of PFAS in various media are used to model human PFAS intake (i.e., nanograms [ng] per day or ng/kilogram [kg] per day). If sufficient measurement data are available, it may be possible to estimate the relative importance of various pathways of exposure.

3. *Is the available research sufficiently robust such that recommendations for modifications to behavior can be made?* Studies differ in terms of their quality and generalizability (i.e., how well the results translate from the studied population and conditions to other populations and conditions). They therefore also differ in terms of the confidence one can place in the results. Factors impacting confidence in studies can include sample size, quality assurance, inter- and intrastudy consistency in results, and completeness of reporting. In addition, for recommendations applicable to the United States, it is important to consider whether the studies reflect conditions that apply to behaviors and exposures in this country.

The remainder of this paper first describes our approach to identifying and reviewing the literature. We then provide results by medium. Next, we discuss efforts to model relative contributions of media and products to overall PFAS intakes in the United States.

We note that there are more than 9,000 PFAS compounds (NASEM, 2021), and any given study generally examines only a very small subset of these. We focus here on the 16 PFAS[4] chemicals included in the Centers for Disease Control and Prevention's (CDC's) *National Report on Human Exposure to Environmental Chemicals* (CDC, 2009).

---

[4] PFBS: perfluorobutane sulfonic acid; PFDA: perfluorodecanoic acid; PFDoA: perfluorododecanoic acid; PFHpA: perfluoroheptanoic acid; PFHxS: perfluorohexane sulfonic acid; PFNA: perfluorononanoic acid; PFOA: perfluorooctanoic acid; n-PFOA: n-perfluorooctanoic acid; Sb-PFOA: branched perfluorooctanoic acid; PFOS: perfluorooctane sulfonic acid; n-PFOS: n-perfluorooctane sulfonic acid; Sm-PFOS: perfluoromethylheptane sulfonic acid; PFOSA or FOSA: perfluorooctane sulfonamide; EtFOSAA: 2-(n-ethyl-perfluorooctane sulfonamido) acetic acid; MeFOSAA: 2-(n-methyl-perfluorooctane sulfonamido) acetic acid; PFUnDA: perfluoroundecanoic acid. The various PFAS abbreviations used in this paper are defined in the listing at the beginning of the paper. For brevity, these abbreviations are not spelled out further in the text of this paper.

E-2

Methods

This review is a scoping review that aims to (1) "identify the types of available evidence in a given field," (2) "report on the types of evidence that address and inform practice in the field and the way the research has been conducted," (3) "examine how research is conducted on a certain topic," and (4) "identify and analyze gaps in the knowledge base" (Munn et al., 2018).

LITERATURE IDENTIFICATION

Online data sources, including PubMed, EMBASE, and Google Scholar, were used to conduct the initial literature searches. We used such keywords as "(PFBS OR PFDA OR PFDoA OR PFHpA OR PFHxS OR PFNA OR PFOA OR n-PFOA OR sb-PFOA OR PFOS OR n-PFOS OR Sm-PFOS OR PFOSA OR FOSA OR EtFOSAA OR MeFOSAA OR PFUnDA OR PFAS)," "PFAS," "perfluoroalkyl," "human," "exposure," "cooking," "dust," "fish," "shellfish," "water," "nail polish," "cleaning," "consumer products," "filter," "water filter," "filtration," "intervention," "determinant," "reduction," "diet," "vacuum," "popcorn," "biomonitor," "breast milk," "breastfeeding," "infant formula," "milk powder," "carpeting," "packaging," "indoor," "bottled water," "air conditioning," "fabrics," "well water," "water treatment," "apparel," "inhalation," "ventilation," "cosmetics," "dental floss," and "personal care products," as well as various combinations of these and related keywords. We selected articles describing interventions designed to reduce human exposure to PFAS, specifically interventions that could be carried out by individuals (e.g., excluding site cleanups, modifications of publicly owned treatments works [POTWs]). Secondary references of retrieved articles were reviewed to identify publications not identified by the electronic search. Additional literature searches were conducted to identify reviews that contained estimates of human PFAS intakes using the following keywords in different combinations: "(PFBS OR PFDA OR PFDoA OR PFHpA OR PFHxS OR PFNA OR PFOA OR n-PFOA OR sb-PFOA OR PFOS OR n-PFOS OR Sm-PFOS OR PFOSA OR FOSA OR EtFOSAA OR MeFOSAA OR PFUnDA)," "PFAS," "exposure," "review," "biomonitor," "PFOA," and "human." The final search date was March 5, 2021.

The criteria for inclusion in the review were as follows: studies of interventions related to personal modifiable behavior and English-language publications. Exclusion criteria included the following: occupational studies and those interventions requiring professional activities, such as modifications to POTWs. For publications on modeled intake estimates, we were interested primarily in PFAS intakes in the United States. Research focused on clinical interventions is outside the scope of this review (e.g., Ducatman et al., 2021; Genuis et al., 2014).

LITERATURE REVIEW AND DATA EXTRACTION

Each study that met the above inclusion criteria was examined by both authors. The data from each intervention study were tabulated. Information extracted from each study included (where available[5]) the following:

- description of the study population: size, composition, source, and location;
- study design: laboratory and population;
- type of specimen and number of samples;

---

[5] Because of the disparate nature of the identified studies, it was often possible to include only a portion of the elements.

- PFAS and concentrations or changes in concentrations; and
- results: percent decrease/increase, concentration decrease/increase, and a measure of precision (e.g., 95% confidence interval [CI], standard deviation [SD]). (For publications with results reported qualitatively, the text was extracted and reproduced verbatim.)

ASSESSMENT OF EVIDENCE

We evaluated the overall strength of evidence for each medium, considering such elements as overall study design, participant selection, sample size, and exposure assessment (LaKind et al., 2014; Vandenbroucke et al., 2007). For studies using biomonitoring data to assess the efficacy of an intervention, we considered whether sampling intervals were designed to capture potential effects of the intervention. To assess the utility of the available evidence for providing recommendations for behavior modifications to reduce PFAS exposures, we examined such factors as (1) the number of available studies for each medium/PFAS chemical/intervention type combination, (2) the quality of the individual studies, (3) the intra- and interstudy consistency in results, and (4) the generalizability of the information to U.S. populations.

E-3

Results

Studies with the potential to provide information on methods that could be used by individuals for reducing exposure to PFAS were identified. The intervention areas include preparation of fish and other foods, reduction of exposure to PFAS in drinking water either via water filtration at point of entry (POE) or point of use (POU) or via consumption of bottled water, selection of cookware, minimization of indoor dust exposure through modification of indoor products, and use of personal care products or dietary fiber. Literature on breast milk/infant formula and both mother and infant PFAS exposure reduction is discussed. Finally, studies on source contributions to overall PFAS intake in the United States were identified. Each of these bodies of literature is described in the following sections.

FOOD PREPARATION: FISH, SHELLFISH, AND MOLLUSKS

Fish, shellfish, and mollusks have been studied for their potential as a source of PFAS exposure in humans. Nine studies examining the effect of various fish and seafood preparation methods on PFAS levels were identified. The studies included different species, cooking practices, and PFAS. We briefly summarize these studies here.

Alves and colleagues (2017) measured PFOS and PFUnA levels in mackerel and flounder purchased in markets in Spain, Italy, and the Netherlands (25 samples per species/location). PFAS were measured in raw samples and from samples steamed at 105°C for 15 minutes. No significant differences in concentrations were found between the steamed and raw samples. Mean concentrations of PFOS in flounder for raw and steamed samples were 24±1.5 nanograms per gram (ng/g) wet weight (ww) and 22±1.5 ng/g ww, respectively. Mean concentrations of PFUnA in mackerel for raw and steamed samples were 3.1±0.2 ng/g ww and 2.9±0.1 ng/g ww, respectively.

The 16 PFAS compounds identified previously (PFBA, PFPeA, PFHxA, PFHpA, PFOA, PFNA, PFDcA, PFUnA, PFDoA, PFTrA, PFTeA, PFBS, PFHxS, PFHpS, PFOS, and PFDS) were measured in raw and steamed tuna, hake, plaice (n = 25 each; fillets), and mussels (n = 50) purchased in European markets (Barbosa et al., 2018). Steaming was performed at 105°C for 15 minutes (fish) or 5 minutes (mussels). The effects of steaming varied considerably by both species and compound. Selected results shown in Figure E-2 illustrate these differences.

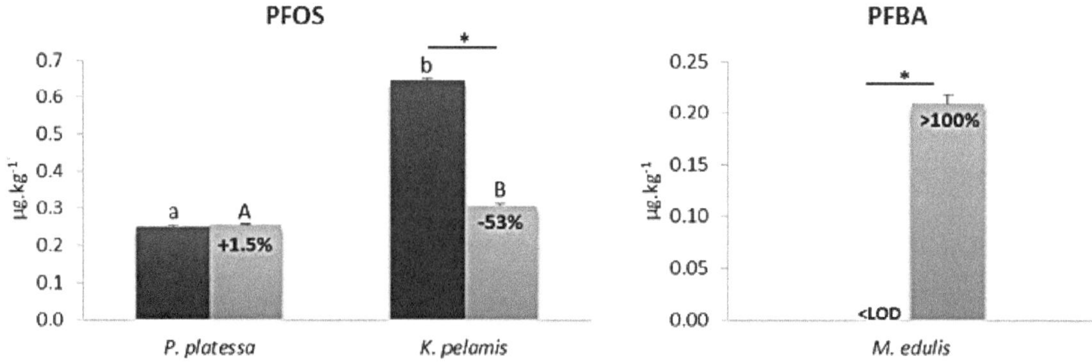

FIGURE E-2 Examples of PFAS content (micrograms per kilogram [µg/kg] wet weight [ww]) in raw and steamed seafood samples and percentages of PFAS content increase (+) and decrease (−) after steaming (mean ± standard deviation [SD]).
NOTE: * = significant differences (p <0.05) between raw and steamed samples.
SOURCE: Partially reprinted from Barbosa et al., 2018.

Bhavsar and colleagues (2014) measured various PFAS (including perfluoroalkyl carboxylic acids, perfluoroalkane sulfonic acids, perfluoroalkyl phosphonic acids, perfluoroalkyl phosphinic acids, and polyfluoroalkyl phosphoric acid diesters) in fresh-caught Chinook salmon, lake trout, common carp, and walleye obtained in Ontario, Canada (fillets from four or five fish per species). PFAS levels in the cooking oil were below the limit of detection (LOD). Fillet samples were placed on a layer of canola oil and fried, baked, or broiled for 10–15 minutes. PFOS was by far the dominant PFAS in each species; thus, the results were focused on the effects of cooking methods on PFOS fish concentrations. All cooking methods resulted in mostly statistically significant increases in concentrations of PFOS in salmon, trout, and walleye (the increase in walleye after frying did not appear to be statistically significant). For example, PFOS concentrations in raw, baked, broiled, and fried chinook salmon (mean ± standard deviation [SD]) were, respectively, 12.70±12.61 ng/g ww, 16.56±18.00 ng/g ww, 16.45±15.63 ng/g ww, and 16.03±15.19 ng/g ww. For carp, broiling and frying resulted in no significant changes. The authors also assessed the change in amount of PFOS (in contrast with the concentration change) to account for loss in mass due to cooking. With this approach, results were mixed in directionality and mostly nonsignificant.

Del Gobbo and colleagues (2008) analyzed fish and shellfish species purchased in Canadian markets for PFOA, PFNA, PFDA, PFUA, PFDoDA, PFTeDA, and PFOS. Species included catfish, cuttlefish, grey mullet, grouper, monkfish, octopus, red snapper, sea squirt, skate, whiting, and yellow croaker. Composites from each species were comprised of at least three individuals from three different sources or markets (total of nine). The fillets (sometimes with skin) were baked, boiled, or fried in water, sesame oil, rice wine, or vegetable oil. PFAS levels in the cooking oil were below the LOD. Baking and frying times were 15 minutes at 163°C or 325°C. All cooking interventions were found to decrease PFAS concentrations, with boiling and frying reducing total PFAS concentrations by an average of 79 percent and 54 percent, respectively. Baking reduced all measured PFAS to below the LOD (0.03–10 ng/g ww). Boiling appeared to increase concentrations of PFOS in octopus (from nondetect to 0.23 ng/g ww) but not in red snapper, skate, or yellow croaker. The authors hypothesize that large loss of mass in the boiled octopus (87 percent) may have resulted in an increase in levels above the LOD.

Hu and colleagues (2020) measured 13 PFAS compounds (PFBA, PFPeA, PFHxA, PFHpA, PFOA, PFNA, PFDA, PFUnDA, PFDoDA, PFBS, PFHxS, PFOS, and FOSA) in grass carp taken from Tangxun Lake, China (n = 5). The fillets were either steamed, boiled, fried, or grilled, with cooking temperatures for the various processes ranging from 100°C to 210°C. The authors note that in the "cooking blank juice samples all PFAS were below the MLQs [Method Quantitation Limit] except PFBS and PFOS. The concentrations of PFBS and PFOS in cooking blank samples were 1.31–2.43 ng/g and 0.131–0.169 ng/g, respectively" (Hu et al., 2020, p. 4). The effects of the cooking methods on concentrations of various PFAS compounds were found to be inconsistent. For example, median PFOS concentrations increased from 71.3 ng/g ww in uncooked fish to 146 ng/g ww in fried fillets. In contrast, median PFBS concentrations decreased from 20.3 ng/g ww in raw fish to 8.08 ng/g ww after grilling. These results exemplify the difficulty in characterizing the directionality and magnitude of the effectiveness of cooking fish as a proposed type of intervention.

Kim and colleagues (2020) measured 19 PFAS (PFOS, PFDS, PFHxS, PFTeDA, PFTrDA, PFDoA, PFUnDA, PFBS, PFDA, PFNA, PFOA, PFHpA, PFHxA, PFPeA, PFOSA, N-EtPFOSAA, N-MePFOSAA, L-PFHpS, and PFBA) in mackerel bought from a market in Korea (n = 10). Composited fillets from three mackerels each were prepared with various washing, soaking, and cooking (grill, braise, steam, or fry) methods (two composites each). The fish were cooked between 6 and 25 minutes with various ingredients including oil, water, potato, soy sauce, pepper paste, sugar, garlic, and ginger. Preparation methods included such traditional Korean practices as soaking the fillets in sake or rice-washed solutions. These soaking practices reduced PFAS levels by 51 to 80 percent. Washing the mackerel with water resulted in a reduction in PFAS of 74 percent. Similarly, all cooking methods reduced total PFAS content compared with the raw samples: grilling—91 percent, steaming—75 percent, frying—58 percent, and braising—47 percent. While cooking with potatoes further reduced PFAS levels in the mackerel, it also increased the levels in the potatoes.

Luo and colleagues (2019) assessed the effect of pretreatments and cooking on levels of 19 PFAS (PFOA, PFOS, PFBA, PFPeA, PFHxA, PFHpA, PFNA, PFDA, PFUnDA, PFDoDA, PFTrDA, PFTeDA, PFBS, PFHxS, L-PFHpS, PFDS, PFOSA, N-MePFOSAA, and N-EtPFOSAA) in fish cakes (n = 4 brands) and swimming crabs (n = 100) purchased in markets in Korea. The effects of a wide range of cooking methods were evaluated. Additional foods used in the cooking processes included soybean oil, two types of soy sauce, and Korean radish. Blanching, commonly used as a pretreatment before cooking fish cakes, did not yield significant changes in PFAS levels. Significant reductions in total PFAS were observed after boiling, frying, and stir-frying fish cakes (total PFAS in control, boiled, fried, and stir-fried fish cakes, respectively, were as follows: 2.96 ± 0.6 ng/g, 1.60 ± 0.16 ng/g, 1.93 ± 0.19 ng/g, and 1.94 ± 0.07 ng/g). For the crabs, presoaking reduced PFAS levels. PFAS in the crabs were significantly decreased after steaming and stewing.

Taylor and colleagues (2019) collected school prawn, blue swimmer crab, and dusky flathead from contaminated or reference estuaries in New South Wales and analyzed them for 20 PFAS compounds (PFAS above the limit of quantitation [LOQ] were PFHxA, PFHpA, PFOA, PFNA, PFDA, PFUnDA, PFDoDA, PFBS, PFHxS, PFOS, PFDS, FOSA, and NEtFOSE) before and after cooking. Dusky flathead fillets were baked or pan-fried in olive oil, while the crab and prawn were boiled in salted water. PFAS were below LODs in the cooking water and oil. Five or six replicates were used for each species and each cooking treatment (for prawns, each replicate was comprised of a composite of 10 individuals). The effect on PFAS concentrations varied with species, cooking method, and chemical. For instance, PFOS concentrations showed no change in crab following boiling, whereas PFHxS and PFOA concentrations were reduced. However, PFOS, PFHxS, and PFOA concentrations increased in prawns after cooking. For the dusky flathead, PFOS levels did not change significantly after frying, but baking resulted in a small but significant increase. The authors conclude that "cooking does not consistently reduce PFAS concentrations, and cannot mitigate dietary exposure" (Taylor et al., 2019, p. 280).

Vassiliadou and colleagues (2015) obtained several species of fish (anchovy, bogue, hake, picarel, sardine, sand smelt, and striped mullet) and shellfish (Mediterranean mussel, shrimp, and squid) from local markets in Greece and mussels from a mariculture farm. Twelve PFAS compounds (PFBA, PFPeA, PFHxA, PFHpA, PFOA, PFNA, PFDA, PFUnDA, PFDoA, PFBS, PFHxS, and PFOS) were measured in raw or washed samples that were then fried (in virgin olive oil at 170°C) or grilled (at 180°C). Total PFAS concentrations were found to be mostly higher after grilling and after frying, but changes in individual PFAS were inconsistent.

In general, there appear to be numerous factors related to preparation of fish, shellfish, and mollusks that can impact changes (increases or decreases) in concentrations of PFAS. Taylor and colleagues (2019) observe that whether cooking reduces PFAS concentrations in fish depends on the physicochemical properties of the chemical, the cooking method used, and the species. They note several processes that could impact changes in PFAS levels in fish from food preparation: losses to the cooking medium (e.g., cooking oil), moisture loss during cooking, PFAS precursors in fish tissues transforming during cooking to PFAAs, protein loss, or protein increase. Additional factors possibly affecting changes in concentration are the size, shape, and thickness of fish fillets (Hu et al., 2020).

Overall, robust recommendations for fish preparation interventions would ideally be based on consistent intra- and interstudy results; use of fish species and preparation methods common to the United States; and well-powered, replicated studies. The results from the studies reviewed here indicate that the effects of preparation of fish and shellfish on PFAS levels are inconsistent (examples are shown in Figure E-3). It is therefore difficult to use this information to inform recommendations regarding the efficacy of fish and shellfish preparation for reducing PFAS intake.

FIGURE E-3 Examples of inconsistent changes in selected PFAS concentrations after fish and shellfish preparation (arrows indicate direction of concentration change).
SOURCES: Blue swimmer crab information is from Taylor et al., 2019. Carp information is from Hu et al., 2020. Crab photo: Judy S. LaKind, CC BY SA 3.0, https://commons.wikimedia.org/w/index.php?curid=4963391. Carp photo: Dezidor—Self-photographed, CC BY 3.0, https://commons.wikimedia.org/w/index.php?curid=12661115.

Furthermore, some of the fish species and preparation methods included in this body of research are not common to the United States, limiting the generalizability of results for U.S. communities. Additionally, because very few of these studies used the same cooking methods and species, it is difficult to corroborate even the instances in which a particular intervention appears to have been effective (within this small group of studies, various cooking preparation approaches included soaking followed by stewing, steaming, boiling, frying, or stir-frying; grilling, steaming, baking, boiling, or frying without presoaking; and baking in rice wine or vegetable oil). Furthermore, several of the studies included very small sample sizes. Finally, while some studies considered the effect of moisture or mass variations on PFAS concentration changes, others reported only concentration data. A more relevant metric in terms of human exposure would be the mass of PFAS remaining in the samples after preparation.

FOOD PREPARATION: OTHER

Two studies were identified that examined the effect of preparation on PFAS levels in foods other than fish (Binnington et al., 2017; Jogsten et al., 2009). Jogsten and colleagues (2009) measured several PFAS in various uncooked and cooked foods (see Table E-1).[6] Specifically, composite samples (n = 2) of beef, pork, or chicken were cooked in an oil mixture using nonstick cookware. Only PFHxS, PFOS, PFHxA, and PFOA were detected in at least one of two composite samples. PFOS levels increased in grilled pork, grilled chicken, and fried chicken compared with the raw samples. In contrast, levels did not increase in cooked veal or fried pork. The results from this study are not directly relevant for intervention recommendations. First, foods were purchased outside of the United States, and it is not known whether PFAS in these foods are similar to those in foods found in the United States. Second, only two samples per food type were included; there was no information on brands and limited information on cooking procedures; and results were inconsistent regarding the efficacy of cooking in reducing PFAS concentrations. Finally, it is not clear whether the effects of cooking can be disentangled from those of the Teflon-coated cookware used in this study.

---

[6] Jogsten and colleagues (2009) also measured PFAS in foods wrapped in different types of packaging. The foods were purchased in Spain, and the brands were not identified, so the relevance to the U.S. population is unclear. Egeghy and Lorber (2011) note that while fluorochemical-treated food packaging can be a source of PFAS in food, it appears that PFAS levels in packaging such as fast-food wrappers have decreased over time.

Binnington and colleagues (2017) studied the effects of preparation of beluga whale blubber on nutrients and environmental chemicals, including PFAS. They collected samples from two male whales (aged 24 and 37 years) from the Northwest Territories and prepared them using traditional approaches. Measured PFAS (detected in ≥50 percent of the samples) were PFNA, PFDA, PFUnDA, PFDoDA, and PFOS. PFAS were measured in raw and prepared (boiled, roasted, and aged) samples. Roasting increased concentrations of some of the PFAS compared with certain other treatments (e.g., air-drying, hang-drying, and boil pot), but were reduced in oil (Table E-1 includes concentration results for PFOS). According to the authors, issues with sample preparation may prevent these results from being considered representative of the overall mixture.

TABLE E-1 Summary of Results of Studies Examining the Effect of Food Preparation on PFAS Levels

Source	PFAS	Study Location	Food Type	Number	Concentration
Jogsten et al., 2009	PFBuS PFHxS PFOS PFHxA PFHpA PFOA PFNA PFDA PFUnDA PFDoDA	Spain	Beef, pork, chicken	Two composite samples from at least six subsamples for each food type from each of two sampling locations	PFOS, ng/g fresh weight (standard deviation) *Veal* raw: <0.015 grilled: <0.008 fried: <0.018 *Pork* raw: <0.008 grilled: 0.011 (0.009) fried: <0.008 *Chicken* raw: <0.008 grilled: 0.012 (0.01) fried: 0.010 (0.007)
Binnington et al., 2017	PFNA PFDA PFUnDA PFDoDA PFOS	Canadian Arctic	Male beluga whale blubber	Blubber sample divided into portions for different preparation processes	nanograms per gram wet weight (ng/g ww)[a] PFOS: nondetect (ND) to <1 PFNA: ND to <4 PFDA: ND to <4 PFUnDA: ND to <4 PFDoDA: ND to <4

[a] Estimated from publication figures.

In summary, only two studies of changes in PFAS levels associated with cooking foods were identified. These studies included small sample sizes and yielded inconsistent results.

LOCAL FOOD CONSUMPTION ADVISORIES

State advisories for guidance on consumption of locally grown foods could be a source of information to inform exposure reduction decisions. Eleven states have advisory guidelines for consumption of fish, wildlife, and other foods (California [seafood], Connecticut, Hawaii [in process], Maine [fish, beef, and milk], Michigan [fish and deer], Minnesota, New Hampshire, New Jersey, New York, Washington [in process], Wisconsin [fish and deer]) to protect human health from exposure to PFAS.[7] These advisories offer guidance on limiting the quantity of these foods consumed. Depending on the state-specific PFAS and concentrations, different consumption levels are indicated, ranging from do not eat (e.g., fish or deer in Michigan with PFOS concentrations of more than 300 parts per billion [ppb])

---

[7] See https://www.ecos.org/wp-content/uploads/2021/04/Updated-Standards-White-Paper-April-2021.pdf (accessed June 30, 2022).

to unlimited consumption (e.g., fish in New Jersey with 0.56 ng/g PFOS). While fish consumption has a role in a healthy diet (Mozaffarian and Rimm, 2006), weighing the risks of PFAS exposure from fish consumption against the benefits of fish consumption is a complex process, and no intervention studies were identified that evaluated the impact of reduced consumption of fish on PFAS levels in blood and urine.

DRINKING WATER

Ingestion of drinking water is thought to be a major pathway for PFAS exposure (Domingo and Nadal, 2019). Research on two types of drinking water interventions is described in this section. The first addresses whether—and the extent to which—the use of water filters at POE into the home, under the sink (POU), or in water pitchers reduces PFAS exposure. The second addresses whether the use of purchased bottled water results in lower PFAS exposure compared with the use of tap water. Because PFAS levels in water can vary widely, we focus on studies that measured PFAS in tap and bottled water obtained from the same geographic area.

Six publications and one agency report evaluating possible drinking water interventions were identified. Four (Ao et al., 2019; Iwabuchi and Sato, 2021; MDH, 2008; Patterson et al., 2019) evaluated use of POE, POU, and water pitcher filtration devices; and three (Ao et al., 2019; Gellrich et al., 2013; Heo et al., 2014) evaluated differences in PFAS concentrations between tap water and bottled water. These studies are summarized here. Also discussed is one association paper assessing the relationship between drinking water source and serum levels in a highly contaminated area (Emmett et al., 2006).

POE, POU, and Water Pitcher Filtration

Ao and colleagues (2019) (also discussed in the next section) measured six PFAS compounds (PFOA, PFOS, PFNA, PFBS, PFHpA, and PFHxS) in tap (n = 9), filtered (n = 9), and bottled (n = 9) water in Shanghai, China. Paired tap and filtered water samples were collected from each of nine homes served by three different water sources. The filtered water samples were collected from the effluent of the home's water purification device. No further information on sampling or quality control in the field was given, nor was information on the type or brand of filter provided. ΣPFAS median concentrations in tap water and filtered water were 4.44 nanograms per Liter (ng/L) and 3.13 ng/L, respectively, but the differences were not statistically significant.

Herkert and colleagues (2020) tested municipal, well, and filtered (n = 89) and unfiltered (n = 87) tap water in residences (n = 73) in North Carolina for 11 PFAS compounds (GenX, PFBS, PFBA, PFHxS, PFOS, PFPA, PFHxA, PFHpA, PFOA, PFNA, and PFDA). The 89 POE and POU filters tested varied in both type (e.g., pitcher, under sink, faucet, whole house) and filtration method (reverse osmosis [RO], granular activated carbon [GAC], single-stage, two-stage). Notably, RO filters and dual-stage filters were found to consistently remove most measured compounds (except PFNA and GenX) at an average of ≥90 percent efficiency. On the other hand, GAC filters had more variable performance and were far less effective in removing short-chain PFAS compounds, with an average removal efficiency of just 41 percent for those chemicals. Whole-house activated carbon POE systems resulted in increased levels of PFAS in half of the tests (n = 4). The authors did not observe any correlations between removal efficiency and brand, source water, loading, or filter age.

Iwabuchi and Sato (2021) tested pitcher-type water filters for their ability to reduce concentrations of six PFAS compounds (PFOA, PFOS, PFHxA, PFDA, PFDoA, and PFHxS). Four different models from four manufacturers were evaluated—two with a carbon, ceramic, and hollow fiber membrane design; and two with an activated carbon (AC) and ion exchange design (brand names not given). One liter of the test water was applied to the water filters 200 times, with filtrate analyzed after 10 L, 100 L, and 200 L had been passed through the filters. For each model, filtration effectiveness decreased with prolonged use, but three of the four models were effective in removing the majority of all PFAS compounds. Removal efficiency did not appear to be related to filter material type. Removal efficiency varied by carbon-chain length (i.e., more efficient removal was observed for longer-carbon-chain PFAS) and the PFAS functional group (PFOS >PFOA, and PFHxS >PFHxA). All tests were performed with initial PFAS concentrations of 50 ng/L. It is possible that filter effectiveness may vary with initial concentration levels. Iwabuchi and Sato (2021) conclude that household water purifiers are effective at reducing PFAS levels in drinking water.

Patterson and colleagues (2019) tested five commercially available POU/POE water treatment systems for six PFAS (PFOS, PFOA, PFHpA, PFHxS, PFBS, and PFNA). These included three RO systems (iSpring RCS5T, HydroLogic Evolution RO1000, and Flexeon LP-700) and two GAC systems (Calgon Filtrasorb 600 AR+ and Evoqua 1230CX). The filters were tested at various flow rates. The authors found that both RO and GAC systems had the potential to remove PFAS to below the LOD under their experimental water quality and operational conditions. They note, though, that performance was variable and that the long-term performance of the systems was not tested.

In addition to the publications described above, the Minnesota Department of Health (MDH, 2008) conducted a survey of POU water filtration devices for PFAS. Fourteen filters were lab-tested, and 11 of these that passed initial testing were field-tested using water from municipal wells. The brands were AC (n = 8)—Aquion Rainsoft Hydrefiner P-12 9878, Kinetico MACguard 7500, and Sears Kenmore Elite 625.385010; and RO (n = 6)—GE Smartwater GXRM10GBL and Watts Premier WP-4V. The four AC devices removed PFAS compounds to below the analytical reporting level (0.2 micrograms [µg]/L). The RO devices also removed PFAS to below the reporting limit. In terms of use by consumers, MDH notes, AC filters, RO membranes, and other filter elements have a limited service life and must be periodically replaced. Manufacturer recommendations vary, but many suggest replacing filters after 500 gallons of treatment or every 6 months (MDH, 2008).

MDH also tested a small, inexpensive, faucet-mounted carbon filter (PUR models FM-2000B, FM-3333B) using chlorinated and unchlorinated water with levels of PFAS that exceeded the U.S. Environmental Protection Agency (EPA) Lifetime Health Advisory Levels. The filter removed all PFAS from the unchlorinated water to below the LODs (which ranged from 5 to 10 ng/L). Some breakthrough of PFBA occurred for the chlorinated water test, but even at the filter capacity recommended by the manufacturer (100 gal), the filter was still removing 73 percent of the PFBA.[8]

In the course of their work sampling private well water in Washington County, Minnesota, MDH staff collected incidental samples of water treated by homeowner-installed carbon and RO systems. While many of these systems performed quite well, others achieved only partial removal of PFAS or none at all. MDH did not have the capacity to investigate further regarding the reasons for poor performance, but suspected that inadequate system maintenance may have been the main cause. When conveying the results of such tests, whether the system was performing well or not, MDH cautions homeowners that no guarantee can be made regarding long-term effectiveness of the system and recommends that they work with a qualified water treatment company to ensure proper maintenance of their system (Virginia Yingling, personal communication, July 14, 2021).

For those homes that exceed the state's drinking water guidance values, the Minnesota Pollution Control Agency installs and maintains whole-house GAC filter systems consisting of two 90 lb carbon canisters in series. As there are currently more than 1,000 of these systems in place, the state does not manage them individually, but changes the carbon out annually (which is more frequent than needed for the levels of PFAS present in the groundwater). Testing shows that the state's GAC systems remove all PFAS below reporting limits (4.4 ng/L). However, some trace levels, especially of PFBA, have been detected above the method detection limits (which range from 0.5 to 1.0 ng/L). In a few instances, the filter systems were found not to be working, but these cases were related to the homeowners having accidentally switched the bypass valve or altered the plumbing in a way that bypassed the filter. As a result, the state's contractor now inspects every system and plumbing at the annual filter changeout (Virginia Yingling, personal communication, July 14, 2021).

---

[8] The information in this and the subsequent two paragraphs was provided by Virginia Yingling, Environmental Health Division, Minnesota Department of Health, Saint Paul, Minnesota.

A study of residents living in an area served by the Little Hocking water system in Ohio—with water PFOA levels in the low ppb range at the time of the study—examined whether a community relying on highly contaminated public water could significantly reduce exposure through the use of a carbon water filter (Emmett et al., 2006). Serum PFOA was measured in a random sample of study participants (n = 324), who also provided information on their drinking water habits. Those who used only water from the Little Hocking water system in their homes were categorized as using a home carbon water filtration system (n = 64) versus no home water filtration system, a system not known to remove PFOA, or a system of unknown type (n = 209). Participants with home carbon water filters were shown to have statistically significantly (p = 0.008) lower median serum PFOA levels compared with those who did not (318 nanograms per milliliter [ng/ml] versus 421 ng/ml, respectively). The difference in serum PFOA levels in these two groups of participants was not as large as that seen for individuals using bottled, spring, or cistern water (see the next section for more information). The authors ascribe this finding to the limited effectiveness of water filters, as well as reliability issues associated with filter maintenance. They do not recommend use of home filtration systems that were available at the time.

All but one of the studies reviewed here found that various filtration methods showed evidence of their potential effectiveness. These studies suggest that pitcher-type, POE, and POU filtration systems can reduce PFAS levels in drinking water under the conditions tested. It is worth noting that optimal filtration depends on the user's actively maintaining these devices, and no study has yet looked at the effectiveness of these interventions in real-world circumstances.[9]

Bottled Water (Versus Tap Water)

Here we summarize three studies that included measurements of PFAS in bottled, tap, or filtered water from the same area in order to compare levels across drinking water sources. This research was conducted in China, Germany, and Korea.

Ao and colleagues (2019) measured six PFAS compounds (PFOA, PFOS, PFNA, PFBS, PFHpA, and PFHxS) in tap (n = 9), filtered (n = 9), and bottled (n = 9) water in Shanghai, China. Tap and filtered water samples were collected from each of nine families served by three different water sources. Paired tap and filtered water samples were taken from each home. The filtered water samples were collected from the effluent of the home's water purification device. Bottled water was purchased from local markets and represented nine best-selling brands. No further information on sampling or quality control in the field was given. ΣPFAS median concentrations in tap water, filtered water, and bottled water were 4.44 ng/L, 3.13 ng/L, and 2.36 ng/L, respectively; the differences were not statistically significant.

Mineral water (n = 119), tap water (n = 26), and spring water (n = 18) samples were measured for 10 or 19 (tap water only) PFAS compounds (PFBA, PFBS, PFPeA, PFHxA, PFHxS, HPFHpA, PFHpA, PFOA, H4PFOS, PFOS, FOSA, PFNA, H2PFDA, PFDA, PFDS, H4PFUnDA, PFUnDA, PFDoDA, and PFTeDA) (Gellrich et al., 2013). The mineral water samples were from Germany; spring water samples from Switzerland, the Czech Republic, and Germany; and tap water samples from homes in unidentified locations. No further information on water sampling was given. The highest ΣPFAS concentration was in tap water (42.7 ng/L). The proportions of individual PFAS differed across water type. For example, PFOS was below the LOD in all of the spring water samples but was detected in 9 percent of all of the mineral water samples. The authors note that the PFAS concentrations in the three water types were similar and described the concentrations as "low." For example, the median PFOA levels in mineral, spring, and tap water samples were 1.6 ng/L, 1.4 ng/L, and 2.6 ng/L, respectively.

Heo and colleagues (2014) measured 16 PFAS compounds (PFBA, PFPeA, PFHxA, PFHpA, PFOA, PFNA, PFDA, PFUnDA, PFDoDA, PFTrDA, PFTeDA, PFBS, PFHxS, PFHpS, PFOS, and

---

[9] Information on certified water filters can be found at the following website: https://www.nsf.org/knowledge-library/perfluorooctanoic-acid-and-perfluorooctanesulfonic-acid-in-drinking-water (accessed May 12, 2021). Note that the certification applies only to PFOA and PFOS, and the water filter must be able to reduce these chemicals to under 70 parts per trillion (ppt).

PFDS) in tap (n = 34) and bottled (n = 8) water from Busan, Korea. The bottled water samples were purchased in markets, and the tap water samples were collected from 16 districts in Busan (no further information on types of bottled water or sampling methods for tap water was provided). PFAS concentrations and detection frequencies were higher in the tap water than in the bottled water samples. For example, mean ΣPFAS levels in bottled and tap water were 0.48 ng/L and 41.3 ng/L, respectively. The authors do not provide information on whether the observed differences were statistically significant.

A study of residents living in an area served by the Little Hocking water system in Ohio suggests that a community relying on highly contaminated public water could significantly reduce exposure through the use of bottled water (Emmett et al., 2006). Serum PFOA was measured in a random sample of study participants (n = 324), who also provided information on their drinking water habits. Residents who reported drinking primarily bottled/spring/cistern water had a median serum PFOA level of 71 ng/mL, compared with a statistically significantly higher level of 374 ng/mL for those who drank only Little Hocking system water. Overall, the authors observed a strong relationship between serum PFOA levels and PFOA concentrations in the drinking water source.

For communities with highly contaminated water supplies, the use of alternative drinking water sources has been shown to be associated with significantly reduced exposures. However, none of the intervention studies reviewed here provide robust evidence for the effectiveness of replacing tap water with bottled water for U.S. locations with background levels of PFAS. The studies reviewed here were conducted in Europe and Asia, and the water PFAS concentrations there may not be generalizable to the United States. Lack of brand information and small sample sizes present additional challenges for evaluating this intervention.

Use of bottled water as a replacement for tap water can be expensive and inconvenient. To be confident that the use of bottled water will result in a reduction in PFAS exposure, an understanding of local water conditions in comparison with PFAS levels in specific types of bottled water is needed. As described above, while PFAS levels in bottled water tend to be approximately between <LOD to <100 ng/L, it cannot be assumed that levels in bottled water are always lower than those in tap water. Outside of highly contaminated areas, the degree of spatial granularity for concentration data required to ensure that replacing tap water with bottled water will reduce PFAS exposure is not well understood. However, based on public data POTWs (EPA, 2017), PFAS levels in treated water can vary widely within regions or states (e.g., PFOS water concentrations in Delaware ranged from <MRL [Minimum Reporting Level] of 0.04 to 1.8 μg/L, while PFOA water concentrations in Pennsylvania ranged from <MRL of 0.02 to 0.349 μg/L).

The EPA's Third Unregulated Contaminant Monitoring Rule (UCMR) (EPA, 2017) includes monitoring data for PFOS, PFOA, PFNA, PFHxS, PFHpA, and PFBS (data from 2013 to 2015 from a representative sample of public water systems serving ≤10,000 people). Figure E-4 shows varying concentrations of these six PFAS across the United States for PFAS levels above the method reporting limits. Levels in drinking water can vary across the United States by at least an order of magnitude. The concentrations are in the low ng/L range, similar to reported levels in bottled water. Also note that for much of the United States, PFAS levels in drinking water are below the method reporting limit (see Figure E-5, green symbols) and so would possibly be similar to those in bottled water.

BREAST MILK AND INFANT FORMULA

In considering the idea of "intervention" as it pertains to breast milk, there are two underlying concepts. The first is whether there are interventions that could reduce exposure to the breastfeeding infant, and the second is whether lactating can be an effective method for reducing the mother's levels of PFAS compounds. We discuss each of these concepts in this section.

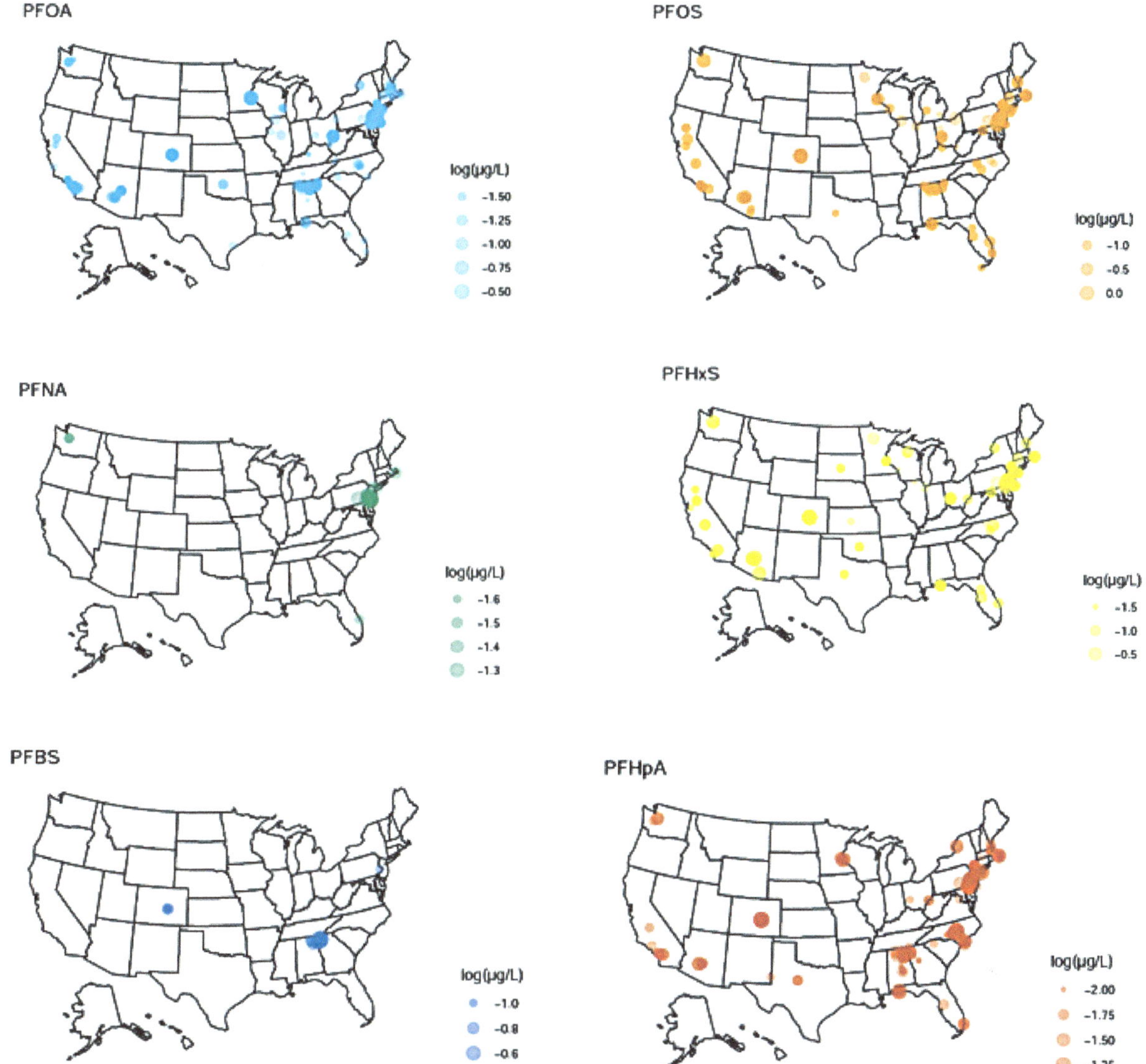

FIGURE E-4 Variation in PFAS levels above the method reporting limit in drinking water in the United States. NOTES: The relative sizes of the symbols correspond to the actual measured water concentrations. The legends show exemplar concentrations. Plotted: $\log_{10}$(concentration), excluding data below minimum reporting levels. μg/L = microgram per Liter.
SOURCE: Data were extracted from the U.S. Environmental Protection Agency's (EPA's) Third Unregulated Contaminant Monitoring Rule (UCMR 3) (2013–2015) database (EPA, 2017). Individual concentration data for each point on the graphics can be found at http://lakindassociates.com/interactive-map (accessed June 30, 2022).

The Infant's Exposure: Does Formula Feeding in Place of Breastfeeding Reduce Exposure?

Many mothers choose to breastfeed rather than use infant formula. Breastfeeding confers various health advantages to the infant and mother (AAP, 2012; WHO, 2020). At the same time, breast milk includes environmental chemicals (LaKind et al., 2001, 2018; Lehmann et al., 2018) that derive from the mother's body and are transferred to the infant via breastfeeding. Thus, concern has been expressed about the possible health effects on the infant from those environmental chemical exposures. This concern raises the question of whether formula feeding would serve as a method for "intervening," or reducing infant exposure to PFAS.

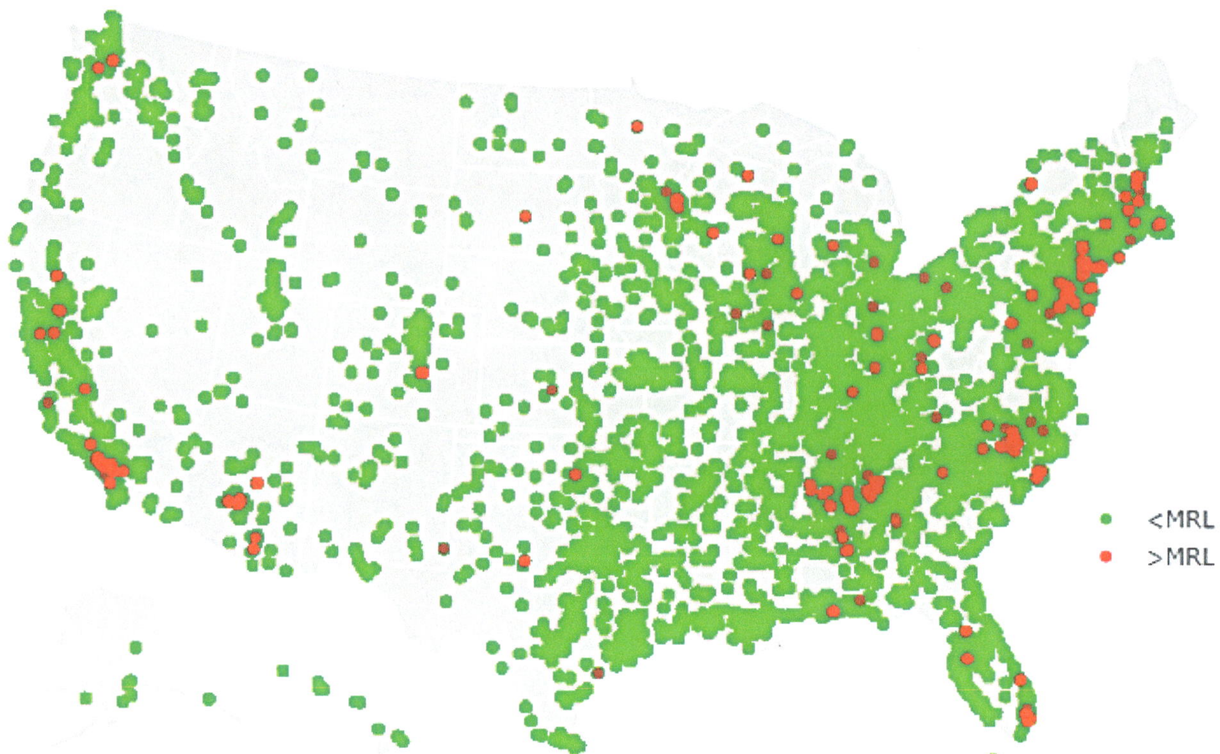

FIGURE E-5 PFAS levels above (red) and below (green) the method reporting limit in drinking water in the United States.
NOTES: The symbols represent a total of 215,963 measurements from the U.S. Environmental Protection Agency's (EPA's) Third Unregulated Contaminant Monitoring Rule (UCMR 3) for PFOA, PFOS, PFHxS, PFBS, PFHpA, and PFNA. MRL = Minimum Reporting Level.
SOURCE: Data were extracted from the EPA's UCMR 3 (2013–2015) database (EPA, 2017).

At least two factors need to be taken into consideration. The first is that infants' exposures begin in utero. Some studies suggest that despite environmental chemical exposures associated with breastfeeding, infants who are breastfed may do better in a number of health-related aspects compared with formula-fed infants (LaKind et al., 2008). The World Health Organization has stated, "in studies of infants, breastfeeding was associated with beneficial effects, in spite of the contaminants present. The subtle effects noted in the studies were found to be associated with transplacental, rather than lactational, exposure" (WHO, 2000, p. 237). These conclusions are drawn from research on persistent chemicals, such as dioxins; to our knowledge, there is no PFAS-specific literature on health effects of breastfed versus formula-fed infants. At present, there does not appear to be sufficient PFAS-related research that would allow for a definitive conclusion regarding infant health and the choice between breastfeeding and use of infant formula. Thus, it is not clear that formula feeding is a scientifically supported "intervention" that would prevent adverse health outcomes.

The second factor is that in choosing formula feeding to reduce infant exposure to PFAS, the assumption is being made that infant formulas do not themselves contain PFAS compounds. While studies have reported on detectable levels of PFAS in infant formula in other countries (e.g., Llorca et al., 2010; Macheka et al., 2021), it is instructive to review available data on levels of PFAS in breast milk versus infant formula in the United States to assess whether levels in formula are lower than those found in breast milk. As an additional complication, it is not uncommon to purchase powdered formula and reconstitute it with drinking water. Therefore, we provide here a synopsis of U.S. levels of PFAS in breast milk, formula, and drinking water.

PFAS in Breast Milk in the United States

Reviews of PFAS in breast milk (Lehmann et al., 2018; Liu et al., 2020) identify three studies reporting measurements of PFAS levels in breast milk in the United States. One of these studies (Kuklenyik et al., 2004) is an analytical methods study. No information on either the milk donors or the sampling procedures is reported; the information from this study is not relevant to this discussion. In a second study, von Ehrenstein and colleagues (2009) collected milk samples from 34 breastfeeding women in North Carolina at 2–7 weeks and 3–4 months postpartum. Nine PFAS were measured (PFOS, PFOA, PFNA, PFHxS, PFOSA, MeFOSAA, EtFOSAA, PFBS, and PFDA). Measurements below the LOD were assigned a value of LOD/sqrt2. PFAS levels were below the LOQ in most of the 34 milk samples collected at both sample times (note that the LOQs for PFAS in milk ranged from 0.15 to 0.60 ng/ml, or ppb). Specifically, PFAS were detected in samples from only 4 of the 34 women, and of these only three PFAS were above the LOQ: Et-PFOSA-AcOH (1.0 ng/ml) and Me-PFOSA-AcOH (0.7 ng/ml) in one woman, and PFOSA in three women (0.3 ng/ml, 0.5 ng/ml, and 0.6 ng/ml). The remainder of the milk samples from both collections were found to have concentrations <LOQ.

In contrast, Tao and colleagues (2008a) measured PFAS (PFOS, PFOA, PFHxS, PFNA, PFHpA, PFDA, PFUnDA, and PFDoDA) in milk samples collected from 45 primiparous and multiparous women in Massachusetts. Levels that were below the detection limit were assigned a value of zero, while those detected but below the LOQ were assigned a value of one-half the LOQ. Mean levels of PFOS and PFOA were 131±103 parts per trillion (ppt) (mean±SD) and 43.8±33.1 ppt, respectively. Mean PFHxS and PFNA levels were 14.5±13.7 ppt and 7.26±4.70 ppt, respectively. The remainder of the PFAS were detected in only ≤4 samples (<24 ppt).

PFAS in Infant Formula in the United States

In a 2018 review of environmental chemicals in breast milk and infant formulas (Lehmann et al., 2018), only one publication is identified with measurements of PFAS in infant formula in the United States (Tao et al., 2008b). Tao and colleagues (2008b) measured PFAS in 21 formula samples purchased in Washington, DC, and Boston, Massachusetts. The brands represented >99 percent of the U.S. market. Most of the samples were organic or nonorganic milk- or soy-based powders and ready-to-use or concentrated liquids. PFOS was detected in one sample (11.3 ppt; LOQ = 11.0 ppt). PFHxS was detected in two samples (1.36 ppt and 3.59 ppt; LOQ = 1.35 ppt). No other PFAS (PFOA, PFNA, PFBS, PFHpA, PFDA, PFUnDA, and PFDoDA) were detected in any samples.

PFAS in Drinking Water in the United States

An assessment of infant exposure to environmental chemicals would not be complete without considering exposures via drinking water used to reconstitute infant formula (LaKind et al., 2005). ATSDR (2019) has noted that a source of PFAS exposure to infants and toddlers is "formula mixed with PFAS contaminated water." It is important to be able to provide information on what is meant by "contaminated," as PFAS levels in tap water vary widely. For example, Andrews and Naidenko (2020), using national and state databases, estimated that 18–80 million people in the United States use tap water containing at least 10 ng/L (ppt) PFOA and PFOS combined, and more than 200 million people may have water with a combined PFOA and PFOS level of at least 1 ng/L.

Formula may also be reconstituted with bottled water. There is a paucity of data on PFAS levels in bottled water in the United States. Akhbarizadeh and colleagues (2020) reviewed the international literature on PFAS levels in bottled water (with none from the United States) and report levels of various PFAS in the low ng/L range, with some levels as high as the low 100s ng/L depending on the type and number of PFAS included in the reporting. They note that researchers have attributed the PFAS in bottled water to several possible sources, including PFAS from the plastic bottles themselves; introduction of PFAS to the water before bottling or during bottle closure; or contamination of contact materials during bottling, handling, and storage of the bottles. In a study of tap, spring, and mineral water from Germany, Gellrich and colleagues (2013) found PFAS levels to be generally in the low ng/L range.

Looking beyond the peer-reviewed literature, *Consumer Reports* (Felton, 2020) conducted a study on PFAS levels in bottled water purchased in stores in the United States and from online retailers. The noncarbonated water levels in 31 brands were less than 1 ppt, and two other brands had levels of 1.21 ppt and 4.64 ppt. These results derive from the averages of two to four samples of each product, but it is not clear whether these were replicates from the same bottle or different bottles. The report does not include information on detection limits or specific PFAS compounds detected, but merely describes "total PFAS as the sum of average concentrations of all PFAS detected in the samples tested of a product."

The U.S. Food and Drug Administration (FDA) measured PFOA and PFOS in carbonated and noncarbonated bottled water (n = 30; brands not identified) and found that levels were below the lower LOQ in all samples (0.004 µg/L).[10]

Studies have also detected PFAS in bottled water from other countries (see, e.g., Le Coadou et al., 2017). As brand names are not provided by these authors, it is not clear how this information could be used for intervention recommendations for the United States.

Based on these very limited data, it is possible that ready-to-use formulas may have lower PFAS levels than formulas reconstituted with tap or bottled water (see Figure E-6). However, the following caveats must be noted: data comparisons across studies are complicated by differences in total PFAS included; measurements may include more legacy PFAS compounds and not include PFAS compounds used to replace the older chemistries; studies use varying detection limits and approaches for assigning values to measurements below the LOD; and sampling in these studies is not representative of regions within the United States.

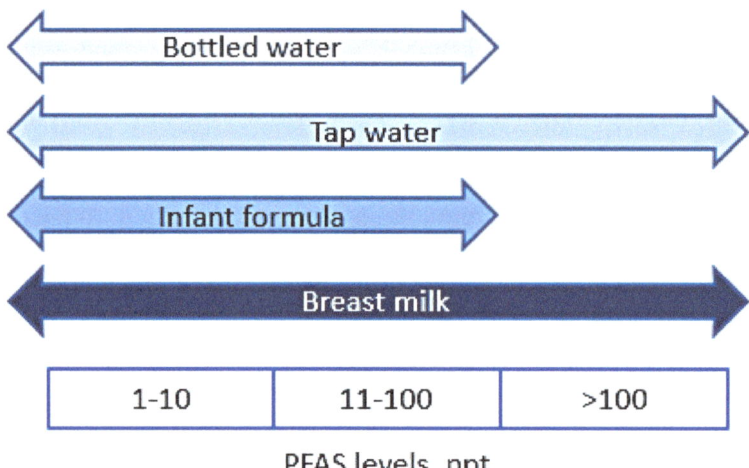

FIGURE E-6 Limited data on PFAS levels in breast milk and infant formula in the United States show general overlapping concentrations, which also overlap with PFAS concentrations in drinking water that could be used to reconstitute formula.
NOTES: Double-headed arrows indicate that these bars could extend in either direction as new data are obtained. ppt = parts per trillion.
SOURCE: Author generated.

---

[10] See https://www.fda.gov/media/130564/download (accessed May 12, 2021).

The Mother's Exposure: Does Lactation Reduce Internal Exposure?

There are two likely sources of persistent chemicals in breast milk: legacy stores in the mother from her lifetime exposure, and current sources of exposure, such as diet (LaKind, 2007). So the question arises: If a mother lactates, can she lower the stores of chemicals in her body, thus reducing her internal exposures? (If she disposes of her milk rather than breastfeeding, a process referred to as pumping-and-dumping, can she similarly reduce her exposure?) The reduction in chemical levels in the body via lactation, referred to as depuration, has been studied for such persistent chemicals as dioxins, polychlorinated biphenyls (PCBs), polybrominated diphenyl ethers (PBDEs), and chlorinated organic compounds (LaKind, 2007). Those studies have yielded mixed results (reviewed by LaKind et al., 2001), with some showing substantial declines in breast milk levels over time (Klein et al., 1986; Yakushiji et al., 1978) and others showing either minimal declines, no changes, or increasing levels over the course of lactation (Hooper et al., 2007; LaKind et al., 2009).

If lactation is an effective process for reducing the legacy stores of PFAS in the mother, one would expect to see levels of PFAS in serum and breast milk decline as lactation progressed and the mother's stores were depleted. We review here the literature on this topic.

In a review of the literature on breastfeeding and serum levels in mothers, VanNoy and colleagues (2018) conclude that the published studies support an association between breastfeeding and serum PFAS concentrations among women. However, they also observe that key aspects of breastfeeding, including duration, exclusivity, and timing of sample collection, should influence the breastfeeding–serum relationship, yet only one study in their review included all three exposure variables. They further describe the importance of parity for predicting maternal serum PFAS levels and note that most of the studies reviewed were unable to disentangle the effects of breastfeeding versus those of parity on PFAS exposure. The ability to fully assess the impact of breastfeeding on maternal PFAS levels is therefore limited.

Another approach to assessing the impact of breastfeeding on maternal PFAS exposure is to examine changes in PFAS levels in breast milk over the course of lactation. If stores of PFAS predominate over current exposures via diet and other sources, and if lactation resulted in mobilization and excretion of those stores, then lactation (with either breastfeeding or "pump-and-dump") could result in lower internal exposures for the mother. Three studies were identified that followed individual women over the course of lactation and collected multiple breast milk samples, which were analyzed for PFAS.

Lee and colleagues (2018) collected breast milk samples (n = 293) from 127 mothers at four different times postpartum (<7, 15, 30, and 90 days) as part of the Children's Health and Environmental Chemicals in Korea Cohort study. Samples were measured for 16 PFAS. Of the 127 women, only 15 provided samples across the lactation period, and samples from these women were pooled by sampling time. The concentrations of PFOS, PFOA, PFNA, and ΣPFAS 30 days postpartum were statistically significantly higher than in milk from sampling at <7 days postpartum.

In a study in Norway with nine women (Thomsen et al., 2010), milk samples were collected monthly from approximately 2 weeks postpartum up to 1 year postpartum (n = 3–10 per woman). The authors found a consistent decrease in concentrations of PFOS and PFOA, except for PFOS in one woman (these were the only PFAS >LOQ). The modeled depuration rates were reported as 7.7 percent and 3.1 percent reduction per month for PFOA and PFOS, respectively. Regarding generalizing these results, the authors note that because they did not have information on changes in the mothers' body weight or diet during the sampling timeframe, they could not evaluate the influence of these factors on depuration rates.

Fromme and colleagues (2010) also measured PFAS in breast milk samples collected monthly over 5 months from seven women. No significant differences in PFOS levels over the 5 months were observed.

In summary, our understanding of PFAS depuration based on breast milk measurements rests on three studies with between 7 and 15 women that yielded conflicting results. This is not necessarily surprising as depuration rates likely depend on numerous factors, such as current exposures, volume of breast milk excreted, and initial levels of PFAS in the body. Thus, the value of lactation as an intervention is unknown and requires additional study.

Would "Real-Time" Testing of Milk Help Guide Decision Making Regarding Infant Feeding?

It is unclear whether "real-time" testing of breast milk is an advantageous approach to guiding new mothers in decision making regarding infant feeding. First, testing requires sampling, shipping to laboratories, and conducting the analyses, which comes with a monetary cost. But even with unlimited resources, the time component is a critical consideration, as results may not be available until well into the infant's first few weeks or months. Second, a single measurement may not capture the infant's actual exposure as PFAS levels in milk may change over the duration of lactation, and the direction and rate of change are not well understood.

INDOOR DUST

Dust is a potential exposure pathway for PFAS compounds (Trudel et al., 2008). One dust-related PFAS intervention study was identified (Young et al., 2021). Dust samples were collected from floors by vacuum in "PFAS-free" refurbished rooms (7 with a full intervention and 28 with a partial intervention) and 12 control rooms at a university in the northeastern United States. Fifteen PFAS compounds were measured (PFOS, PFOA, PFHxA, PFHxS, FOSA, PFHpA, PFPeA, PFNA, PFBS, PFDS, PFBA, PFDA, PFUnDA, PFDoDA, and N-MeFOSAA), with detection limits ranging from 0.06 to 1.5 ng/g. PFAS levels in field blanks were either below the LOD or substantially lower than levels in the samples.

The geometric mean ΣPFAS levels were 481 ng/g (225–1,140 ng/g) in rooms with no intervention, 252 ng/g (18.1–8,310 ng/g) in rooms with partial interventions, and 108 ng/g (43.6–243 ng/g) in rooms with full interventions. Use of PFAS-free furnishing resulted in a statistically significant (78%, 95% CI: 38–92) ΣPFAS reduction in dust compared with control rooms. The results from this study suggest a possible intervention for reducing exposure to PFAS in dust.

While not an intervention study, research conducted by Scher and colleagues (2019) examined associations between PFAS-contaminated soil outside of homes and dust concentrations inside of the homes. The authors collected dust samples from the interior of the house and entryways to determine whether entryway dust levels indicate that "track-in" is an important contributor to house dust PFAS levels. They observed higher PFAS levels in the interior of the homes compared with soil levels, and suggest that soil track-in was not an important source of PFAS in interior dust.

Interpretation of PFAS dust studies is complicated by the lack of standardized sampling techniques for dust collection; the impact of use of different sampling methods on PFAS concentrations is unknown (Scher et al., 2019). Furthermore, for interventions that reduce PFAS in dust, the extent to which overall human intakes of PFAS would be reduced is not known. Estimates of the PFAS intake via dust have varied considerably (Nadal and Domingo, 2014).

OTHER POTENTIAL INTERVENTIONS

Other possible actions that could be considered interventions include limiting the use of PFAS-containing household goods and personal care products and introducing substances to the diet specifically intended to remove PFAS from the body. No intervention studies for these behavioral changes were identified, but here we describe one study on nonstick cookware, one study related to use of dental floss, and one dietary modification approach in the context of future exploration.

Nonstick Cookware

Nonstick cookware has been studied as a source of PFAS exposure. In terms of interventions, at issue is whether replacing nonstick cookware with items that do not contain PFAS would result in a measurable decrease in human PFAS exposures. While no intervention studies were identified, we briefly describe one study conducted in the United States that compared the release of PFAS into air and water from nonstick versus stainless steel frying pans. Sinclair and colleagues (2007) purchased four brands of domestic and imported nonstick frying pans and one brand of stainless steel frying pan (three to five of each brand) in New York. The pan brand names are not identified. All pans were precleaned with hot, soapy water; rinsed with Milli-Q water; and dried with a towel. The stainless steel pans were used as controls. The authors report that under normal cooking conditions (179°C to 233°C surface temperature), PFOA in the gas phase was measured at 11–503 picograms per square centimeter (pg/cm^2) from the nonstick frying pans. (Fluorotelemer alcohols were also detected, but these are not chemicals of focus in this paper and are not discussed further.) Gas-phase PFOA decreased after repeated use of one brand of pan but not the others (n = 1 for each brand). The authors also measured PFOA in Milli-Q water boiled for 10 minutes in selected pans and found inconsistent results (certain pans resulted in measurable levels of PFOA in the water, while others did not).

Since brand names are not included in the publication, this information cannot be used as the basis for specific intervention recommendations. Even if brand names were included, given the small sample size and the lack of study replication, it would be difficult to use this information as the basis for general recommendations. Finally, as neither air nor water concentrations are provided, the extent to which this exposure source contributes to overall intake is unclear.

Dental Floss

One study examined the association between use of dental floss and serum PFAS (PFOA, PFNA, PFDeA, PFHxS, PFOS, and Me-PFOSA-AcOH) levels (Boronow et al., 2019). Serum PFAS levels were measured in 178 middle-aged women. An administered questionnaire included one question on use of dental floss: "In the last month, how often did you use Oral-B Glide dental floss?" Response choices were as follows: "Never or almost never, Several times a month, 2 or more times a week, Every day." Only "ever" and "never" were used in the regression analysis. While five of the PFAS compounds did not show significant associations, a 24.9 percent (95% CI: 0.2–55.7) higher level of PFHxS was found in subjects who used Oral-B Glide floss. It is worth noting that the questionnaires were administered several years after blood sample collection.

The authors also analyzed 18 dental floss products (only one or two samples per brand were analyzed, except the Oral-B Glide brand [five samples]) for total fluorine (as an indicator for polytetrafluoroethylene or PTFE), 6 of which yielded detectable levels. Given the lag in time between blood collection and questionnaire administration, the small number of floss samples analyzed, and the mostly nonsignificant results, it is not clear whether dental floss is an important route of human exposure to PFAS.

Fiber Intake

Studies have observed relationships between higher fiber intake and lower serum PFAS levels (e.g., PFOA, PFOS, and PFNA [Dzierlenga et al., 2021]; PFOS and PFOA [Halldorsson et al., 2008]; PFOS, PFOA, PFHxS, EtFOSAA, MeFOSAA, and PFNA [Lin et al., 2020]). This finding is based on studies examining associations between dietary recall information and serum levels in cross-sectional study designs. Thus, it is unknown whether other factors (e.g., whether diets with high fiber generally have lower levels of PFAS) influence this relationship.

MODELED INTAKES AS THE BASIS FOR RECOMMENDATIONS FOR REDUCING EXPOSURE TO PFAS

Well-conducted and generalizable intervention studies can be considered the gold standard for recommending approaches to reductions in personal exposure to PFAS. In the absence of such studies, it may be informative to use results from studies designed to model the relative contributions of exposure pathways to overall PFAS exposure. This approach would focus attention and intervention strategies on the most important exposure pathways. In this section, we discuss results from studies using concentration data from several media and products (e.g., soil, water, and food concentration data) in combination with generic intake factors (e.g., g intake/kg body weight per day) to model human intakes of PFAS. We explore whether any common, generalizable themes emerge from such modeling studies regarding dominant intake pathways.

Because of geographic differences in both environmental media and product concentrations, the emphasis here is on studies conducted for the U.S. population. First, however, we provide a brief synopsis of reviews on studies conducted for populations outside of the United States. Although some of the modeling efforts included less well-studied PFAS (e.g., PFHxS), the most commonly assessed PFAS were PFOA and PFOS. Sunderland and colleagues (2019) recently compiled data from the literature on percent source contribution in adults. The results are summarized in Figures E-7 and E-8. The modeling approaches represented by the underlying papers included different concentration data and intake assumptions. Regardless of the approach used, diet appears to be the major pathway of exposure to PFOS, with the percent contribution ranging from 65 to 96 percent. For PFOA, the range of percentages for dietary contributions is wider (6–86 percent). While these results point to the importance of diet, the location of a study and the presence or absence of point sources will influence the relative contributions of various pathways to overall PFAS intake. Behavioral differences across countries can also impact the results and their generalizability to communities in the United States. We therefore focus on two studies that modeled source contributions for PFOA (Lorber and Egeghy, 2011) and PFOS (Egeghy and Lorber, 2011) in the United States.

For PFOA intake modeling, Lorber and Egeghy (2011) used the following data, assumptions, and approach. Their intakes were developed for adults and 2-year-old children. PFOA concentrations in various media from the published literature were used in combination with EPA exposure contact rates. The authors generated intake distributions by inputting different exposure media concentrations into the model. They assumed that PFOA concentrations in indoor air were 20 times higher than those in outdoor air, with outdoor air data being derived from a study in Albany, New York. Whether these air concentrations represent more general U.S. exposures is not discussed. The authors also used dietary data from the Canadian Total Dietary Survey (Tittlemier et al., 2007); thus, it is not known whether this aspect of the model is representative of exposures to PFOA in the United States. PFOA levels in drinking water were estimated from surface water concentrations in various parts of the United States (New York, North Carolina, New Jersey [drinking water], Great Lakes, Tennessee, and Florida). Dust concentration data were obtained from dust samples from homes and day care centers in Ohio and North Carolina.

Lorber and Egeghy (2011)[11] discuss numerous limitations around their estimates of intakes, including the following: no estimates of intake via direct contact with consumer products (e.g., treated carpets or cosmetics), no inclusion of PFOA precursors, and a lack of PFOA data for food in the United States. However, Lorber and Egeghy (2011) also used a simple one-compartment, first-order pharmacokinetic model in combination with National Health and Nutrition Examination Survey (NHANES) 2003–2004 data and back-calculated intakes. Their central tendency intake estimates for adults and children (70 ng/day and 26 ng/day, respectively) are not dissimilar to the intakes back-calculated from the NHANES data (56 ng/day and 37 ng/day for males and females, respectively). These models would benefit from the use of improved measurement data in food and other media specific to— and representative of—the United States (or at least areas or regions of interest within the United States) and the use of more recent NHANES data.

---

[11] This text has been changed since the release of the pre-publication version of the report to remove reference to a figure that cannot be reproduced due to lack of copyright permissions.

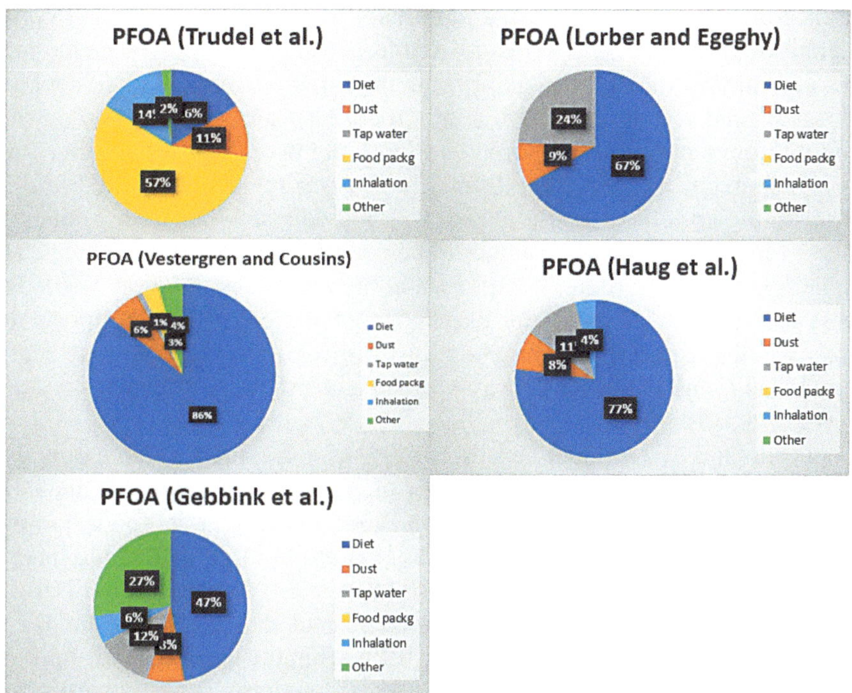

FIGURE E-7 Relative contribution percentiles for various pathways of exposure to PFOA.
NOTE: Values less than 1 percent were assigned a value of zero.
SOURCE: Data from Sunderland et al., 2019.

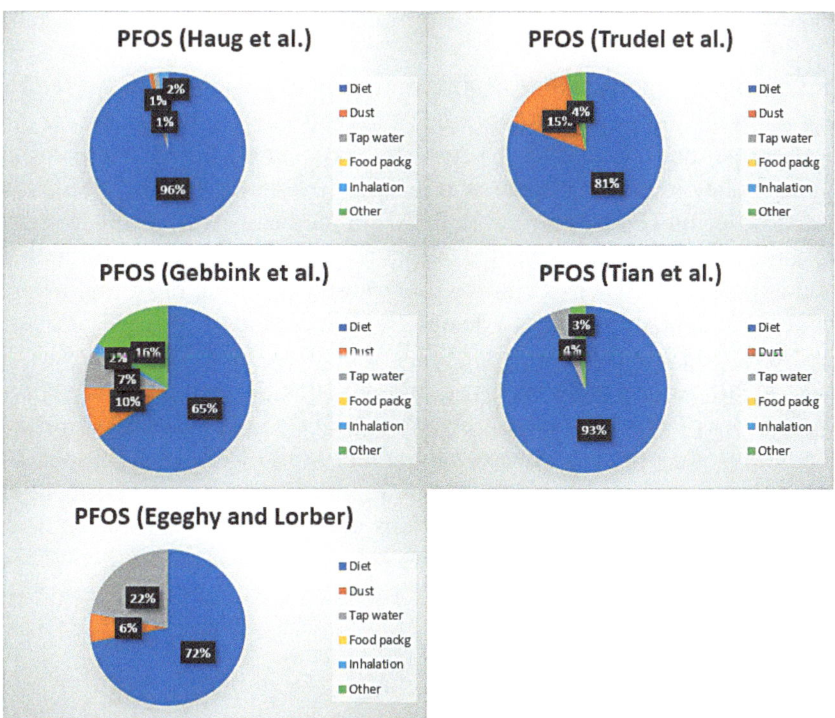

FIGURE E-8 Relative contribution percentiles for various pathways of exposure to PFOS.
NOTE: Values less than 1 percent were assigned a value of zero.
SOURCE: Data from Sunderland et al., 2019.

Egeghy and Lorber (2011) used a similar approach to model intakes of PFOS and its precursors. In addition to modeling of precursors, this effort differs from the PFOA models described above in that the models were developed for both "typical" and "contaminated" scenarios. (The contaminated scenario was similar to the background scenario except that much higher water concentrations were used.) In terms of available data on environmental concentrations of PFOS and precursors, the authors note that "measurement of PFCs in exposure media in North America was relatively sparse for all media compared with European data with the possible exception of dust. Food data are sparse for both continents. No measurements of PFOS in uncontaminated soil could be found" (Egeghy and Lorber, 2011, p. 158). The authors found that diet was the main route of exposure in the general population, while for children, dust ingestion was almost as important a contributor to intake as diet. Perhaps most important in the context of using modeled uptakes as the basis for recommendations for modifying behaviors to reduce exposure, Egeghy and Lorber (2011) found that the pathway-specific contributions spanned several orders of magnitude and overlapped substantially.[12]

Use of model estimates of dominant pathways of PFAS exposure for making recommendations to individuals regarding exposure reduction faces a number of challenges. First, while diet appears to be a major pathway of exposure, there is little information on PFAS in commercial foods commonly consumed in the United States. The FDA has released PFAS data for certain foods, which could be used in future studies involving modeling of source contributions to PFAS intake.[13] However, the FDA observed that its data for produce, meat, dairy, and grain products are based on a small sample size, and the results "cannot be used to draw definitive conclusions about the levels of PFAS in the general food supply" (para. 3). It is not currently well understood whether the data on commercial foods from other countries used in the intake models are representative of levels in the United States. Wu and colleagues (2015) state, "Information on dietary predictors in U.S. is still limited" and "more data are needed to determine the relative contributions of food and dust to serum PFCs for both adult and child populations" (p. 265). A review of PFAS in foods by Domingo and Nadal (2017) identifies only two recent studies for the United States: one focused on PFAS uptake by lettuce and strawberries irrigated with reclaimed water (Blaine et al., 2014) and the other on concentrations of PFAS in freshwater fish samples from urban rivers and the Great Lakes (Stahl et al., 2014).

Furthermore, the relative importance of different sources varies by study (see Figures E-7 and E-8) and by demographic group and population (Sunderland et al., 2019). In describing the findings of their recent review on nonoccupational intakes via background exposures, De Silva and colleagues (2021) observe that the inconsistency in the relative importance of different exposure sources from one study to the next may be due to differing concentrations of PFAS in media, as well as the assignment of different values for exposure intake factors (e.g., exposure frequency and duration). They conclude, "Without rigorously conducted exposure studies it is challenging to rank order the most important human exposure pathways and without these data, our ability to design evidence-based exposure intervention strategies will be limited" (De Silva, 2021, p. 644).

Within even a small geographic area, one could envision varying exposure characteristics that could, in turn, affect sources of PFAS intakes. Vestergren and Cousins (2009) explored this possibility by estimating relative intakes for those with exposure to background PFAS levels, exposure to higher levels in drinking water or drinking water impacted by a PFAS point source, or occupational exposures; they found substantial differences in the relative contributions to overall intakes (see Figure E-9).

---

[12] This text has been changed since the release of the pre-publication version of the report to remove reference to a figure that cannot be reproduced due to lack of copyright permissions.

[13] See https://www.fda.gov/food/cfsan-constituent-updates/update-fdas-continuing-efforts-understand-and-reduce-exposure-pfas-foods (accessed May 12, 2021).

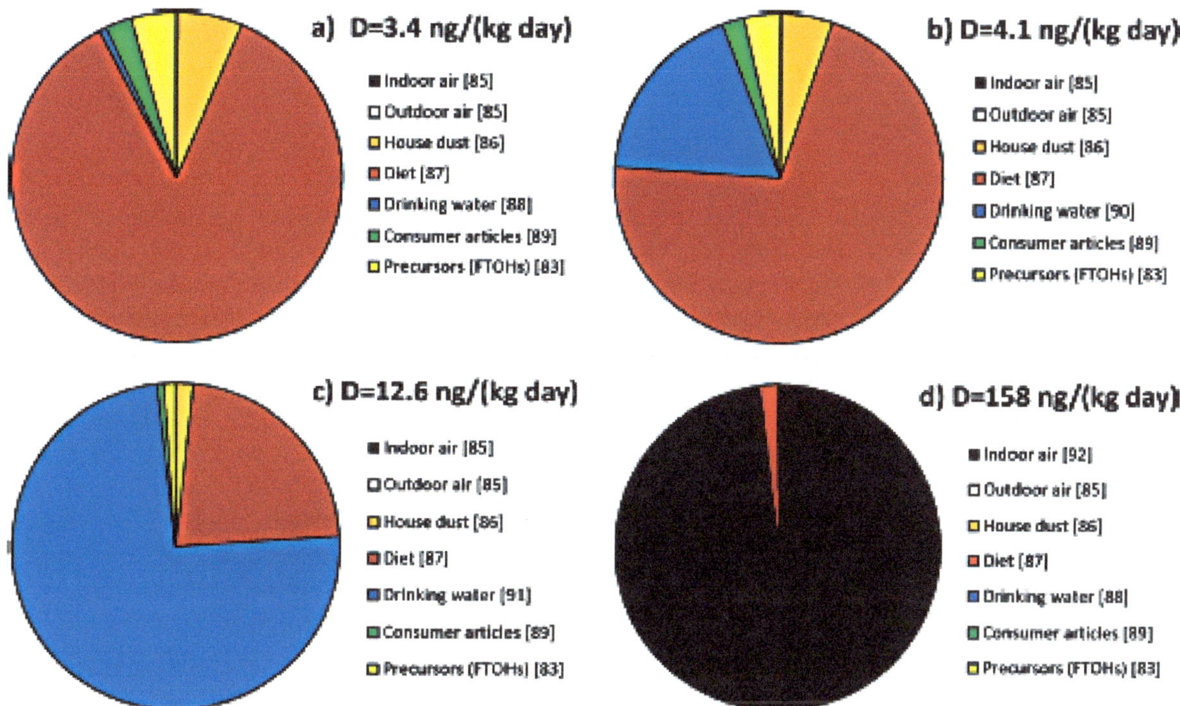

FIGURE E-9 Estimated daily intakes for male adults (D) and relative source contributions.
NOTES: Data for sources are derived from different studies and different countries. Bracketed numbers refer to citations in Vestergren and Cousins, 2009. (a) = background concentrations in drinking water (1.3 [nanograms per Liter [ng/L]); (b) = elevated concentrations in drinking water (40 ng/L); (c) = point sources of drinking water contamination (519 ng/L); and (d) = occupationally exposed individuals (indoor air concentrations 1 micrograms per cubic meter [μg/m^3]). FTOH = fluorotelomer alcohols; ng/(kg day) = nanograms per kilogram per day .
SOURCE: Reprinted with permission from Vestergren and Cousins, 2009. Copyright 2009 American Chemical Society.

Another factor related to modeling principal sources of exposure is the changes in production and use of individual PFAS over time. Sunderland and colleagues (2019) describe the changes in serum PFAS levels following the phase-out of production of PFOS and its precursors, with PFOS declining, but other PFAS, such as PFHxS, increasing. They observe that while exposures to PFOS and PFOA "have been successfully reduced by product phase-outs for many populations, exposures to C-9-C-11 PFCAs have not followed the same trends" (Sunderland et al., 2019, p. 138). Thus, it is important to use recent environmental and consumer product and dietary data to develop robust estimates of current dominant pathways of exposure.

E-4

Discussion

In this review, we have sought to address the following question: Based on current research, are there interventions or *personal* modifiable behavioral changes that individuals can make to *effectively* reduce their PFAS exposure? We have described research on potential ways in which an individual could reduce exposure to PFAS. It is important to acknowledge that communities across the United States have received guidance from state and federal agencies regarding PFAS exposure reduction, including advisories around consumption of drinking water and fish. While it may seem obvious that avoiding exposure to sources of PFAS would result in reduced intake of PFAS and, in turn, lower internal PFAS levels, some caution in assuming that exposure and risk reduction would ensue is warranted. For example, if one is advised to avoid locally caught fish because of known PFAS contamination in that fish, such avoidance could result in reduced exposure. However, if dietary fish is replaced by another food that is also high in PFAS, avoiding the fish may not result in lower PFAS exposure. Another issue to consider is that avoiding one group of chemicals by changing diet or other behaviors can result in increased exposures to other chemicals. Finally, if a certain PFAS source is related to a small portion of overall exposure, then avoidance of that source may not result in appreciable reductions in internal levels of PFAS. Thus, while avoiding known sources of PFAS exposure may be a useful approach to lowering overall exposure, these other considerations should be taken into account; research on the efficacy of various interventions could help shed light on these complexities.

The number of intervention studies available to address the question of whether there are interventions or *personal* modifiable behavioral changes that individuals can make to *effectively* reduce their PFAS exposure ranged from 1 to 11, depending on the source of exposure; for some pathways, no studies were found (see Figure E-10). Overall, only the water filtration studies provided relatively consistent evidence of an effective reduction in PFAS levels. No studies were identified that confirmed reduction in human exposure via biomonitoring. To fully demonstrate the efficacy of an intervention, a study would need to be conducted over a timeframe sufficient to account for the long half-lives of PFAS.

Demonstration that PFAS exposures are reduced in a meaningful way through biomonitoring confirmation is an important step. It is possible that an intervention may reduce PFAS levels in a particular medium, but if this medium is not a major source of overall exposure, then that intervention may not contribute significantly to reduction in human exposures. Overall, the intervention literature is sparse and has many limitations. Thus, the committee may have to rely on assumptions and other bodies of evidence to make recommendations to individuals and communities about exposure reduction.

We also considered whether intake models could assess with some degree of confidence the relative source contributions to overall intake and whether this approach could be used to inform decisions regarding community recommendations. While diet appears to be a major contributor to overall intake, dietary data from the United States were not used in these models. It is not known whether the Canadian data used are representative of levels in the United States. As observed by Macheka-Tendenguwo and colleagues (2018), "numerous investigations have attempted to establish the main exposure pathway of PFAS in humans, but differing viewpoints make the results inconclusive" (p. 36066).

In conclusion, other than using certified POE or POU water filters, the available intervention studies generally do not appear to be sufficiently robust to support recommendations for personal behavioral modifications for communities in the United States. The studies reviewed here were limited by small sample sizes, generally inconsistent study designs and results, and a possible lack of generalizability to the United States. It is also important to consider additional factors when developing recommendations for changes in personal behavior, such as the ease and cost of an intervention, associated trade-offs (i.e., If an intervention lowers concentrations of PFAS, does it increase exposures to other chemicals, including other PFAS?), and whether enough is known about varying PFAS levels in the environment and consumer products to understand the necessary scale of an intervention recommendation (e.g., individual, local, or regional). Information will be needed on local levels of PFAS in drinking water, as well as levels in breast milk, for any recommendations regarding infant nutrition to be well supported.

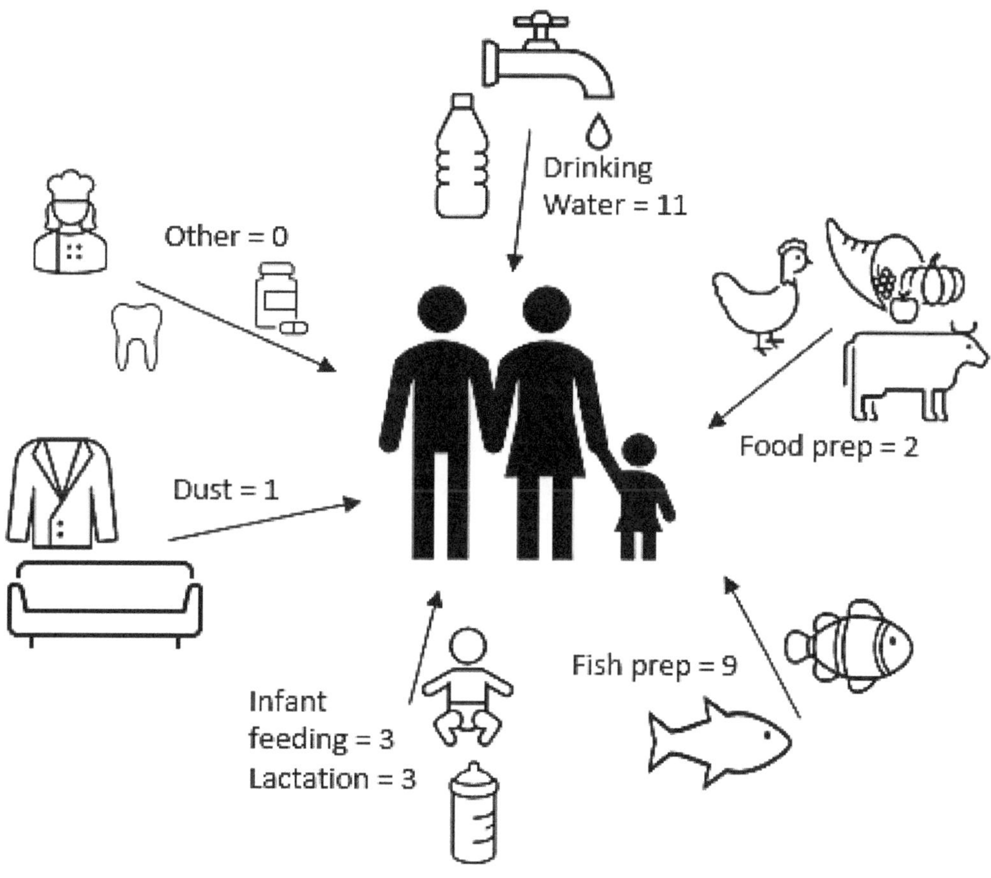

FIGURE E-10 Summary of numbers of studies identified for each PFAS exposure source.
NOTE: "Other" includes nonstick cookware, dental floss, and fiber intake, which these sources are described in the text but for which no intervention studies were identified.
SOURCE: Author generated.

References

AAP (American Academy of Pediatrics). 2012. Policy statement. Breastfeeding and the use of human milk. Section on breastfeeding. *Pediatrics* 129(3):e827–e841. https://doi.org/10.1542/peds.2011-3552.

Ahrens, L., and M. Bundschuh. 2014. Fate and effects of poly- and perfluoroalkyl substances in the aquatic environment: A review. *Environmental Toxicology and Chemistry* 33(9):1921–1929. https://doi.org/10.1002/etc.2663. PMID: 24924660.

Akhbarizadeh, R., S. Dobaradaran, T. C. Schmidt, I. Nabipour, and J. Spitz. 2020. Worldwide bottled water occurrence of emerging contaminants: A review of the recent scientific literature. *Journal of Hazardous Materials* 392:122271.

Alves, R. N., A. L. Maulvault, V. L. Barbosa, S. Cunha, C. J. A. F. Kwadijk, D. Álvarez-Muñoz, S. Rodríguez-Mozaz, Ò. Aznar-Alemany, E. Eljarrat, D. Barceló, M. Fernandez-Tejedor, A. Tediosi, and A. Marques. 2017. Preliminary assessment on the bioaccessibility of contaminants of emerging concern in raw and cooked seafood. *Food and Chemical Toxicology* 104:69–78. https://doi.org/10.1016/j.fct.2017.01.029. PMID: 28202359.

Andrews, D. Q., and O. V. Naidenko. 2020. Population-wide exposure to per- and polyfluoroalkyl substances from drinking water in the United States. *Environmental Science & Technology Letters* 7(12):931–936. https://doi.org/10.1021/acs.estlett.0c00713.

Ao, J., T. Yuan, H. Xia, Y. Ma, Z. Shen, R. Shi, Y. Tian, J. Zhang, W. Ding, L. Gao, X. Zhao, and X. Yu. 2019. Characteristic and human exposure risk assessment of per- and polyfluoroalkyl substances: A study based on indoor dust and drinking water in China. *Environmental Pollution* 254(Pt A):112873. https://doi.org/10.1016/j.envpol.2019.07.041. PMID: 31369910.

ATSDR (Agency for Toxic Substances and Disease Registry). 2019. *PFAS: An overview of the science and guidance for clinicians on per- and polyfluoroalkyl substances (PFAS).* https://www.atsdr.cdc.gov/pfas/docs/clinical-guidance-12-20-2019.pdf.

ATSDR. 2020. *Per- and polyfluoroalkyl substances (PFAS) and your health.* https://www.atsdr.cdc.gov/pfas/health-effects/exposure.html (accessed March 10, 2021).

Barbosa, V., A. L. Maulvault, R. N. Alves, C. Kwadijk, M. Kotterman, A. Tediosi, M. Fernández-Tejedor, J. J. Sloth, K. Granby, R. R. Rasmussen, J. Robbens, B. De Witte, L. Trabalón, J. O. Fernandes, S. C. Cunha, and A. Marques. 2018. Effects of steaming on contaminants of emerging concern levels in seafood. *Food and Chemical Toxicology* 118:490–504. https://doi.org/10.1016/j.fct.2018.05.047. PMID: 29787848.

Bhavsar, S. P., X. Zhang, R. Guo, E. Braekevelt, S. Petro, N. Gandhi, E. J. Reiner, H. Lee, R. Bronson, and S. A. Tittlemier. 2014. Cooking fish is not effective in reducing exposure to perfluoroalkyl and polyfluoroalkyl substances. *Environment International* 66:107–114. https://doi.org/10.1016/j.envint.2014.01.024. PMID: 24561272.

Binnington, M. J., Y. D. Lei, L. Pokiak, J. Pokiak, S. K. Ostertag, L. L. Loseto, H. M. Chan, L. W. Y. Yeung, H. Huang, and F. Wania. 2017. Effects of preparation on nutrient and environmental contaminant levels in Arctic beluga whale (Delphinapterus leucas) traditional foods. *Environmental Science: Processes & Impacts* 19(8):1000–1015. https://doi.org/10.1039/c7em00167c. PMID: 28752885.

Blaine, A. C., C. D. Rich, E. M. Sedlacko, K. C. Hyland, C. Stushnoff, E. R. V. Dickenson, and C. P. Higgins. 2014. Perfluoroalkyl acid uptake in lettuce (Lactuca sativa) and strawberry (Fragaria ananassa) irrigated with reclaimed water. *Environmental Science & Technology* 48:14361–14368.

Boronow, K. E., J. G. Brody, L. A. Schaider, F. Graham, G. F. Peaslee, L. Havas, A. Barbara, and B. A. Cohn. 2019. Serum concentrations of PFASs and exposure-related behaviors in African American and non-Hispanic white women. *Journal of Exposure Science & Environmental Epidemiology* 29:206–217. https://doi.org/10.1038/s41370-018-0109-y.

Butenhoff, J. L., G. W. Olsen, and A. Pfahles-Hutchens. 2006. The applicability of biomonitoring data for perfluorooctanesulfonate to the environmental public health continuum. *Environmental Health Perspectives* 114(11):1776–1782. https://doi.org/10.1289/ehp.9060. PMID: 17107867; PMCID: PMC1665413.

Calafat, A. M., K. Kato, K. Hubbard, T. Jia, J. C. Botelho, and L. Y. Wong. 2019. Legacy and alternative per- and polyfluoroalkyl substances in the U.S. general population: Paired serum-urine data from the 2013–2014 National Health and Nutrition Examination Survey. *Environment International* 131:105048. https://doi.org/10.1016/j.envint.2019.105048. PMID: 31376596.

CDC (Centers for Disease Control and Prevention). 2009. *National report on human exposure to environmental chemicals.* https://www.cdc.gov/exposurereport/pdf/fourthreport.pdf.

De Silva, A. O., J. M. Armitage, T. A. Bruton, C. Dassuncao, W. Heiger-Bernays, X. C. Hu, A. Kärrman, B. Kelly, C. Ng, A. Robuck, M. Sun, T. F. Webster, and E. M. Sunderland. 2021. PFAS exposure pathways for humans and wildlife: A synthesis of current knowledge and key gaps in understanding. *Environmental Toxicology and Chemistry* 40(3):631–657. PMID: 33201517.

Del Gobbo, L., S. Tittlemier, M. Diamond, K. Pepper, B. Tague, F. Yeudall, and L. Vanderlinden. 2008. Cooking decreases observed perfluorinated compound concentrations in fish. *Journal of Agricultural and Food Chemistry* 56(16):7551–7559. https://doi.org/10.1021/jf800827r. PMID: 18620413.

D'Hollander, W., P. de Voogt, W. De Coen, and L. Bervoets. 2010. Perfluorinated substances in human food and other sources of human exposure. *Reviews of Environmental Contamination and Toxicology* 208:179–215. https://doi.org/10.1007/978-1-4419-6880-7_4. PMID: 20811865.

Domingo, J. L., and M. Nadal. 2017. Per- and polyfluoroalkyl substances (PFASs) in food and human dietary intake: A review of the recent scientific literature. *Journal of Agricultural and Food Chemistry* 65(3):533–543. https://doi.org/10.1021/acs.jafc.6b04683. PMID: 28052194.

Domingo, J. L., and M. Nadal. 2019. Human exposure to per- and polyfluoroalkyl substances (PFAS) through drinking water: A review of the recent scientific literature. *Environmental Research* 177:108648. https://doi.org/10.1016/j.envres.2019.108648. PMID: 31421451.

Ducatman, A., M. Luster, and T. Fletcher. 2021. Perfluoroalkyl substance excretion: Effects of organic anion-inhibiting and resin-binding drugs in a community setting. *Environmental Toxicology and Pharmacology* 85:103650. https://doi.org/10.1016/j.etap.2021.103650. https://www.ncbi.nlm.nih.gov/pubmed/33819618.

Dzierlenga, M. W., D. R. Keast, and M. P. Longnecker. 2021. The concentration of several perfluoroalkyl acids in serum appears to be reduced by dietary fiber. *Environment International* 146:106292. https://doi.org/10.1016/j.envint.2020.106292. PMID: 33395939.

EFSA (European Food Safety Authority). 2020. Risk to human health related to the presence of perfluoroalkyl substances in food. *EFSA Journal* 18(9):6223.

Egeghy, P. P., and M. Lorber. 2011. An assessment of the exposure of Americans to perfluorooctane sulfonate: A comparison of estimated intake with values inferred from NHANES data. *Journal of Exposure Science & Environmental Epidemiology* 21(2):150–168. https://doi.org/10.1038/jes.2009.73. PMID: 20145679.

Eichler, C. M. A., and J. C. Little. 2020. A framework to model exposure to per- and polyfluoroalkyl substances in indoor environments. *Environmental Science: Processes & Impacts* 22(3):500–511. https://doi.org/10.1039/c9em00556k. PMID:32141451.

Emmett, E. A., F. S. Shofer, H. Zhang, D. Freeman, C. Desai, and L. M. Shaw. 2006. Community exposure to perfluorooctanoate: Relationships between serum concentrations and exposure sources. *Journal of Occupational and Environmental Medicine* 48(8):759–770. https://doi.org/10.1097/01.jom.0000232486.07658.74. PMID: 16902368; PMCID: PMC3038253.

EPA (U.S. Environmental Protection Agency). 2017. *The Third Unregulated Contaminant Monitoring Rule (UCMR 3): Data summary, January 2017.* https://www.epa.gov/dwucmr/occurrence-data-unregulated-contaminant-monitoring-rule#3 (accessed March 12, 2021).

EWG (Environmental Working Group). 2016. *How can I avoid PFAS chemicals?*
    https://www.ewg.org/avoidpfas (accessed March 10, 2021).
Felton, R. 2020, September 24. What's really in your bottled water? Consumer reports found toxic PFAS
    chemicals in several popular water brands, especially carbonated ones. *Consumer Reports.*
    https://www.consumerreports.org/bottled-water/whats-really-in-your-bottled-water (accessed
    March 22, 2021).
Fromme, H., S. A. Tittlemier, W. Völkel, M. Wilhelm, and D. Twardella. 2009. Perfluorinated
    compounds—Exposure assessment for the general population in Western countries. *International
    Journal of Hygiene and Environmental Health* 212(3):239–270.
    https://doi.org/10.1016/j.ijheh.2008.04.007. PMID: 18565792.
Fromme, H., C. Mosch, M. Morovitz, I. Alba-Alejandre, S. Boehmer, M. Kiranoglu, F. Faber, I.
    Hannibal, O. Genzel-Boroviczény, B. Koletzko, and W. Völkel. 2010. Pre- and postnatal
    exposure to perfluorinated compounds (PFCs). *Environmental Science & Technology*
    44(18):7123–7129. https://doi.org/10.1021/es101184f.
Gellrich, V., H. Brunn, and T. Stahl. 2013. Perfluoroalkyl and polyfluoroalkyl substances (PFASs) in
    mineral water and tap water. *Journal of Environmental Science and Health, Part A
    Toxic/Hazardous Substances and Environmental Engineering* 48(2):129–135.
    https://doi.org/10.1080/10934529.2013.719431. PMID: 23043333.
Genuis, S. J., Y. Liu, Q. I. Genuis, and J. W. Martin. 2014. Phlebotomy treatment for elimination of
    perfluoroalkyl acids in a highly exposed family: A retrospective case-series. *PLoS One*
    9(12):e114295. https://doi.org/10.1371/journal.pone.0114295.
    http://www.ncbi.nlm.nih.gov/pubmed/25504057.
Göckener, B., T. Weber, H. Rüdel, M. Bücking, and M. Kolossa-Gehring. 2020. Human biomonitoring of
    per- and polyfluoroalkyl substances in German blood plasma samples from 1982 to 2019.
    Environment International 145:106123. https://doi.org/10.1016/j.envint.2020.106123. PMID:
    32949877.
Halldorsson, T. I., C. Fei, J. Olsen, L. Lipworth, J. K. McLaughlin, and S. F. Olsen. 2008. Dietary
    predictors of perfluorinated chemicals: A study from the Danish National Birth Cohort.
    Environmental Science & Technology 42(23):8971–8977. https://doi.org/10.1021/es801907r.
    PMID: 19192827.
Health Canada. 2019, November. *Fifth report on human biomonitoring of environmental chemicals in
    Canada.* Results of the Canadian Health Measures Survey Cycle 5 (2016–2017).
    https://www.canada.ca/content/dam/hc-sc/documents/services/environmental-workplace-
    health/reports-publications/environmental-contaminants/fifth-report-human-biomonitoring/publ-
    eng.pdf (accessed March 21, 2021).
Heo, J. J., J. W. Lee, S. K. Kim, and J. E. Oh. 2014. Foodstuff analyses show that seafood and water are
    major perfluoroalkyl acids (PFAAs) sources to humans in Korea. *Journal of Hazardous Materials*
    279:402–409. https://doi.org/10.1016/j.jhazmat.2014.07.004. PMID: 25093550.
Herkert, N. J., J. Merrill, C. Peters, D. Bollinger, S. Zhang, K. Hoffman, P. L. Ferguson, D. R. U. Knappe
    and H. M. Stapleton. 2020. Assessing the effectiveness of point-of-use residential drinking water
    filters for perfluoroalkyl substances (PFASs). *Environmental Science & Technology Letters*
    7(3):178–184.
Herrick, R. L., J. Buckholz, F. M. Biro, A. M. Calafat, X. Ye, C. Xie, and S. M. Pinney. 2017.
    Polyfluoroalkyl substance exposure in the Mid-Ohio River Valley, 1991–2012. *Environmental
    Pollution* 228:50–60. https://doi.org/10.1016/j.envpol.2017.04.092. PMID: 28505513; PMCID:
    PMC5540235.
Hooper, K., J. She, M. Sharp, J. Chow, N. Jewell, R. Gephart, and A. Holden. 2007. Depuration of
    polybrominated diphenyl ethers (PBDEs) and polychlorinated biphenyls (PCBs) in breast milk
    from California first-time mothers (primiparae). *Environmental Health Perspectives*
    115(9):1271–1275. https://doi.org/10.1289/ehp.10166. PMID: 17805415; PMCID: PMC1964891.

Hu, Y., C. Wei, L. Wang, Z. Zhou, T. Wang, G. Liu, Y. Feng, and Y. Liang. 2020. Cooking methods affect the intake of per- and polyfluoroalkyl substances (PFASs) from grass carp. *Ecotoxicology and Environmental Safety* 203:111003. https://doi.org/10.1016/j.ecoenv.2020.111003. PMID: 32678765.

Iwabuchi, K., and I. Sato. 2021. Effectiveness of household water purifiers in removing perfluoroalkyl substances from drinking water. *Environmental Science and Pollution Research International* 28(9):11665–11671. https://doi.org/10.1007/s11356-020-11757-1. PMID: 33410030.

Jogsten, I. E., G. Perelló, X. Llebaria, E. Bigas, R. Martí-Cid, A. Kärrman, and J. L. Domingo. 2009. Exposure to perfluorinated compounds in Catalonia, Spain, through consumption of various raw and cooked foodstuffs, including packaged food. *Food and Chemical Toxicology* 47(7):1577–1583. https://doi.org/10.1016/j.fct.2009.04.004. PMID: 19362113.

Kannan, K., S. Corsolini, J. Falandysz, G. Fillmann, K. S. Kumar, B. G. Loganathan, M. A. Mohd, J. Olivero, N. Van Wouwe, J. H. Yang, and K. M. Aldoust. 2004. Perfluorooctanesulfonate and related fluorochemicals in human blood from several countries. *Environmental Science & Technology* 38(17):4489–4495. https://doi.org/10.1021/es0493446. PMID: 15461154.

Kim, M. J., J. Park, L. Luo, J. Min, J. H. Kim, H. D. Yang, Y. Kho, G. J. Kang, M. S. Chung, S. Shin, and B. Moon. 2020. Effect of washing, soaking, and cooking methods on perfluorinated compounds in mackerel (*Scomber japonicus*). *Food Science & Nutrition* 8(8):4399–4408. https://doi.org/10.1002/fsn3.1737. PMID: 32884720; PMCID: PMC7455985.

Klein, D., J. C. Dillon, J. L. Jirou-Najou, M. J. Gagey, and G. Debry. 1986. The kinetics of the elimination of organochlorine compounds during the 1st week of breast feeding. *Food and Chemical Toxicology* 24(8):869–874.

Kuklenyik, Z., J. A. Reich, J. S. Tully, L. L. Needham, and A. M. Calafat. 2004. Automated solid-phase extraction and measurement of perfluorinated organic acids and amides in human serum and milk. *Environmental Science & Technology* 38(13):3698–3704. https://doi.org/10.1021/es040332u. PMID: 15296323.

LaKind, J. S. 2007. Recent global trends and physiologic origins of dioxins and furans in human milk. *Journal of Exposure Science and Environmental Epidemiology* 17:510–524.

LaKind, J. S., C. Berlin, and D. Q. Naiman. 2001. Infant exposure to chemicals in breast milk in the United States: What we need to learn from a breast milk monitoring program. *Environmental Health Perspectives* 109:75–88.

LaKind, J. S., R. L. Brent, M. L. Dourson, S. Kacew, G. Koren, B. Sonawane, A. J. Tarzian, and K. Uhl. 2005. Human milk biomonitoring data: Interpretation and risk assessment issues. *Journal of Toxicology and Environmental Health* 68(20):1713–1770.

LaKind, J. S., C. M. Berlin, Jr., and D. R. Mattison. 2008. The heart of the matter on breastmilk and environmental chemicals: Essential points for healthcare providers and new parents. *Breastfeeding Medicine* 3(4):251–259.

LaKind, J. S., C. M. Berlin, Jr., A. Sjödin, W. Turner, R. Y. Wang, L. L. Needham, I. M. Paul, J. L. Stokes, D. Q. Naiman, and D. G. Patterson, Jr. 2009. Do human milk concentrations of persistent organic chemicals really decline during lactation? Chemical concentrations during lactation and milk/serum partitioning. *Environmental Health Perspectives* 117(4):500–507.

LaKind, J. S., M. Goodman, and D. R. Mattison. 2014. Bisphenol A and indicators of obesity, glucose metabolism/type 2 diabetes and cardiovascular disease: A systematic review of epidemiologic research. *Critical Reviews in Toxicology* 44(2):121–150.

LaKind, J. S., M. Davis, G. M. Lehmann, E. Hines, S. A. Marchitti, C. Alcala, and M. Lorber. 2018. Infant dietary exposures to environmental chemicals and infant/child health: A critical assessment of the literature. *Environmental Health Perspectives* 126(9):96002.

Le Coadou, L., K. Le Ménach, P. Labadie, M. H. Dévier, P. Pardon, S. Augagneur, and H. Budzinski. 2017. Quality survey of natural mineral water and spring water sold in France: Monitoring of hormones, pharmaceuticals, pesticides, perfluoroalkyl substances, phthalates, and alkylphenols at the ultra-trace level. *Science of the Total Environment* 603–604:651–662. https://doi.org/10.1016/j.scitotenv.2016.11.174. PMID: 28343692.

Lee, S., S. Kim, J. Park, H. J. Kim, G. Choi, S. Choi, S. Kim, S. Y. Kim, S. Kim, K. Choi, and H. B. Moon. 2018. Perfluoroalkyl substances (PFASs) in breast milk from Korea: Time-course trends, influencing factors, and infant exposure. *Science of the Total Environment* 612:286–292. https://doi.org/10.1016/j.scitotenv.2017.08.094. PMID: 28865262.

Lehmann, G. M., J. S. LaKind, M. Davis, E. Hines, S. A. Marchitti, C. Alcala, and M. Lorber. 2018. Environmental chemicals in breast milk and formula: Exposure and risk assessment implications. *Environmental Health Perspectives* 126(9):96001.

Li, Y., T. Fletcher, D. Mucs, K. Scott, C. H. Lindh, P. Tallving, and K. Jakobsson. 2018. Half-lives of PFOS, PFHxS and PFOA after end of exposure to contaminated drinking water. *Occupational and Environmental Medicine* 75(1):46–51. https://doi.org/10.1136/oemed-2017-104651. PMID: 29133598; PMCID: PMC5749314.

Lin, P. D., A. Cardenas, R. Hauser, D. R. Gold, K. P. Kleinman, M. F. Hivert, A. F. Fleisch, A. M. Calafat, M. Sanchez-Guerra, C. Osorio-Yáñez, T. F. Webster, E. S. Horton, and E. Oken. 2020. Dietary characteristics associated with plasma concentrations of per- and polyfluoroalkyl substances among adults with pre-diabetes: Cross-sectional results from the Diabetes Prevention Program Trial. *Environment International* 137:105217. https://doi.org/10.1016/j.envint.2019.105217. PMID: 32086073; PMCID: PMC7517661.

Liu, Y., A. Li, S. Buchanan, and W. Liu. 2020. Exposure characteristics for congeners, isomers, and enantiomers of perfluoroalkyl substances in mothers and infants. *Environment International* 144:106012. https://doi.org/10.1016/j.envint.2020.106012. PMID: 32771830.

Llorca, M., M. Farré, Y. Picó, M. L. Teijón, J. G. Alvarez, and D. Barceló. 2010. Infant exposure of perfluorinated compounds: Levels in breast milk and commercial baby food. *Environment International* 36(6):584–592. https://doi.org/10.1016/j.envint.2010.04.016. PMID: 20494442.

Lorber, M., and P. P. Egeghy. 2011. Simple intake and pharmacokinetic modeling to characterize exposure of Americans to perfluorooctanoic acid, PFOA. *Environmental Science & Technology* 45(19):8006–8014. https://doi.org/10.1021/es103718h. PMID: 21517063.

Loria, K. 2019. To reduce PFAS levels in food, cook at home. *Consumer Reports.* https://www.consumerreports.org/food-safety/to-reduce-pfas-levels-in-food-cook-at-home (accessed March 10, 2021).

Luo, L., M. J. Kim, J. Park, H.-D. Yang, Y. Kho, M.-S. Chung, and B. Moon. 2019. Reduction of perfluorinated compound content in fish cake and swimming crab by different cooking methods. *Applied Biological Chemistry* 62:44. https://doi.org/10.1186/s13765-019-0449-x.

Macheka, L. R., J. O. Olowoyo, L. L. Mugivhisa, and O. A. Abafe. 2021. Determination and assessment of human dietary intake of per and polyfluoroalkyl substances in retail dairy milk and infant formula from South Africa. *Science of the Total Environment* 755(Pt 2):142697. https://doi.org/10.1016/j.scitotenv.2020.142697. PMID: 33065506.

Macheka-Tendenguwo, L. R., J. O. Olowoyo, L. L. Mugivhisa, and O. A. Abafe. 2018. Per- and polyfluoroalkyl substances in human breast milk and current analytical methods. *Environmental Science and Pollution Research International* 25(36):36064–36086. https://doi.org/10.1007/s11356-018-3483-z. PMID: 30382519.

MDH (Minnesota Department of Health). 2008, January. *MDH evaluation of point-of-use water treatment devices for perfluorochemical removal interim report.* Fact Sheet. https://wrl.mnpals.net/islandora/object/WRLrepository%3A1862/datastream/PDF/view (accessed March 21, 2021).

Mozaffarian, D., and E. B. Rimm. 2006. Fish intake, contaminants, and human health: Evaluating the risks and the benefits. *Journal of the American Medical Association* 296(15):1885–1899. https://doi.org/10.1001/jama.296.15.1885. Erratum in *Journal of the American Medical Association* 2007; 297(6):590. PMID: 17047219.

Munn, Z., M. D. J. Peters, C. Stern, C. Tufanaru, A. McArthur, and E. Aromataris. 2018. Systematic review or scoping review? Guidance for authors when choosing between a systematic or scoping review approach. *BMC Medical Research Methodology* 143. https://doi.org/10.1186/s12874-018-0611-x (accessed March 21, 2021).

Myers, A. L., P. W. Crozier, P. A. Helm, C. Brimacombe, V. I. Furdui, E. J. Reiner, D. Burniston, and C. H. Marvin. 2012. Fate, distribution, and contrasting temporal trends of perfluoroalkyl substances (PFASs) in Lake Ontario, Canada. *Environment International* 44:92–99. https://doi.org/10.1016/j.envint.2012.02.002. PMID: 22406021.

Nadal, M., and J. L. Domingo. 2014. Indoor dust levels of perfluoroalkyl substances (PFASs) and the role of ingestion as an exposure pathway: A review. *Current Organic Chemistry* 18(17):2200–2208. https://doi.org/10.2174/1385272819666140804230713.

NASEM (National Academies of Sciences, Engineering, and Medicine). 2021. *Federal government human health PFAS research workshop: Proceedings of a workshop—in brief.* Washington, DC: The National Academies Press. https://doi.org/10.17226/26054.

ODH (Ohio Department of Health). 2020. *How to reduce your exposure to PFAS. Health assessment section.* https://epa.ohio.gov/Portals/28/documents/pfas/PFASHowtoReduceExposure.pdf (accessed March 10, 2021).

Oliaei, F., D. Kriens, R. Weber, and A. Watson. 2013. PFOS and PFC releases and associated pollution from a PFC production plant in Minnesota (USA). *Environmental Science and Pollution Research International* 20(4):1977–1992. https://doi.org/10.1007/s11356-012-1275-4. PMID: 23128989.

Patterson, C., J. Burkhardt, D. Schupp, E. R. Krishnan, S. Dyment, S. Merritt, L. Zintek, and D. Kleinmaier. 2019. Effectiveness of point-of-use/point-of-entry systems to remove per- and polyfluoroalkyl substances from drinking water. *AWWA Water Science* 1(2):1–12. https://doi.org/10.1002/aws2.1131. PMID: 31338490; PMCID: PMC6650157.

Sajid, M., and M. Ilyas. 2017. PTFE-coated non-stick cookware and toxicity concerns: A perspective. *Environmental Science and Pollution Research International* 24(30):23436–23440. https://doi.org/10.1007/s11356-017-0095-y. PMID: 28913736.

Scher, D. P., J. E. Kelly, C. A. Huset, K. M. Barry, and V. L. Yingling. 2019. Does soil track-in contribute to house dust concentrations of perfluoroalkyl acids (PFAAs) in areas affected by soil or water contamination? *Journal of Exposure Science & Environmental Epidemiology* 29:218–226. https://doi.org/10.1038/s41370-018-0101-6.

Sinclair, E., S. K. Kim, H. B. Akinleye, and K. Kannan. 2007. Quantitation of gas-phase perfluoroalkyl surfactants and fluorotelomer alcohols released from nonstick cookware and microwave popcorn bags. *Environmental Science & Technology* 41(4):1180–1185. https://doi.org/10.1021/es062377w. PMID: 17593716.

Stahl, L. L., D. D. Snyder, A. R. Olsen, T. M. Kincaid, J. B. Wathen, and H. B. McCarty. 2014. Perfluorinated compounds in fish from U.S. urban rivers and the Great Lakes. *Science of the Total Environment* 499:185–195. https://doi.org/10.1016/j.scitotenv.2014.07.126. PMID: 25190044.

Sunderland, E. M., X. C. Hu, C. Dassuncao, A. K. Tokranov, C. C. Wagner, and J. G. Allen. 2019. A review of the pathways of human exposure to poly- and perfluoroalkyl substances (PFASs) and present understanding of health effects. *Journal of Exposure Science & Environmental Epidemiology* 29(2):131–147. https://doi.org/10.1038/s41370-018-0094-1. PMID: 30470793; PMCID: PMC6380916.

Tao, L., K. Kannan, C. M. Wong, K. F. Arcaro, and J. L. Butenhoff. 2008a. Perfluorinated compounds in human milk from Massachusetts, U.S.A. *Environmental Science & Technology* 42(8):3096–3101. https://doi.org/10.1021/es702789k. PMID: 18497172.

Tao, L., J. Ma, T. Kunisue, E. L. Libel, S. Tanabe, and K. Kannan. 2008b. Perfluorinated compounds in human breast milk from several Asian countries, and in infant formula and dairy milk from the

United States. *Environmental Science & Technology* 42(22):8597–8602.
https://doi.org/10.1021/es801875v. PMID: 19068854.

Taylor, M. D., S. Nilsson, J. Bräunig, K. C. Bowles, V. Cole, N. A. Moltschaniwskyj, and J. F. Mueller.
2019. Do conventional cooking methods alter concentrations of per- and polyfluoroalkyl
substances (PFASs) in seafood? *Food and Chemical Toxicology* 127:280–287.
https://doi.org/10.1016/j.fct.2019.03.032. PMID: 30905867.

Thomsen, C., L. S. Haug, H. Stigum, M. Frøshaug, S. L. Broadwell, and G. Becher. 2010. Changes in
concentrations of perfluorinated compounds, polybrominated diphenyl ethers, and
polychlorinated biphenyls in Norwegian breast-milk during twelve months of lactation.
Environmental Science & Technology 44(24):9550–9556. https://doi.org/10.1021/es1021922.
Erratum in *Environmental Science & Technology* 2011; 45(7):3192. PMID: 21090747.

Tittlemier, S. A., K. Pepper, C. Seymour, J. Moisey, R. Bronson, X. L. Cao, and R. W. Dabeka. 2007.
Dietary exposure of Canadians to perfluorinated carboxylates and perfluorooctane sulfonate via
consumption of meat, fish, fast foods, and food items prepared in their packaging. *Journal of
Agricultural and Food Chemistry* 55:3203–3210.

Trudel, D., L. Horowitz, M. Wormuth, M. Scheringer, I. T. Cousins, and K. Hungerbühler. 2008.
Estimating consumer exposure to PFOS and PFOA. *Risk Analysis* 28(2):251–269.
https://doi.org/10.1111/j.1539-6924.2008.01017.x. Erratum in *Risk Analysis* 2008; 28(3):807.
PMID: 18419647.

Vandenbroucke, J. P., E. von Elm, D. G. Altman, P. C. Gøtzsche, C. D. Mulrow, S. J. Pocock, C. Poole, J.
J. Schlesselman, and M. Egger for the STROBE Initiative. 2007. Strengthening the reporting of
observational studies in epidemiology (STROBE): Explanation and elaboration. *Epidemiology*
18:805–835.

VanNoy, B. N., J. Lam, and A. R. Zota. 2018. Breastfeeding as a predictor of serum concentrations of
per- and polyfluorinated alkyl substances in reproductive-aged women and young children: A
rapid systematic review. *Current Environmental Health Reports* 5(2):213–224.
https://doi.org/10.1007/s40572-018-0194-z. PMID: 29737463.

Vassiliadou, I., D. Costopoulou, N. Kalogeropoulos, S. Karavoltsos, A. Sakellari, E. Zafeiraki, M.
Dassenakis, and L. Leondiadis. 2015. Levels of perfluorinated compounds in raw and cooked
Mediterranean finfish and shellfish. *Chemosphere* 127:117–126.
https://doi.org/10.1016/j.chemosphere.2014.12.081. PMID: 25676497.

Vestergren, R., and I. T. Cousins. 2009. Tracking the pathways of human exposure to perfluorocarboxylates.
Environmental Science & Technology 43(15):5565–5575. https://doi.org/10.1021/es900228k.
PMID: 19731646.

von der Trenck, K. T., R. Konietzka, A. Biegel-Engler, J. Brodsky, A. Hädicke, A. Quadflieg, R.
Stockerl, and T. Stahl. 2018. Significance thresholds for the assessment of contaminated
groundwater: Perfluorinated and polyfluorinated chemicals. *Environmental Sciences Europe*
30(1):19. https://doi.org/10.1186/s12302-018-0142-4. PMID: 29930891; PMCID: PMC5992233.

von Ehrenstein, O. S., S. E. Fenton, K. Kato, Z. Kuklenyik, A. M. Calafat, and E. P. Hines. 2009.
Polyfluoroalkyl chemicals in the serum and milk of breastfeeding women. *Reproductive
Toxicology* 27(3–4):239–245. https://doi.org/10.1016/j.reprotox.2009.03.001. PMID: 19429402.

WHO (World Health Organization). 2000. Consultation on assessment of the health risks of dioxins; reevaluation of the tolerable daily intake (TDI): Executive summary. *Food Additives &
Contaminants* 17:223–240.

WHO. 2020. *Infant and young child feeding.* https://www.who.int/en/news-room/fact-sheets/detail/infant-
and-young-child-feeding (accessed May 12, 2021).

Wu, X. M., D. H. Bennett, A. M. Calafat, K. Kato, M. Strynar, E. Andersen, R. E. Moran, D. J. Tancredi,
N. S. Tulve, and I. Hertz-Picciotto. 2015. Serum concentrations of perfluorinated compounds
(PFC) among selected populations of children and adults in California. *Environmental Research*
136:264–273. https://doi.org/10.1016/j.envres.2014.09.026. PMID: 25460645; PMCID:
PMC4724210.

Yakushiji, T., I. Watanabe, K. Kuwabara, S. Yoshida, K. Koyama, I. Hara, and N. Kunita. 1978. Long-term studies of the excretion of polychlorinated biphenyls (PCBs) through the mother's milk of an occupationally exposed worker. *Archives of Environmental Contamination and Toxicology* 7(4):493–504.

Young, A. S., R. Hauser, T. M. James-Todd, B. A. Coull, H. Zhu, K. Kannan, A. J. Specht, M. S. Bliss, and J. G. Allen. 2021. Impact of "healthier" materials interventions on dust concentrations of per- and polyfluoroalkyl substances, polybrominated diphenyl ethers, and organophosphate esters. *Environment International* 150:106151. https://doi.org/10.1016/j.envint.2020.106151. PMID: 33092866; PMCID: PMC7940547.